Lecture Notes in Control and Information Sciences

Edited by M. Thoma and A. Wyner

For information about Vols. 1-96 please contact your bookseller or Springer-Verlag

Lecture Notes in Control and Information Sciences

Edited by M. Thoma and A. Wyner

159

X. Li, J. Yong (Eds.)

Control Theory of Distributed Parameter Systems and Applications

Proceedings of the IFIP WG 7.2 Working Conference
Shanghai, China, May 6–9, 1990

Springer Science+Business Media, LLC

Editors
Xunjing Li
Jiongmin Yong

Dept. of Mathematics
Fudan University
Shanghai 200433
China

ISBN 978-3-540-53894-3 ISBN 978-3-540-46372-6 (eBook)
DOI 10.1007/978-3-540-46372-6

Originally published by Springer-Verlag Berlin Heidelberg New York in 1991

61/3020-543210 Printed on acid-free paper

FORWORD

The IFIP-TC7 Conference on Control Theory of Distributed Parameter Systems and Applications was held at Fudan University , Shanghai, China on May 6–9, 1990. More than thirty scholars from seven countries attended the meeting. There were five invited talks and about thirty contributed talks. This proceeding gethers most papers presented at the conference. The topics of this conference involves the following areas of distributed parameter systems: optimal control, identification, stability, numerical optimization, stochastic control, etc.

We would like to express our thanks to the following organizations which sponsored this conference:

State Education Commission of the People's Republic of China
National Science Fundation of China
International Federation for Information Processing (IFIP)
Fudan University
Institute of System Sciences, Chinese Academy of Sciences

We also would like to extend our gratitude to all the authors for their real interests in the conference and all members of the local organizing committee for their suggestions and supports. Our thanks also go to Professor I. Lasiecka, the Chairman of the IFIP-TC7, for her consistent helps in organizing the meeting and our colleagues at Fudan University for their cooperation which made the meeting really happen.

Xunjing Li and Jiongmin Yong

Department of Mathematics
Fudan University
Shanghai
China

The IFIP-TC7 International Program Committee:

A. Bermudez, Fac. De Ciencias, Santiago de Compostelo, Spain
A. Butkovski, Control Institute, Moscow
R. Curtain, Univ. of Groningen, Netherlands
G. Da Prato, Scoula Normale, Pisa, Italy
R. Glowinski, INRIA, Paris, France
K. Hoffman, Univ. of Augsburg, Germany
G. Krabs, Technische Hochschule, Darmstadt, Germany
A. Kurzhanskij, IIASA, Laxenburg, Austria
I. Lasiecka (Chairman), Univ. of Virginia, USA
J. L. Lions, College de France and CNES, Paris, France
U. Mosco, Univ. of Rome, Rome, Italy
O. Pironneau, INRIA, Paris, France
P. Yvon, INRIA, Paris, France
J. P. Zolesio, Univ. de Nice, Nice, France

The Local Organizing Committee:

Dexing Feng, Institute of System Science, Chinese Academy of Sciences
Guangyuan Huang, Shandong Univ.
Xunjing Li (Chairman), Fudan Univ.
Yongzai Lu, Zhejiang Univ.
Laixiang Sun, Fudan Univ.
Jingyuan Yu, Beijing Institute of Information and Control

LIST OF PARTICIPANTS

Banks, H.T.
——Center for Applied Mathematical Sciences, DRB-306, University of South California, Los Angeles, CA 90089-1113, USA
Butkovskiy, A.G.
——Institute of Control Sciences, Moscow, USSR
Caffarelli, G.V.
——Dipartimento di Matematica, Univerita Degli Studi di Trento, 38050 Povo (Trento), Italy
Chavent, G
——INRIA, Domaine de Voluceau, Rocquencourt, B.P.105, 78153 Le Chesnay Cedex, France
Chen, Shuping
——Department of Mathematics, Zhejiang University, Hangzhou, China
Deng, Shaomei
——Nanjing Insititute of Hydrology Ministry of Water Conservancy, Nanjing, Jiangsu 210024, China
Le Dimet, F.-X.
——Department of Applied Mathematics, University Blaisc Pascal Clermont-Ferrand, B.P. 45-63170 Aubiere, France
Gao, Hang
——Department of Mathematics, Northeastern Normal University, Changchun, Jilin 130024, China
Gao, Lin
——Institute of Population Research, Chinese People's University, Beijing 100872, China
Huang, Shaoyun
——Department of Mathematics, Beijing University, Beijing 100871, China
Huang, Yu
——Department of Mathematics, Zhongshan University, Guangzhou 510275, China
Kappel, F.
——Institut fur Mathematik, Karl-Franzens-Universitat Graz, A-8010 Graz, Elisabethstrasse 16, Austria
Li, Chengzhi
——Xiamen University, Xiamen, Fujian 361005, China
Li, Ping
——Beijing College of Technology, Beijing, China
Li, Xunjing
——Department of Mathematics, Fuadn University, Shanghai 200433, China
Lu, Yongzai
——Reesearch Institute of Industrial Process Control, Zhejiang University, Hangzhou, China

Luce, R
——Department of Applied Mathematics, University of Technology of Compiegne, 60206 Compiegne, France
Nakagiri, Shin-ichi
——Department of Applied Mathematics, Faculty of Engineering, Kobe University, Kobe, Nada 657, Japan
Pan, Liping
——Institute of Mathematics, Fudan University, Shanghai 200433, China
Peng, Shige
——Department of Mathematics, Shandong University, Jinan, Shandong 250100, China
Sakawa, Yoshyuki
——Department of Control Engineering, Faculty of Engineering Sciences, Osaka Univ., Toyonaka, Osaka, Japan
Simon, Jacques
——Department of Applied Mathematics, Univeristy Blaise Pascal Clermont-Ferrand, B.P. 45-63170 Aubiere, France
Situ, Rong
——Department of Mathematics, Zhongshan University, Guangzhou 510275, China
Song, Wen
——Department of Mathematics, Harbin Normal University, Harbin 150080, China
Sun, Haiwei
——Department of Mathematics, Zhongshan University, Guangzhou 510275, China
Wang, Miansen
——Department of Mathematics, Xi'an Jiaotong University, Xi'an, Sanxi 710049, China
Wang, Yun
——Center for Applied Mathematical Sciences, DRB-306, University of South California, Los Angeles, CA 90089-1113, USA
Wang, Yuwen
——Department of Mathematics, Harbin Normal University, Harbin 150080, China
Wu, Jingbo
——Department of Computer & System Sciences, Nankai University, Tianjin 300071, China
Xu, Yanqing
——Department of Computer & System Sciences, Nankai University, Tianjin 300071, China
Yong, Jiongmin
——Department of Mathematics, Fudan University, Shanghai 200433, China
Zhang, Weitao
——Institute of System Sciences, Academica Sinica, Beijing 100080, China
Zhao, Yaowen
——Department of Mathematics, Shandong University, Jinan, Shandong 250100, China
Zhou, Hongxin
——Department of Mathematics, Shandong University, Jinan, Shandong 250100, China

CONTENTS

METODS AND MODELS TO DESIGN MOBILE CONTROLS ON SURFACE

A.G.Butkovskiy, V.A.Kubyshkin, V.I.Finyagina
Institute of Control Sciences
Profsojuznaja 65, 117345 Moscow, USSR

The objectes with a mobile heat source which periodically varies its position along an assigned trajectory on an object surface are considered. The object state is described by two-dimentional heat transfer equation. The problem to obtain and maintain an object state closed to the assigned one is stated. Two types of models are used to solve the problem. These are stationary models with distributed control and nonstationary ones in which heat source movement is taken into account. The calculation method of controls making use of above two types of models has been developed. The paper contains the calculation examples of the source movement laws along the linewise object surface trajectories, power of the mobile source, dynamics of temperature field and grafical result representations.

1. INTRODUCTION

The systems with mobile source, such as electronic, ion or laser beams, possess some features complicating their modelling, design and analysis [1]. The main of them are nonlinearity of controls and fast movement of a source with respect to an object.

At present some publications highlight the developments on the choice of models and designing of mobile controls [1-3]. They present in sufficient details the investigations of the cases with one-dimensional approximation of real objects. However in practice a mobile power source is most frequently surfacing the object along a curvilinear trajectory. In this case one-dimensional models are too warse approximation to be applied in practice.

The paper is concerned with the choice and validation of models of objects with mobile action as well as the design of methods and algorithms for calculating the source movement laws along a trajectory on the object surface.

2. PROBLEM STATEMENT

Object whose state $Q(x_1,x_2,t)$ is described by heat-transfer equations (nonlinear in general case) with mobile heat source are considered

$$\frac{\partial Q}{\partial t} = a(Q)\left(\frac{\partial^2 Q}{\partial x_1^2} + \frac{\partial^2 Q}{\partial x_2^2}\right) - q(Q) + F(x_1,x_2,t),\ (x_1,x_2)\in D,\ t>0, \qquad (2.1)$$

$$Q(x_1,x_2,0) = Q_0(x_1,x_2),\quad (x_1,x_2) \in D, \qquad (2.2)$$

$$\left[\alpha Q + \lambda\frac{\partial Q}{\partial n}\right]_{(x_1,x_2)\in\Gamma} = \alpha Q_\Gamma. \qquad (2.3)$$

Here t is time, $x=(x_1,x_2)$ is a spatial coordinate, D is a bounded domain of object determining, Γ is a domain boundary, $a(Q)$ is a coefficient of thermal conductivity, α, λ are constant coefficients, $Q_0(x_1,x_2)$ is an assigned function, $\partial Q/\partial n$ is a derivative of the external normal direction to Γ, Q_Γ is an assigned number, $q(Q)$ is a nonnegative function, determining heat removal from the object surface, $F(x_1,x_2)$ is a mobile heat source having the form of

$$F(x_1,x_2) = u(t)\cdot\Phi\left[x_1 - s_1(t),\ x_2 - s_2(t)\right]. \qquad (2.4)$$

Here $u(t)$ is power of heat source, $\Phi(x_1,x_2)$ is an assigned source power distribution on the object relative to its centre, $\Phi(x_1,x_2)\geq 0$,

$$\iint_{-\infty}^{\infty} \Phi(x_1,x_2)dx_1dx_2 = 1,$$

(usually $\Phi(x_1,x_2)$ has the form of the Gaussian distribition),

$s(t)=(s_1(t), s_2(t))$ is a position of the source centre in the assigned domain $G\in D$.

Assume that in the domain G the trajectory $\mathfrak{L}$ with the length S is assigned by parametrical equations $x_1=X_1(s_1)$, $x_2=X_2(s_1)$ and as the parameter s_1 we have chosen the trajectory arc length measured from the arc beginning - the point l_0 to its end - in the point l_1. The source centre moves along the trajectory from l_0 to l_1 and inversely periodically with the period T. Then the law of source movement is fully determined by the trajectory equations $X_1(s_1)$ and $X_2(s_1)$ and by the position of the source centre $s_1(t)$ on the trajectory in each moment of time

$$F(x_1,x_2) = u(t)\cdot\Phi\left[x_1- X_1(s_1(t)),\ x_2- X_2(s_1(t))\right].$$

In genegal form the control problem is stated as follows. A desired object state $Q^*(x_1,x_2)$, $(x_1,x_2)\in G$ is assigned. It is required to find the trajectory $\mathfrak{L}\subset G$, the law of a periodical source movement along the trajectory $s_1(t)$ and the source power $u(t)$ with the constraints $0\leqslant ut)\leqslant U_{max}$, which provide the object state $Q(x_1,x_2,t)$ in the steady mode (with $t\to\infty$), whose deviation from an assigned state is minimal or it does not exeed an acceptable value δ. The measure of such a deviation can be assumed, for example, as the functional

$$\mathfrak{J} = \max_{\substack{t\in[t_1+T] \\ t_1\to\infty}} \int_G \left[Q^*(x_1,x_2) - Q(x_1,x_2,t)\right]^2 dx_1 dx_2.$$

3. STATIONARY MODEL OF AN OBJECT

The problem solution in a general form is complicated by the fact that over and above the object equation nonlinearity the controls $X_1(s_1)$, $X_2(s_1)$, $s_1(t)$ entering in the equation are nonlinear too. Therefore one has to use simple models for the problem solution.

It is known that under the condition of periodical heat source movement along the trajectory $\mathfrak{L}$ an object state also becomes a periodical function of time with $t \to \infty$ i.e. the condition $Q(x_1, x_2, t+T) = Q(X_1, x_2, t)$ is fulfilled. Then the object state can be represented in the form of a sum of two componentes: an averaged one $\bar{Q}(x_1, x_2)$ and an ocsillating one $Q_k(x_1, x_2, t)$. It can be shown [3] that an ocsillating component tends to zero with $T \to 0$ under the norm $L_2(D)$. Then, with $t \to \infty$ the object equation (2.1)-(2.3) can be approximately replaced by a corresponding stationary equation for the averaged component $\bar{Q}(x_1, x_2)$

$$a(Q) \cdot \left(\frac{\partial^2 \bar{Q}}{\partial x_1^2} + \frac{\partial^2 \bar{Q}}{\partial x_2^2} \right) - q(\bar{Q}) + \bar{F}(x_1, x_2) = 0, \quad (x_1, x_2) \in D, \tag{3.1}$$

with boundary conditions (2.2) and (2.3).

In equation (3.1) $\bar{Q}(x_1, x_2)$ is the averaged object state:

$$Q(x_1, x_2) = \frac{1}{T} \cdot \int_t^{t+T} Q(x_1, x_2, \tau) d\tau$$

When $t \to \infty$ it does not depend on time due to periodicity of $Q(x_1, x_2, t)$. The averaged control $\bar{F}(x_1, x_2)$ has the form of

$$\bar{F}(x_1, x_2) = \frac{1}{T} \cdot \int_t^{t+T} u(\tau) \cdot \Phi\left[x_1 - X_1(s_1(\tau)),\ x_2 - X_2(s_1(\tau))\right] d\tau =$$

$$= \frac{1}{T} \cdot \int_0^S \Phi\left[x_1 - X_1(s_1),\ x_2 - X_2(s_1)\right] \cdot u(t(s_1)) \cdot t(s_1) ds_1 \tag{3.2}$$

It is implemented in (3.2) the substitution $\tau = arc\left[s_1(t)\right] = t(s_1)$, $t(s_1)$ is a function inverse to $s_1(t)$. Thus in a steady mode an averaged object state can be determined by means of a stationary model (3.1), (2.2), (2.3).

For a stationary model the control problem is transformed as follows. Find the source power $0 \leqslant u(t) \leqslant U_{max}$, the trajectory $\left\{X_1(s_1), X_2(s_1)\right\} = = X(s_1) \subset G$ and the source centre movement law $s_1(t)$ along the trajectory $\mathfrak{L}$ on the object surface. All of them are the controls which by virtue of equation (3.1) provide a minimal value of the functional

$$\mathfrak{I} = \int_G \left[Q^*(x_1,x_2) - \bar{Q}(x_1,x_2)\right]^2 dx_1 dx_2 \tag{3.3}$$

4. METHODS FOR CALCULATING THE TWO-DIMENSIONAL MOBILE CONTROLS

To solve the problems it is suggested to decompose it into two simpler ones. 1. To find the distributed control $F^*(x_1,x_2)$ which being substituted into equation (3.1) for $\bar{F}(x_1,x_2)$ of the form (3.2) provides a minimal value or the one not exceeding the value δ of the functional (3.3). 2. To find the controls $X_1(s_1)$, $X_2(s_1)$, $s_1(t)$, $u(t)$ which provide a minimal value or the one not exceeding an acceptable value δ of the functional

$$\mathfrak{I} = \| F^*(x_1,x_2) - \bar{F}(x_1,x_2 \|_{L_2(G)} \tag{4.1}$$

Solution of the first problem by traditional methods (the principle of maximum, dynamical programming, etc.) proves to be difficult due to two-dimentionality of the problem and nonlinearity of the object equation. Therefore to solve the problem we propose to use the substitution method.

The method is to substitude the chosen in a special way function $Q_A(x_1,x_2)$, $(x_1,x_2)\in D$ which approximates an assigned state and the following requirements must be satisfied: a) it has to be a second piecewise-continuous derivative; b) it has to satisfy boundary conditions (2.3); c) being substituted in equation (3.1) for $\bar{Q}(x_1,x_2)$ it has to determine the function $F^*(x_1,x_2)$ satisfying the constraints $F^*(x_1,x_2)\geqslant 0$, $\int_G F^*(x_1,x_2)dx_1dx_2\leqslant U_{max}$ in domain G and it has to be equal to zero beyond G.

This method is rather visualized. It makes possibility to find acceptable solutions for practice by simple and unvieldy computations, and moreover the nonlinearity of equations slightly affects the solution process.

The function $Q_A(x_1,x_2)$ is conveniently assigned in the form of three functions "sewn" together $Q_A(x_1,x_2) = \sum_{i=1}^{3} Q_i(x_1,x_2)\cdot\chi_i(x_1,x_2)$ where $\chi_i(x_1,x_2)$, $i=1,2,3$ are characteristic functions of the domains respectively $\Omega_1=\backslash G$, $\Omega_2=\Gamma_{1\varepsilon}\backslash\Omega_1$ (Γ_1 is a G-domain boundary, $\Gamma_{1\varepsilon}$ is the ε-nighbourhood of Γ_1) and $\Omega_3=G\backslash\Omega_2$.

The function $Q_1(x_1,x_2)$ is assigned so that $F^*(x_1,x_2)$ will be equal to zero in the domain Ω_1. It is determined as the solution of equation (3.1) with boundary conditions (2.2) on the boundary Γ and $\mu_1(x_1,x_2)$, $(x_1,x_2)\in\Gamma_1$ on the boundary Γ_1. The function $\mu_1(x_1,x_2)$ is assigned following from physical considerations and updated in the course of calculations.

In the domain Ω_3 the function $Q_3(x_1,x_2)$ either coincides with $Q^*(x_1,x_2)$ or approximates it. Approximation becomes necessary when the conditions a) and c) are not observed.

The domain Ω_2 is used for transition from the function Q_1 to the function Q_3. The known methods of two-dimentional interpretation [4] can be used to assign $Q_2(x_1,x_2)$. The value of the domain Ω_3 is chosen so that in matching the constraints on $F^*(x_1,x_2)$ are satisfied.

Now consider in more details the problem of implementing the distributed control by a mobile one. Let be assigned the distributed control $F^*(x_1,x_2)$. It is required to find the trajectory $X_1(s_1)$, $X_2(s_1)$, the source movement law $s_1'(t)$ along the trajectory and source power u(t) all of which provide a minimal value of the functional (3.3).

In the stated problem there are three controls (with the fixed source form $\Phi(x_1,x_2)$): the trajectory $X_1(s_1)$, $X_2(s_1)$, the source movement law $s_1(t)$ along the trajectory and source power u(t). It is known [3] that with the fixed trajectory $\mathfrak{L}$ the wider class of averaged controls $\bar{F}(x_1,x_2)$ can be obtained by the movement control $s_1(t)$ with the fixed power u. That is why hereinafter the power u is to be referred to as a constant value which is to be determined.

In some cases it can be additionally assigned that a source movement occurs periodically at a constant speed in each period. Then, the problem is stated to determin the trajectory $\mathfrak{L}$ of a minimal lenght S with which the (3.3) functional value does not exceed an assigned admissible value δ.

In other cases it is more reasonable to specify the source trajectory. Then the problem consists in finding the power u and the source movement law $s_1(t)$ which provide a minimal value or not exceeding in a admissable value of functional (4.1).

Consider the problem solution for the widely used in practice case, namely for the linewise source movement trajectory in a rectangular domain.

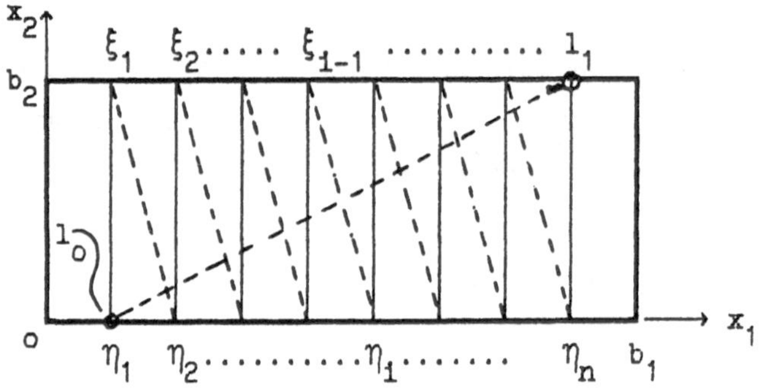

Fig.1

Let in the rectangular object surface domain $G=[0,b_1]x[0,b_2]$ a mobile source moves along the trajectory $\mathfrak{L}$ shown in Fig.1. In this case the source movement law in each line can be calculated by the approximated formulae which are derived from the assumption that a heat source is located on a sufficiently smaller area compared with the total area of the domain G [2] (a heat source is closer to a point one) and the number of lines is sufficiently large. These formulae are

$$\tau_i(s) = \int_0^s F^*(\eta_i,x_2)dx_2, \quad 0 \leqslant s \leqslant b_2, \quad i=1,2,\dots,N,$$

$$t_i(s) = \tau_i(s)\cdot T/\sum_{i=1}^{N} \tau_i(b_2) + t_i(b_2), \quad i=1,2,\dots,N, \quad t_o(b_2)=0. \tag{4.2}$$

Here η_i is a position of each line along the coordinate x_1, N is the total number of lines, T is the time of source movement along the whole trajectory from point l_0 to point l_1 (source movement period), $\tau_i(s)$ is an intermediate variable, $t_i(s)$, $0 \leqslant s \leqslant b_2$ is the function being inverse to the source movement law $s_i(t)$ in each line, $0 \leqslant s_i \leqslant b_2$, $t_{i-1}(b_2) \leqslant t \leqslant t_i(b_2)$. It is assumed that the source centre moves from line to line (from the point ξ_{i-1} to the point η_i, Fig.1) in the infinitesimal time.

Implementation accuracy of distributed control by mobile one with application of the above formulae improves if the line number is increased and if the effective source dimension is decreased (the source dimension is being approached to the point one).

5. CALCULATION EXAMPLES OF TWO-DIMENSIONAL MOBILE CONTROLS

Example 1. *Calculation of the distributed control* $F^*(x_1,x_2)$.

The object is a rectangular plate $D=(0,b_1)x(0,b_2)$, Fig.1. The source is moving along linewise trajectory. The assigned state is

$$Q^*(x_1,x_2) = C = const.,\ (x_1,x_2)\in D,\ q(Q) = \sigma(Q^4- Q_T^4).$$

Here σ is a constant coefficient. Domains D and G coincide. Thus the domain Ω_1 is not available while domain Ω_3 is a rectangle $[\varepsilon,b_1-\varepsilon]x[\varepsilon,b_2-\varepsilon]$.

Find the distributed control by the substitution method. The function $Q_A(x_1,x_2)$ consists of two parts $Q_A(x_1,x_2) = Q_2\chi_2(x_1,x_2)+Q_3\chi_3(x_1,x_2)$. Here $Q_3(x_1,x_2)=C=const.,\ (x_1,x_2)\in\Omega_3$ and $Q_2(x_1,x_2)$ serves for matching with the boundary conditions and it is represented as a product $Q_2(x_1,x_2)=\varphi_1(x_1)\cdot\varphi_2(x_2)$ where

$$\varphi_{1,2}(x_{1,2}) = C\begin{cases} k\cdot(x_{1,2}-\varepsilon)^2+1, & \text{if}\quad 0\leqslant x_{1,2}\leqslant\varepsilon,\\ 1, & \text{if}\quad \varepsilon\leqslant x_{1,2}\leqslant b_{1,2}-\varepsilon,\\ -k\cdot\left[x_{1,2}-(b_{1,2}-\varepsilon)\right]^2+1, & \text{if}\ b_{1,2}-\varepsilon\leqslant x_{1,2}\leqslant b_{1,2}, \end{cases}$$

$$(x_1,x_2)\in\Omega_2 \qquad (5.1)$$

The coefficient k is chosen from the matching conditions between the function $Q_A(x_1,x_2)$ and its derivative on the boundary Γ_1. In particular for boundary conditions (2.3) this coefficient is

$$k = \frac{Q_T - \lambda}{\lambda(\varepsilon)^2 \mp 2\alpha\varepsilon}$$

Fig.2 presents the approximating function $Q_A(x_1,x_2)$ and Fig.3 shows the distributed control corresponding with this function which has been calculated by the substitution method.

Example 2. The object is the same as in Example 1. The assigned state $Q^*(x_1,x_2)$ has the form shown in Fig.4. The approximating function shown in Fig.5 is obtained by smoothing the discontinuities of the function $Q^*(x_1,x_2)$ in the domain Ω_3 and matching with the boundary conditions of the function $Q_2(x_1,x_2)$ in the domain Ω_2. The corresponding with this problem statement distributed control calculated by substituting the approximating function in the object equation is shown in Fig.6.

Example 3. *Implementation of calculated distributed controls.* The source movement laws calculated by the approximated formulae (4.2) for the linewise trajectory are shown in Figs.7 and 8. These laws provide the distributed controls (Figs.3 and 6) described in Examples 1 and 2.

6. TWO-DIMENTIONAL NONSTATIONARY MODEL WITH MOBILE CONTROL

Calculations by approximated formulae require to be checked on digital models taking into account the source movement. That is why the two-dimentional model of a rectangular plate being heated by a mobile heat source has been designed. The method of finite differences is used to calculate the temperature field making use of equations (2.1)-(2.3). The finite difference equations were formed by means of a mesh with constant steps along spatial coordinates and a variable step in time matching with the source movement law along the trajectory.

The trajectory and the movement law along the trajectory were specified separetely.

The model takes much computer time due to a fast heat source movement and, consequently, smaller mesh step size along the time coordinate. As a result a computing time becomes unacceptable when is needed to attain the steady mode. Therefore the temperature field is calculated in two stage. In the first stage it is calculated the temerature field of an object with averaged action (3.2). In approaching the steady mode action (3.2) is replaced by mobile action (2.4) and a temperature field pattern is finally determined.

The designed temperature field is used for determining field pulsations either in time or along statial coordinates. Fig.9 shows the temperatute field in a steady mode. It has been calculated by the model under the source movement law shown in Fig.7.

7. CONCLUSION

Two tipe of models are proposed to compute mobile controls on surface: 1) stationary models with control which represents an action averaged for the period of source movement or a distributed control neither connected with the source movement; 2) nonstationary (full) models taking into account the source movement.

The mobile control calculation consists of the following stages: a) calculation of a distributed control making use of the stationary model; b) calculation of the source power and the source movement law along the trajectory which provide an averaged mobile source action, i.e. the action which is close to the distributed control; c) checking on a nonstationary model deviation of the obtained state from the calculated one, the temperature variations in space and time, updating the trajectory and the choice of a sourse movement period.

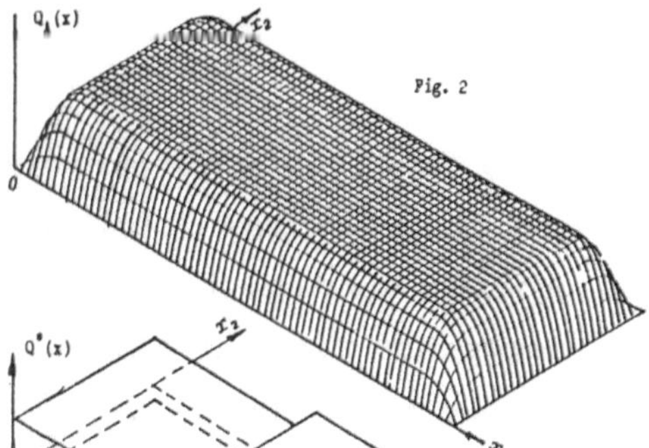

Fig. 2

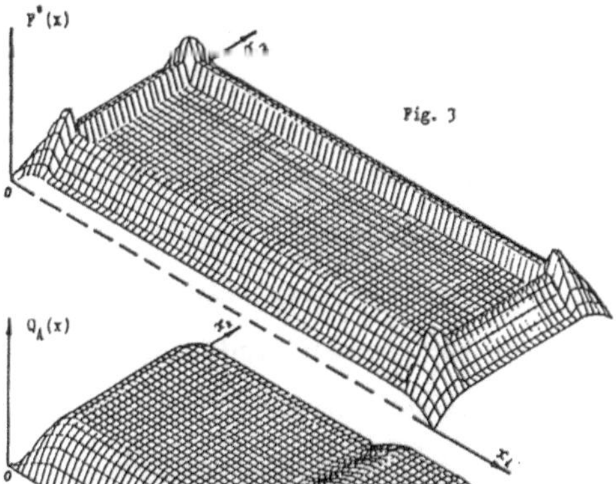

Fig. 3

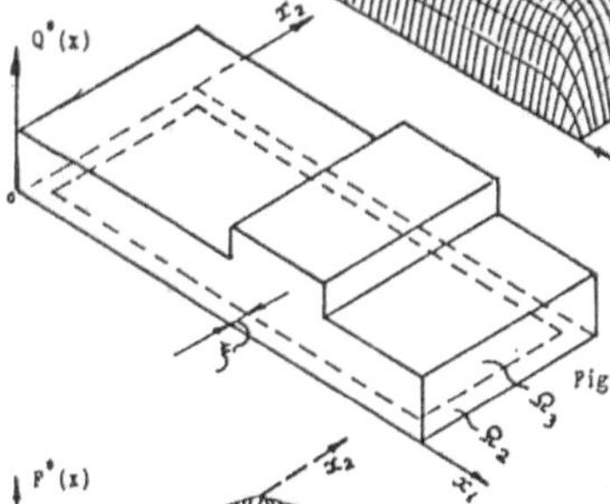

Fig. 4

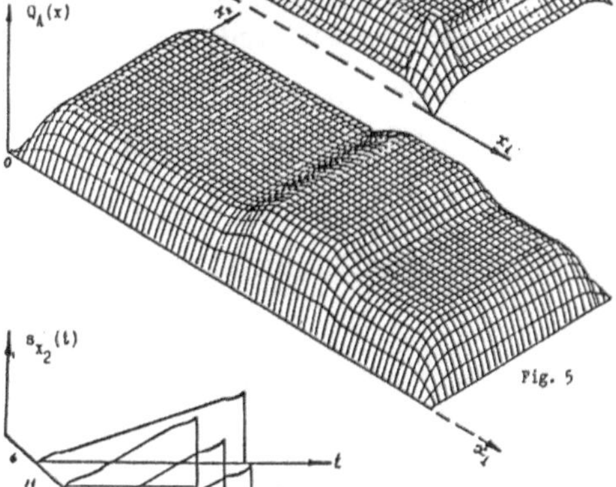

Fig. 5

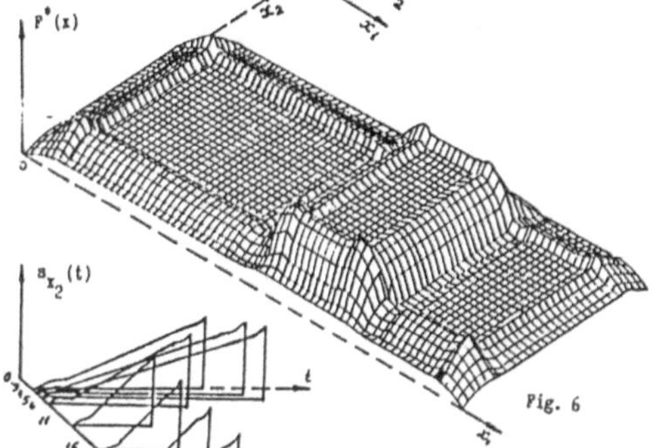

Fig. 6

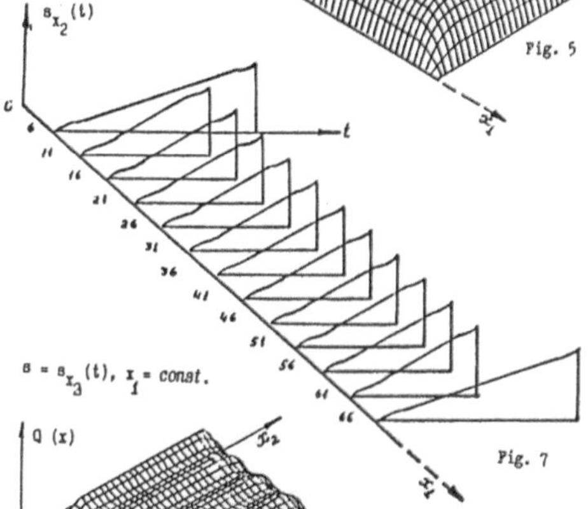

$s = s_{x_3}(t)$, $x_1 = const.$

Fig. 7

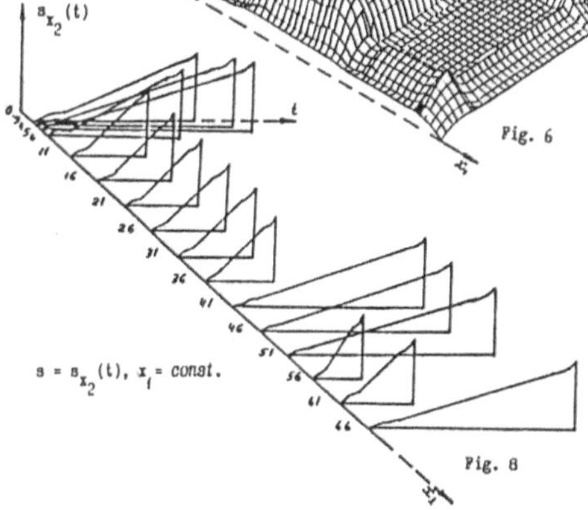

$s = s_{x_2}(t)$, $x_1 = const.$

Fig. 8

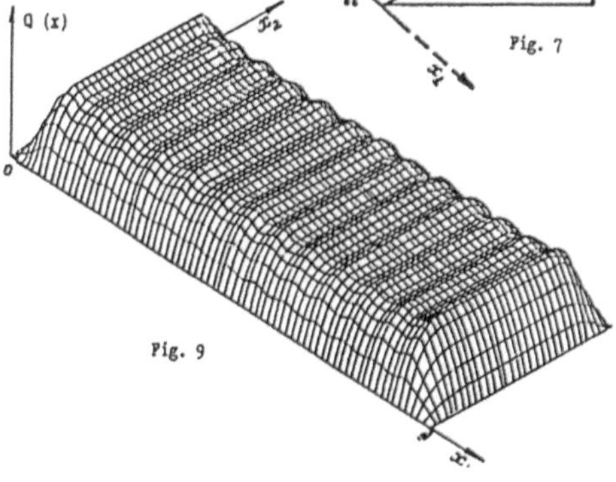

Fig. 9

The distributed control is calculated by substituting a specially chosen function approximating an spesified object state in the object equation. An approximating function is performed by means of "sewing" the functions on the domains joint where an action equals zero and an obtained object state is determined.

This method allows us to find the controls for an object described by nonlinear equations and it is visualized and easy to follow.

REFERENCES

1. Butkovskiy, A.G. and Pustil'nikov, L.M. Mobile Control Distributed Parameter Systems. (Ellis Horwood Limited Publishers, Chichester, 1987).
2. Butkovskiy, A.G., Kubyshkin, V.A., Smirnov, A.G., Tverdochlebov, E.S., Chubarov, E.P. A Method for Computing and Implementing Distributed Control Signals. (Preprints IFAC 9-th World Congress, vol.IX, 1984).
3. Breger, A.M., Butkovskiy, A.G., Kubyshkin, V.A. and Utkin, V.I. Sliding Modes in Control of Distributed Plants Subjected to a Mobile Multicycle Signal. Automation and Remote Control, vol.41, No.3, Part 1, (1980).
4. Foux, I.D. and Prett, M.J. Computational Geometry for Design and Manufacture. (Halsted Press: A Division of John Wiley & Sons, New York - Chichester - Brisbane - Toronto). Ellis Horwood Ltd., 1979.

A GEOMETRICAL THEORY FOR NONLINEAR LEAST SQUARES PROBLEMS

Guy Chavent
CEREMADE, University of Paris-Dauphine,
75775 Paris Cédex 16, FRANCE
and
INRIA, Domaine de Voluceau-Rocquencourt
BP 105, 78153 Le Chesnay Cédex, FRANCE

Summary

1 - Introduction

The motivation for the geometrical tools described in this paper is the study of the non linear least squares problem :

find $\hat{x} \in C$ s.t $J(x) = \|\varphi(x)-z\|^2_F$ = Min over C (1.1)

where

C is a convex subset of some linear vector space (set of admissible parameters)

F is an Hilbert space (data space)

$\varphi : C \rightarrow F$ is a given mapping (parameter $\rightarrow$ output) (1.2)

$z \in F$ is a given point (experimental data).

We have given in (1.2) the interpretation of problem (1.1) in the context of parameter estimation in PDEs or ODEs, but problems like (1.1) arise in various other areas, as approximation theory, control theory, optimum design, inverse problems.

Problem (1.1) has been extensively studied when φ is linear (see for example reference [1]). When φ is non linear, which is most generally the case in the above mentionned contexts, very few results are available : of course, compactness techniques allow in certain cases to proove the existence of a solution $\hat{x}$ to problem (1.1), but give no indication on uniqueness of $\hat{x}$, its continuous dependance on the data z, or the absence of parasitic local minima. This last property is of great practical importance when it comes to the numerical resolution of (1.1) on a computer, as it will guarantee that gradient technique will be able to find the global minimum $\hat{x}$, and will not stop in some local minimum of J. That is why we include this property (absence of local minimum) in our definition of well-posedness :

Definition 1.1. Let

$d(x,y)$ = distance on C (1.3)

be given. The non-linear least squares problem (1.1) is said to be Quadratically-well-posed (Q-well-posed in short) on some open neighborhood $\mathcal{V}$ of $\varphi(C)$ for the distance d on C iff :

i) for any $z \in \mathcal{V}$, (1.1) has a unique solution $\hat{x}$

ii) for any $z \in \mathcal{V}$, J has no parasitic local minimum

iii) for $z \in \mathcal{V}$, any minimizing sequence x_n converges to $\hat{x}$ for the distance d

iv) the mapping $z \to \hat{x}$ is locally Lipschitz continuous from $(\mathcal{V}, \| \|_F)$ to (C,d).

We describe in this paper sufficient conditions for problem (1.1) to be Q-well-posed on some cylindrical neighborhood of $\varphi(C)$, with explicit formula for the size of the neighborhood and for the Lipschitz constant of $z \to \hat{x}$ mapping.

Of course, such result requires that one is able to project z onto $\varphi(C)$ in a unique and stable way. That is why we shall devote paragraph 2 to the definition of strictly quasiconvex sets D of F, which have this nice property for all z of some neighborhood $\mathcal{V}$ of D. Then we shall give in paragraph 3 two sufficient conditions (size × curvature conditions) to recognize strictly quasiconvex sets, based on geometric properties of some family of paths defined on D. Finally we shall apply in paragraph 4 the above size × curvature conditions to $D=\varphi(C)$ and obtain sufficient conditions for the Q-wellposedness of problem (1.1).

The Q-wellposedness result has been used in reference [2] for the study of a plane wave detection problem, in reference [3] for the estimation of a space-varying diffusion coefficient in a one-dimensional elliptic equation and in reference [6] for the determination of the regularization parameter when problem (1.1) is ill posed.

We present in this paper only the main steps of the construction, and refer the reader to references [4] and [5] for more detailed discussion and all proofs.

2 - Strictly quasiconvex sets

We are given in this paragraph

$$F = \text{Hilbert space} \qquad (2.1)$$
$$D \subset F$$

and we want to find conditions on D so that the projection on D is nicely behaved on some neighborhood $\mathcal{V}$ of D. The idea is to mimic the theory of projection on convex sets by replacing the convex set by D and the segments of the convex by paths of D :

Definition 2.1. (path of D)

A mapping $P:[0,L] \to D$ is a path of D iff :

$v \to P(v)$ is in $W^{2,\infty}([0,L])$ (2.2)

$\|P'(v)\|_F = 1$ for a.e. $v \in [0,L]$ (2.3)

Definition 2.2. (attributes of a path)

Let a path P be given. Then :

$v \in [0,L]$ is the arc-length along P (2.4)

$\delta(P) \triangleq L$ is the length of P (2.5)

$v(v) \triangleq P'(v)$ is the unit tangent vector to P at $P(v)$ (2.6)

$a(v) \triangleq P''(v)$ is the accelleration vector to P at $P(v)$ (2.7)

$\rho(v) \triangleq \|a(v)\|^{-1} \in IR^+ \cup \{+\infty\}$ is the radius of curvature of P at $P(v)$ (2.8)

$R(P) \triangleq \underset{\gamma \in [0,\delta(P)]}{\text{Inf Ess}}\ \rho(v) = \|a\|_\infty^{-1} > 0$ is smallest radius of curvature along P (2.9)

We have now to chose, among all possible paths of D, the ones which are going to play for D the same role segments play for a convex ; we need for that :

Definition 2.3. (collection of paths)

A set $\mathcal{P}$ of paths is a collection of paths for D iff :

$\mathcal{P}$ is made of paths of D (2.10)

$\mathcal{P}$ is complete, i.e. $\forall X,Y \in D$, $X \neq Y$, $\exists P \in \mathcal{P}$ such that $P(0)=X$, $P(\delta(P))=Y$ (2.11)

$\mathcal{P}$ is stable with respect to restriction, i.e. $\forall P \in \mathcal{P}$, $\forall v',v'' \in [0,\delta(P)]$, $v'<v''$, (2.12)
the path $\tilde{P} : v \in [0,v''-v'] \to P(v'+v)$ belongs to $\mathcal{P}$.

Of course, if D is convex, the set made of all segments of D is a collection of paths for D ! So we shall suppose in the sequel that one collection of paths $\mathcal{P}$ for the set D has been chosen. This will be emphasized by writing $(D,\mathcal{P})$ instead of D (for example, when $D=\varphi(C)$, one can take for $\mathcal{P}$, provided φ satisfies some conditions, the set of the images by φ of the segments of the convex set C as we shall see in paragraph 4).

As we are going to study the projection of a point $z \in F$ onto D, the paths P joining two points of D located at a distance of z slightly larger than $d(z,D)$ will play a crucial role. Hence we define, for any $z \in F$ and $\eta > 0$:

$$\mathcal{P}(z,\eta) = \{P \in \mathcal{P} \mid \|P(j)-z\| \leq d(z,D)+\eta,\ j=0,\delta(P)\}. \tag{2.13}$$

The key number for the definition of quasiconvex sets is then defined by :

$$k(z,\eta) = \underset{P \in \mathcal{P}(z,\eta)}{\text{Sup}}\ \underset{v \in [0,\delta(P)]}{\text{Sup Ess}}\ \langle z-P(v), a(v) \rangle_F \tag{2.14}$$

We can now give :

Definition 2.4. (quasiconvex and strictly quasiconvex sets).

- A set $(D,\mathcal{P})$ is quasiconvex iff :

  i) $\mathcal{P}$ is a collection of paths for D

ii) There exists a neighborhood $\mathcal{V}$ of D in F and a ℓ.s.c. function $\varepsilon:\mathcal{V}\to]0,+\infty]$ such that :

$$\begin{matrix} z\in\mathcal{V} \\ 0<\eta<\varepsilon(z) \end{matrix} \quad\Rightarrow\quad k(z,\eta)<1 \tag{2.15}$$

- The set $(D,\mathcal{P})$ is strictly quasiconvex if moreover :

$$\begin{matrix} z\in\mathcal{V} \\ P\in\mathcal{P} \\ d(z,P)<d(z,D)+\varepsilon(z) \end{matrix} \quad\Rightarrow\quad \begin{matrix} \text{the "distance to z" function is} \\ \text{strictly quasiconvex} \\ \text{along the path } P \end{matrix} \tag{2.16}$$

The above definitions are quite technical, and we refer to reference [5] for a geometrical interpretation of $k(z,\eta)$. But the sufficient conditions to come in paragraph 3 will have a more natural geometrical interpretation.

Proposition 2.1.

Let $(D,\mathcal{P})$ be strictly quasiconvex. Then there exists a largest open neighborhood $\mathcal{V}$ of D, and a largest ℓ.s.c. function $\varepsilon:\mathcal{V}\to]0,+\infty]$ satisfying (2.15), (2.16).

The motivation for the introduction of strictly quasiconvex sets is the fact that on $\mathcal{V}$, the projection onto D behaves as if D were convex :

Theorem 2.1. (projection on strictly quasiconvex sets)

Suppose that :

$(D,\mathcal{P})$ is strictly quasiconvex (2.17)

Then :

i) uniqueness : for any $z\in\mathcal{V}$, there exists at most one projection $\hat{X}$ of z on D.

ii) local minima : if $z\in\mathcal{V}$ admits a (necessarily unique) projection $\hat{X}$ on D, the "distance to z" function has no parasitic local minima on D distinct from $\hat{X}$.

iii) continuity : if $z_0,z_1\in\mathcal{V}$ admit projections $\hat{X}_0,\hat{X}_1$ on D, and are close enough so that there exists $d\geq 0$ satisfying :

$$\|z_0-z_1\|+\underset{j=0,1}{\mathrm{Max}}\, d(z_j,D)\leq d<\underset{j=0,1}{\mathrm{Min}}\left\{d(z_j,D)+\varepsilon(z_j)\right\} \tag{2.18}$$

then, for any path P going from $\hat{X}_0$ to $\hat{X}_1$ one has :

$$\|\hat{X}_0-\hat{X}_1\|\leq\delta(P)\leq(1-k)^{-1}\|z_0-z_1\|_F. \tag{2.19}$$

where $k<1$ is defined by :

$$\begin{matrix} k=(k(z_0,\eta_0)+k(z_1,\eta_1))/2 \\ 0<\eta_j=d-d(z_j,D)<\varepsilon(z_j) \qquad j=0,1 \end{matrix} \tag{2.20}$$

iv) existence : if we suppose moreover that

D is closed in F (2.21)

then any $z\in\mathcal{V}$ has a (unique) projection $\hat{X}$ on D, and any minimizing sequence X_n satisfies $\|X_n-\hat{X}\|_F\to 0$ and $\delta(P_n)\to 0$, where P_n is any path of $\mathcal{P}$ going from X_n to $\hat{X}$.

In the very simple case where D is an arc of circle of radius R and length L, one checks easily that D is strictly quasiconvex if and only if $L/R<\pi$.

3 - Size × curvature conditions

We present in this paragraph two sufficient conditions for the strict quasiconvexity of a set D, which yield cylindrical neighborhoods $\mathcal{V}$. We call them "size × curvature conditions" because they express the fact that in some sense the size of the set D is not too large with respect to its curvature, i.e. the size × curvature product of the set is not too large. The first condition is based on the notion of global radius of curvature of a path, the second on the deflection of a path.

We begin the definition of the global radius of curvature, which was first introduced in [5].

Definition 3.1

Let a path P be given. Then, for any $v,v' \in [0,\delta(P)]$, $v \neq v'$, we define the affine normal half space $N(v,v')$ to P at v seen from v' by :

$$N(v,v') = \{ z \in F \mid < z-P(v), \lambda v(v) >_F \leq 0 \quad \forall \lambda \in \mathbb{R}, \quad v+\lambda \in [\mathrm{Min}(v,v'),\mathrm{Max}(v,v')] \} \qquad (3.1)$$

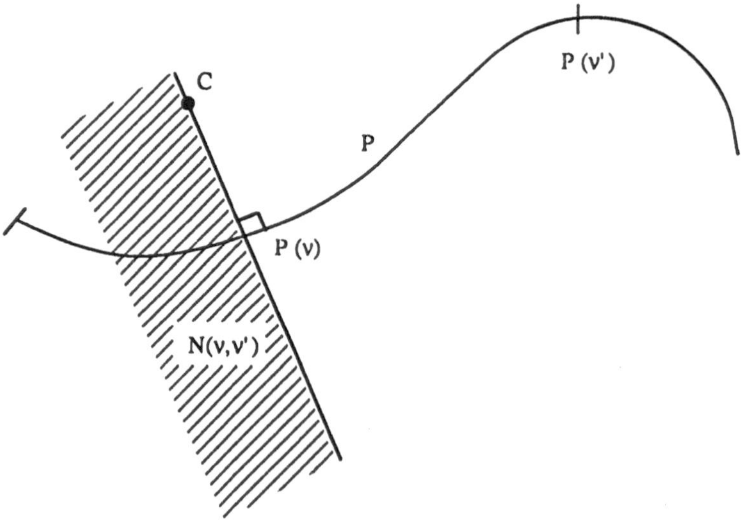

Figure 3.1. The normal half space at v to P seen from v', and the center of curvature C of P at v.

Of course, the center of curvature C of P at P(v), which is defined (cf.(2.8)) by :

$$C = P(v)+a(v) / \|a(v)\|_F^2 \qquad (3.2)$$

belongs to $N(v,v')$ as $< v(v),a(v) > = 0$.

Definition 3.2 (Global radius of curvature)

Let a path P be given. Then, for any $v,v' \in [0,\delta(P)]$, $v \neq v'$, we define the global radius of curvature of P at v seen from v' by :

$$\rho_G(v,v') = d(P(v), N(v,v') \cap N(v',v)) \in [0,+\infty] \qquad (3.3)$$

with the natural convention that $\rho_G(v,v') = +\infty$ if $N(v,v') \cap N(v',v) = \emptyset$.

The global radius of curvature can be easily calculated :

Proposition 3.1.

Let $P \in \mathcal{P}$ and $v,v' \in [0,\delta(P)]$, $v \neq v'$, be given, and define :

$$N = \mathrm{Sgn}(v'-v) < P(v')-P(v), v(v') >_F \qquad (3.4)$$

$$D = (1 - < v(v), v(v') >^2)^{1/2} \qquad (3.5)$$

Then the global radius of curvature $\rho_G(vv')$ is given by :

$$\rho_G(v,v') = \begin{cases} 0 & \text{if } N \leq 0 \\ N & \text{if } N > 0 \text{ and } < v(v),v(v') > \geq 0 \\ N/D & \text{if } N > 0 \text{ and } < v(v)),v(v') >> 0 \end{cases} \qquad (3.6)$$

The global radius of curvature is related, when $v' \to v$, to the usual radius of curvature $\rho(v)$:

Proposition 3.2

Let $P \in \mathcal{P}$ be given. Then, for almost all $v \in [0,\delta(P)]$ one has :

$$\rho_G(v,v') \to \rho(v) \qquad \text{when } v' \to v \qquad (3.7)$$

$$\rho_G(v',v) \to \rho(v) \qquad \text{when } v' \to v \qquad (3.8)$$

We consider now the worst possible case along a path :

Definition 3.3

Let $P \in \mathcal{P}$ be given. We define :

$$R_G(P) = \inf_{v,v' \in [0,\delta(P)]} \rho_G(v,v'). \qquad (3.9)$$

Of course we see from proposition 3.2 that

$$R_G(P) \geq R(P) \qquad \forall P \in \mathcal{P} \qquad (3.11)$$

The motivation for the definition of the global radius of curvature comes from the following result :

Proposition 3.3

Let $P \in \mathcal{P}$ and $z \in F$ be given. Then :

$$0 \leq d(z,P) < R_G(P) \qquad (3.13)$$

implies that the $v \to \|P(v)-z\|_F$ function is strictly quasiconvex.

Coming back to the definition of strictly quasiconvex sets, we see that an immediate consequence of proposition 3.3 is the

Theorem 3.1 (R_G-size×curvature condition)

Let $(D,\mathcal{P})$ be given.

If there exists $R_G > 0$ such that

$R_G(P) \geq R_G > 0 \qquad \forall P \in \mathcal{P}$ (3.14)

Then :

$(D,\mathcal{P})$ is strictly quasiconvex, with a cylindrical neighborhood $\mathcal{V}$ given by :

$\mathcal{V} = \{z \in F \mid d(z,D) < R_G\}$ (3.15)

and an $\varepsilon(z)$ function defined, for any $z \in \mathcal{V}$, by :

$\varepsilon(z) = R_G - d(z,D) > 0$ (3.16)

This sufficient condition is constructive, as it gives simple explicit formul for $\mathcal{V}$ and $\varepsilon(z)$, an can possibly be used as it is when a numerical estimate of R_G is calculated on the the computer using formulas of proposition 3.1. But this approach will often be unpracticable because of the huge amount of computation required, and an analytical determination of R_G using these formula does not seem very easy. Hence we develop now another (less precise) sufficient condition, which will be better suited for analytical calculations, by searching for a lower bound R_G to $R_G(P)$ in term of the deflection of the path P :

Definition 3.4 (deflection along a path)

Let $P \in \mathcal{P}$ be given. For $v,v' \in [0,\delta(P)]$ the deflection of P between v and v' is :

$\theta(v,v') = \mathrm{Arg}\cos < v(v),v(v') > \in [0,\pi],$ (3.17)

and the largest deflection along P is :

$\theta(P) = \underset{v,v' \in [0,\delta(P)]}{\mathrm{Max}} \theta(v,v').$ (3.18)

The deflection can be estimated from the radius of curvature $\rho(v)$ and the arc-length v using the following result :

Theorem 3.2

Let $P \in \mathcal{P}$ be given. Then, for any $v' \in [0,\delta(P)]$, $\partial\theta/\partial v(.,v') \in L^1([0,\delta(P)]$, and :

$$\left| \frac{\partial\theta}{\partial v}(v,v') \right| \leq 1/\rho(v) \qquad \text{for a.e.} \quad v \in [0,\delta(P)] \tag{3.19}$$

so that :

$$\Theta(P) \leq \int_0^{\delta(P)} \frac{dv}{\rho(v)}. \tag{3.20}$$

Hence :

$\Theta(P) \leq \delta(P)/R(P)$ (3.21)

where the equality holds if and only if P is an arc of circle.

Notice that the integrand in the right hand side of (3.20) is the product of a size (arc length dv) by a curvature ($1/\rho(v)$), and so for (3.21). Hence any limitation on the deflection of a path implemented using (3.20) or (3.21) is an actual size × curvature condition ! But paths with small enough deflection should conceivably behave like segments (which are zero deflection paths), or in other terms should exhibit a strictly positive global radius of curvature. This is made precise by :

<u>Theorem 3.3</u>

Let $P \in \mathcal{P}$ be given. Then the following lower bounds for $R_G(P)$ hold, depending on the deflection $\Theta(P)$ of the path :

i) <u>low deflection paths</u>

$$R_G(P) = R(P) \tag{3.22}$$

as soon as :

$$0 \leq \Theta(P) \leq \pi/2 \tag{3.23}$$

ii) <u>large deflection paths</u>

$$R_G(P) \geq R(P) \sin\Theta(P) + \{\delta(P)-R(P)\Theta(P)\} \cos\Theta(P) \tag{3.24}$$

as soon as :

$$\pi/2 \leq \Theta(P) \leq \pi \tag{3.25}$$

We use now the above result to obtain uniform lower bound on R_G for all paths P of $\mathcal{P}$.

Suppose we have found R,Θ,Δ such that :

$$\left.\begin{array}{l} R(P) \geq R > 0 \\ \Theta(P) \leq \Theta < +\infty \\ \Delta(P) \leq \Delta \leq +\infty \end{array}\right\} \quad \text{for all } P \in \mathcal{P}. \tag{3.26}$$

Of course the upper bound Θ on the deflection is supposed to be at least as precise as the always valid upper bound resulting from (3.21), i.e. :

$$\Theta \leq \Delta/R. \tag{3.27}$$

If $\Theta \leq \pi/2$ we find immediately from (3.22), (3.23) that $R_G(P) \geq R \;\; \forall P \in \mathcal{P}$, so that one can take :

$$R_G = R \tag{3.28}$$

in (3.14).

If $\pi/2 \leq \Theta \leq \pi$, we see from (3.26), (3.24) that :

$$R_G(P) \geq R \sin\Theta(P) + \{\Delta-R\Theta(P)\} \cos\Theta(P),$$

whose right-hand side is a decreasing function of $\Theta(P)$ on the $[0,\Delta/R]$ interval, so that :

$$R_G(P) \geq R \sin\Theta + \{\Delta-R\Theta\} \cos\Theta \qquad \forall P \in \mathcal{P}.$$

Hence one can take :

$$R_G = R \sin\Theta + \{\Delta-R\Theta\} \cos\Theta \tag{3.29}$$

In order to reformulate (3.29) in a more convenient way, we notice that (3.27) can be rewritten in an equivalent way :

$$\Theta = \tau\Delta / R \qquad 0 \leq \tau \leq 1, \tag{3.30}$$

where τ can be interpreted as follows :

- $\tau=0$ means that all paths of $\mathcal{P}$ have a zero deflection, i.e. are segments, which implies that D is convex !
- $\tau=1$ means that the deflection of all paths have been estimated very roughly using the second inequality of (3.20), i.e. as if all paths of P were arcs of circles.

Hence we shall <u>call τ the shape coefficient of the estimates R,Θ,Δ</u> : τ close to 1 means that either the paths of $\mathcal{P}$ are close to arcs of circles, or the estimates R,Θ,Δ are very loose ; τ close to 0 means that the paths of P are "close" to segments, i.e. that D is "close" to a convex (think of a piece of granitated wall paper).

We can now rewrite (3.28), (3.29) using τ as :

$$R_G = \begin{cases} R & 0 \le \Theta \le \pi/2 \\ R\left\{\sin\Theta + \left(\tau^{-1} - 1\right)\Theta\cos\Theta\right\} & \pi/2 \le \Theta \le \pi \end{cases} \tag{3.31}$$

We use now (3.31) to find conditions that will enure $R_G > 0$. We have illustrated for that purpose on figure 3.2 R_G as a function of Θ for various values of τ. Clearly, if we define :

Θ_M = unique solution, in $]\pi/2,\pi]$, of the equation : $\tan\theta + \{\tau^{-1} - 1\}\theta = 0$ (3.32)

then :

$R_G > 0$ as soon as $\Theta < \Theta_M$. (3.33)

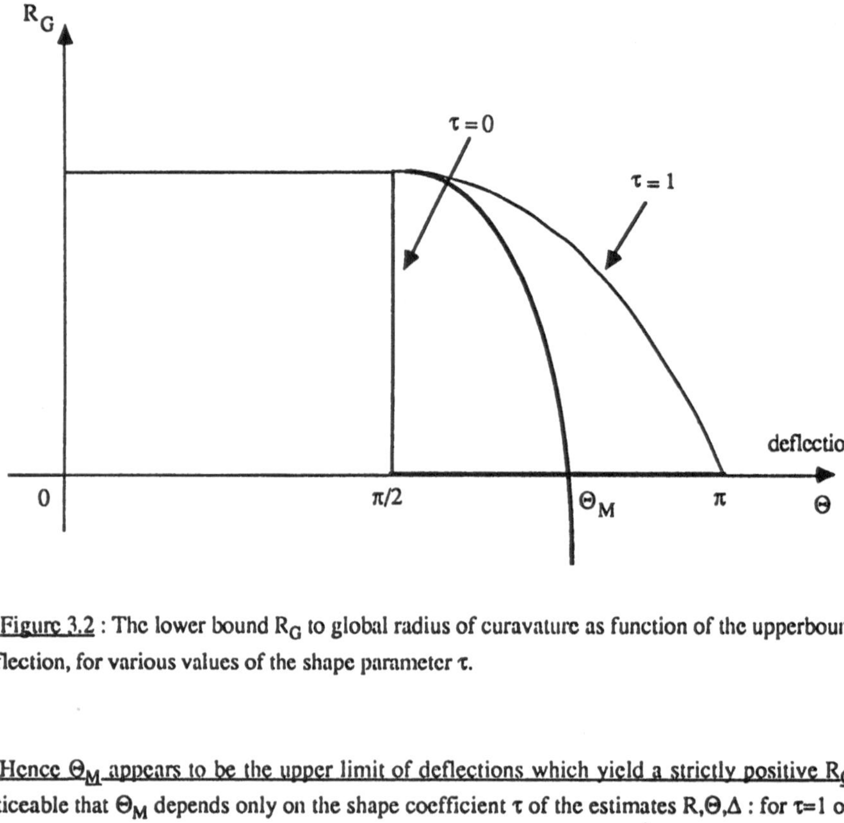

Figure 3.2 : The lower bound R_G to global radius of curavature as function of the upperbound Θ to deflection, for various values of the shape parameter τ.

Hence Θ_M appears to be the upper limit of deflections which yield a strictly positive R_G. It is noticeable that Θ_M depends only on the shape coefficient τ of the estimates R,Θ,Δ : for $\tau=1$ one has $\Theta_M=\pi$, then Θ_M decreases quickly towards $\pi/2$ when τ decreases towards zero (see figure 3.3).

Notice also that the function $\tau \rightarrow \Theta_M$ is more easily expressed through its reciprocal :

$\tau = (1 - \tan\Theta_M / \Theta_M)^{-1}$. (3.34)

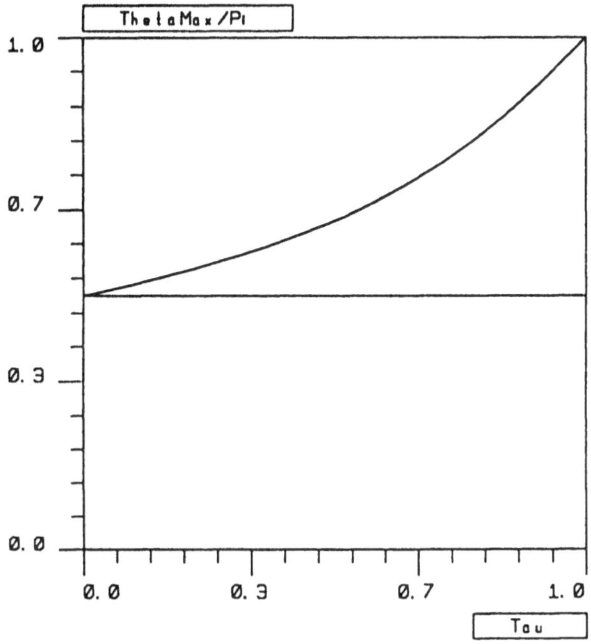

Figure 3.3 : The maximal deviation Θ_M as function τ.

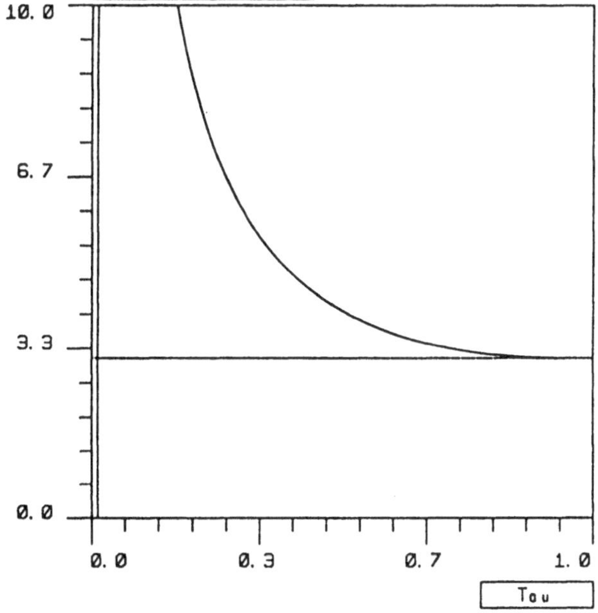

Figure 3.4 : The maximal value $(\Delta/R)_M$ of Δ/R as function of the shape parameter τ.

An equivalent way of writing (3.33) is obtained by imposing an ad-hoc upper bound $(\Delta/R)_M$ on Δ/R using the relation (3.30) between Θ and Δ/R. If we define :

$(\Delta/R)_M = \Theta_M/\tau = \Theta_M - \tan\Theta_M$ (3.35)

then (3.33) is equivalent to :

$R_G > 0$ as soon as $\Delta/R \le (\Delta/R)_M$, (3.36)

which is clearly a size × curvature condition. As shown on figure 3.4 $(\Delta/R)_M$ increases from π to infinity when τ decreases from 1 to 0.

Suppose that, without changing the estimates R and Δ, we refine the estimate Θ on the deflection, i.e. we diminish Θ, and hence τ : as we have seen, we expect a less constraining condition for $R_G > 0$, which is clearly the case for (3.36) (as $(\Delta/R)_M$ increases), but seems apparently not true for (3.33) (as Θ_M decreases !). There is in fact no contradiction with the fact that (3.33) and (3.36) are equivalent : when you refine your estimation Θ of deflection, its not surprising that even a smaller upper bound Θ_M may correspond to a less constraining situation !

We summarize these results in the

Theorem 3.4

Let $(D,\mathcal{P})$ be given, let R,Δ,Θ and τ be real numbers satisfying (3.26), (3.30), and define Θ_M, $(\Delta/R)_M$ by (3.32), (3.35). If the θ-size × curvature condition :

$\Theta < \Theta_M$ (3.37)

or equivalently :

$\Delta/R < (\Delta/R)_M$, (3.38)

holds, then R_G defined by (3.31) satisfies :

$R_G > 0$ (3.39)

and $(D,\mathcal{P})$ is strictly quasiconvex with $\mathcal{V}$ and ε given by (3.15), (3.16).

One checks easily that theorems 3.1 and 3.4 recognize exactly all strictly quasiconvex arcs of circles. More precisely, the R_G-size × curvature condition of theorem 3.1 recognizes exactly all quasiconvex curves, and the Θ-size × curvature condition of theorem 3.4 recognizes exactly all strictly quasiconvex curves made of one arc of circle and one segment.

4 - Application to Q-well posedness of non-linear least squares

We use in this paragraph the Θ-size × curvature condition of theorem 3.4 to study the Q-well-posedness of the non-linear least squares problem (1.1), (1.2). We equip for that the set $D=\varphi(C)$ with the family of paths $\mathcal{P}$ made of the images by φ of the segments of the convex set C :

$$\begin{cases} \text{to any } x_0, x_1 \in C \quad \text{we associate the path} \\ P : t \in [0,1] \to P(t) = \varphi\big((1-t)x_0 + tx_1\big) \end{cases} \tag{4.1}$$

and we suppose the φ is regular enough so that :

$P \in W^{2,\infty}([0,1]) \qquad \forall P \in \mathcal{P}$ (4.2)

Notice that (4.1), (4.2) alone does not imply that $\mathcal{P}$ is a collection of paths in the sense of definition 2.3, as t is not the arc-length ν ! But we may still associate to any path P :

$$V(t) = P'(t) \in F \qquad \text{(velocity along the path)}$$
$$A(t) = P''(t) \in F \qquad \text{(accelleration along the path)} \tag{4.3}$$

which of course are different from v(ν) and a(ν) defined in proposition 2.2 ($\|V(t)\|_F$ is not necessarily equal to 1, and $< V(t),A(t) >$ not necessarily zero !). The relation between V(t),A(t) and v(ν),a(ν) is given, at any point where $V(t) \neq 0$, by :

$$v(\nu) = \frac{V(t)}{\|V(t)\|_F}$$
$$a(\nu) = \frac{A(t)}{\|V(t)\|_F^2} - v(\nu) < v(\nu), \frac{A(t)}{\|V(t)\|_F^2} > \tag{4.4}$$

which implies that :

$$\|v(\nu)\|_F = 1 \qquad \text{(not surprising !)}$$
$$\|a(\nu)\|_F = \frac{\|A(t)\|_F}{\|V(t)\|_F^2} \left\{ 1 - < \frac{V(t)}{\|V(t)\|_F}, \frac{A(t)}{\|A(t)\|_F} >^2 \right\}^{1/2} \leq \frac{\|A(t)\|_F}{\|V(t)\|_F^2} . \tag{4.5}$$

We state now the main hypothesis on C and φ : we suppose that one has been able to find a distance d(x,y) on C such that :

(C,d) is complete (4.6)

there exists $0 < \alpha_m \leq \alpha_M$ such that :

$\forall x_0, x_1 \in C,$ for a.e. $t \in]0,1[$: (4.7)

$\alpha_m \, d(x_o,x_1) \leq \|V(t)\|_F \leq \alpha_M \, d(x_o,x_1)$,

and :

there exists $\Theta \geq 0$ and $R > 0$ such that :

$\forall x_0, x_1 \in C,$ for a.e. $t \in]0,1[$: (4.8)

$\|A(t)\|_F / \|V(t)\|_F \leq \Theta, \|A(t)\|_F / \|V(t)\|^2_F \leq 1/R.$

Hypothesis (4.7) on V(t) implies that $\|V(t)\|_F \neq 0$, $\forall t \in]0,1[$ for any path P of $\mathcal{P}$, so that we see from (4.4) that any path P can be reparametrized as a $W^{2,\infty}([0,\delta(P)])$ function of the arc length ν ; hence $\mathcal{P}$ is now a collection of paths in the sense of definition 2.3.

By analogy with the case where the mapping φ is twice differentiable, we shall say that α_m and α_M are lower and upper bounds to the singular values of φ'(x), x∈ C, and that α_M/α_m is an upper bound to the condition number of the linearized problems.

It is then easy to see using (2.8), (3.20) and (4.5) that Θ and R defined in (4.8) satisfy :

$$\Theta(P) \leq \Theta, \; R(P) \geq R > 0 \quad \forall P \in \mathcal{P}. \tag{4.9}$$

We can also define :

$$\Delta = \alpha_M \operatorname{diam} C, \tag{4.10}$$

which obviously satisfies :

$$\delta(P) \leq \Delta \qquad \forall P \in \mathcal{P}. \tag{4.11}$$

In sight of (4.9), (4.11) and (3.21), we can suppose, without loss of generality, that (3.30) holds, i.e. :

$$\Theta = \tau \, \Delta / R \qquad 0 \leq \tau \leq 1. \tag{4.12}$$

Then, if $\Theta < \Theta_M$ defined in (3.32), all hypothesis of theorem 3.4 are satisfied, so that $(\varphi(C),\mathcal{P})$ is strictly quasiconvex. On the other hand, hypothesis (4.7), part iii) of theorem 2.1 and the completeness of (C,d) imply the closedness of $\varphi(C)$. Hence theorem 2.1 rewrites as :

Theorem 4.1

Let C,d,F and φ be given satisfying (1.2), (4.2), (4.6), (4.7), (4.8), and $\Delta,\tau,R_G,\Theta_M,(\Delta/R)_M$ be defined by (4.10), (4.12), (3.31), (3.32), (3.35).

If the deflection size × curvature condition :

$$\Theta < \Theta_M \tag{4.13}$$

or equivalently :

$$\Delta/R < (\Delta/R)_M = \Theta_M - \tan \Theta_M \tag{4.14}$$

is satisfied, the non-linear least squares problem (1.1) is Q-well posed on the neighborhood :

$$\mathcal{V} = \{z \in F \mid d(z,\varphi(C)) < R_G\} \tag{4.15}$$

for the d(x,y) distance on C, and the following stability estimate holds :

$$\alpha_m d(\hat{x}_1,\hat{x}_0) \leq \int_0^1 \|V(t)\|_F dt \leq (1-d/R)^{-1} \|z_1-z_0\|_F \tag{4.16}$$

(where V(t) is the velocity along the path image by φ of $[\hat{x}_0,\hat{x}_1]$)

as soon as z_0,z_1 satisfy :

$$\|z_1-z_0\|_F + \operatorname*{Max}_{j=0,1} d(z_j,\varphi(C)) \leq d < R_G. \tag{4.17}$$

A special case is given in :

Corollary 4.1.

Theorem 4.1 holds with hypotheis (4.8) replaced by :

there exists $\beta \geq 0$ such that,
$\forall x_0,x_1 \in C$, for a.e. $t \in]0,1[$: (4.18)
$\|A(t)\|_F \leq \beta\, d(x_0,x_1)^2$

provided Θ and R and τ are defined by :

$$\Theta = (\beta/\alpha_m)\text{diam } C \tag{4.19}$$

$$R = \alpha_m{}^2/\beta \tag{4.20}$$

$$\tau = \alpha_m/\alpha_M, \tag{4.21}$$

and the size × curvature condition (4.13) or (4.14) rewrites :

$$(\beta/\alpha_m)\text{diam } C < \Theta_M \tag{4.22}$$

Notice that, when corollary 4.1 applies, τ is the reciprocal of the (upper bound to) condition number α_M/α_m of linearized problems. Hence poorly conditioned linearized problems (α_M/α_m large) will yield a small τ and hence (see figure 3.3) a Θ_M very close (from above) to $\pi/2$. Hence, the practical augmentation of the diam C caused by considering the case $\pi/2 \leq \Theta \leq \pi$ will be negligible as soon as the condition number α_M/α_m is larger than a few units.

Notice also that condition (4.22) can always be satisfied by reducing the size of C : theorem (4.1) and corollary 4.1 yield a "local" non-linear inversion result, but with explicit expressions for the localization.

References :

[1] Campbell S. and Meyer C., "Generalized Inverses of Linear Transformations", London : Pitman, 1979.

[2] Symes W.W., "The Plane Wave Detection Problem", Technical Report 90-1, Department of Mathematical Sciences, Rice University, Houston TX 77251, 1990.

[3] Chavent G., Kunisch K., "L^2-Stability of the Inverse Problem in a 1-D Elliptic Equation from an H^1-Observation", (in preparation).

[4] Chavent G., "Quasiconvex Sets and Size × Curvature Condition, Application to Non-Linear Inversion", to appear in Journal of Applied Mathematics and Optimization, 1991.

[5] Chavent G., "New Size × Curvature Conditions for Strict Quasiconvexity of Sets", to appear in Siam Journal on Optimization and Control.

[6] Chavent G. "A New Sufficient Condition for the Well-Posedness of Non-Linear Least Square Problems Arising in Identification and Control", in "Lectures Notes in Control and Information Sciences", Vol.144, Analysis and Optimization of Systems, A. Bensoussan and J-L. Lions Ed., Springer, pp.452-463, 1990.

DOMAIN VARIATION FOR DRAG IN STOKES FLOW

Jacques SIMON

C.N.R.S. and Université Blaise Pascal (Clermont-Ferrand)

Département de Mathématiques, Université Blaise Pascal,
63177 Aubiere Cedex, France

Introduction.

The drag variations of an uniformly moving body in a Stokes flow, with respect to body variations, or more generally with respect to fluid domain variations, are investigated.

The variations of the initial fluid domain Ω being represented by a vector field u, we are interested in an expansion of $J(\Omega + u)$ with respect to u. We prove the existence of an indefinite expansion, and we calculate the first and second order terms.

First variation value was calculated in [P], and for Navier-Stokes flow first variation was calculated in [MS1] and [S1] and second one in [S2], using general rules for domain variations. However, full proofs were not given.

Proofs for similar problems were given given in [MS2] and [S1], based on usual implicit functions differentiation theorem. For fluid equations, assumptions of this theorem are not satisfied (see comment after (40) in section 6). Here, we give an extension of this implicit function theorem, which allows us to carry on proofs.

Outlines

1. Drag of a body in a Stokes flow
2. Drag variations
3. Velocity and pressure variations
4. Rough calculus of the derivatives
5. Differentiabilty of an implicit equation solution
6. Proof of domain variation results
7. Application to drag minimization for given volume
8. Second variations
9. Rough calculus of second variations
10. Infinite order differentiabilty, and second variations proofs.

1. DRAG OF A BODY IN A STOKES FLOW.

We are interested in a motionless body $\mathcal{B}$ in a viscous incompressible fluid moving at a uniform velocity h. The flow is considered in a bounded region Λ containing $\mathcal{B}$. Thus the fluid occupies the annulus shaped domain $\Omega = \Lambda \setminus \overline{\mathcal{B}}$, whose boundary has an outside part $\partial\Lambda$ and an inside part $\partial\mathcal{B}$.

The velocity $y = (y_1, y_2, y_3)$ and the pressure p satisfy

$$-\nu\Delta y + \nabla p = 0 \quad \text{and} \quad \nabla\cdot y = 0, \quad \text{in } \Omega, \tag{1}$$

$$y = 0 \text{ on } \partial\mathcal{B}, \quad y = h \text{ on } \partial\Lambda, \quad \int_\Omega p\, dx = 0, \tag{2}$$

where the viscosity ν is a positive real number, and $h \in \mathbb{R}^3$.

The energy dissipated by the fluid is

$$J = \tfrac{1}{2}\int_\Omega |Ly|^2\, dx \tag{3}$$

where $(Ly)_{ij} = \partial_i y_j + \partial_j y_i$.

This energy is proportional to the drag, which is $J/|h|$. The velocity and then the drag are uniquely defined by the following result.

Theorem 1. *We assume that*

$$\Omega = \Lambda \setminus \overline{\mathcal{B}}, \quad \Lambda \text{ is open and bounded}, \quad \overline{\mathcal{B}} \subset \Lambda, \quad \partial\Lambda \text{ and } \partial\mathcal{B} \text{ are } Lip^1. \tag{4}$$

Then, there exist a unique pair $y \in H^1(\Omega)^3$, $p \in L^2(\Omega)$, satisfying (1) and (2). Therefore (3) defines a unique $J \in \mathbb{R}$.

Let us recall that $H^k(\Omega) = \{v \mid D^\alpha v \in L^2(\Omega), 0 \le |\alpha| \le k\}$. We say that $\partial\Omega$ is Lip^k if it is locally the graph of a $Lip^k(\mathbb{R}^2)$ function, where

$$Lip^k(\mathbb{R}^2) = \{v \mid D^\alpha v \text{ is Lipschitz continuous on } \mathbb{R}^2, 0 \le |\alpha| \le k-1\}.$$

We will use the following properties of a generalized Stokes problem, which may be found in [L] and [GR].

Theorem 2. (i) *Assume that Ω satisfies (4), and let $f \in H^{-1}(\Omega)^3$, $g \in L^2(\Omega)$, $k \in H^1(\Omega)^3$, $r \in \mathbb{R}$, satisfy*

$$\int_\Omega g\, dx = \int_{\partial\Omega} k \cdot n\, ds. \tag{5}$$

Then, there exists a unique pair $y \in H^1(\Omega)^3, p \in L^2(\Omega)$ such that

$$-\nu\Delta y + \nabla p = f, \quad \nabla \cdot y = g, \quad y = k \text{ on } \partial\Omega, \quad \int_\Omega p\, dx = 0. \tag{6}$$

Conversely, if (6) has a solution, then (5) is satisfied.

(ii) *Assume in addition that $\partial\Omega$ is Lip^2 and $f \in L^2(\Omega)^3$, $g \in H^1(\Omega)$, $k \in H^2(\Omega)^3$.*

Then $y \in H^2(\Omega)^3$, $p \in H^1(\Omega)$, and there exists a real c depending only on Ω such that

$$\|y\|_{H^2} + \|p\|_{H^1} \le c(\|f\|_{L^2} + \|g\|_{H^1} + \|k\|_{H^2} + r).$$

Remark. Theorem 1 is a particular case of theorem 2 since (5) yields $\int_{\partial\Lambda} h \cdot n\, ds = 0$. This is satisfied since $\int_{\partial\Lambda} n\, ds = \int_{\Lambda} \nabla 1\, dx = 0$.

2. DRAG VARIATIONS.

We are interested in the variations of $J(\Omega)$ with respect to small variations of Ω. Let u be a vector field defined in all of $\mathbb{R}^3$ representing variations of Ω. A varying domain is defined as

$$\Omega + u = \{x + u(x) \mid x \in \Omega\}. \tag{7}$$

We assume that $u \in Lip^1(\mathbb{R}^3)^3$, $\|u\|_{Lip^1} < 1$. Then, u is a contraction thus $\Omega + u$ satisfies (4). Therefore, by theorem 1, there exists a unique pair $y_{\Omega+u} \in H^1(\Omega + u)^3$, $p_{\Omega+u} \in L^2(\Omega + u)$ such that

$$-\nu \Delta y_{\Omega+u} + \nabla p_{\Omega+u} = 0 \quad \text{and} \quad \nabla \cdot y_{\Omega+u} = 0 \quad \text{in } \Omega + u, \tag{8}$$

$$y_{\Omega+u} = 0 \text{ on } \partial B + u, \quad y_{\Omega+u} = h \text{ on } \partial\Lambda + u, \quad \int_{\Omega+u} p_{\Omega+u}\, dx = 0. \tag{9}$$

Therefore there exists a unique energy

$$J(\Omega + u) = \tfrac{1}{2} \int_{\Omega+u} |L y_{\Omega+u}|^2\, dx. \tag{10}$$

We are looking for an expansion of $J(\Omega + u)$ with respect to small u. We will get an expansion for u in Lip^2, that is,

$$J(\Omega + u) = J(\Omega) + J'(\Omega; u) + o(\|u\|_{Lip^2}), \tag{11}$$

where $o(t)$ denotes any real number such that $o(t)/t \to 0$ as $t \to 0$.

More precisely, the following result will be proved in section 6.

Theorem 3. *We assume that Ω satisfies (4) and $\partial\Omega$ is Lip^2.*
Then, the expansion (11) holds for all $u \in Lip^2$ such that $\|u\|_{Lip^1} < 1$, with

$$J'(\Omega; u) = -\tfrac{1}{2} \int_{\partial\Omega} u_n \left|\frac{\partial y_\Omega}{\partial n}\right|^2 ds. \tag{12}$$

We denote by $n = n_\Omega$ a unitary vector field on $\partial\Omega$, directed outside Ω. And $u_n = u \cdot n$.

Remark. The expansion yields,

$$\int_{\Omega+u} |L y_{\Omega+u}|^2\, dx = \int_{\Omega} |L y_\Omega|^2\, dx - \int_{\partial\Omega} u_n \left|\frac{\partial y_\Omega}{\partial n}\right|^2 ds. \tag{13}$$

Remark. In the case of a body in a fixed experiment domain, the outside boundary is not supposed to vary. This is modelized by choosing $u = 0$ on Λ.

The varying body $\Omega + u$ depends only on the restriction $u|_{\partial\Omega}$. However, we use a variation $u(x)$ which is defined for all $x \in \mathbb{R}^3$ since it provides a map, $I + u$, from Ω onto $\Omega + u$, which will be used for proofs.

Remark. The integral in (12) is defined, since $y \in H^2(\Omega)^3$ thus $\frac{\partial y}{\partial n} \in L^2(\partial\Omega)^3$. In the case where $\partial\Omega$ is only Lip^1, it is not defined.

Remark. As for any optimization problem, the first variation yields a necessary optimality condition. An example will be given in section 7.

Moreover, based on J', we can use a gradient method to construct a locally optimal domain Ω_0.

Remark. In particular, the Frechet differentiability, that is (11), yields variations depending on a small real parameter t. For a fixed u and for all $0 \le t \le t_u$, $J(\Omega + tu)$ is defined, and

$$J(\Omega + tu) = J(\Omega) + tJ'(\Omega; u) + o(t). \tag{15}$$

This is Gateaux differentiability. However, as it will be seen in section 7, it is not enough for applications. This is the reason why we use Frechet derivative.

3. VELOCITY AND PRESSURE VARIATIONS.

3.1. Local variations y', p'. To get the variation J', we will use the variation y' of the velocity. That is an expansion of the following type,

$$y_{\Omega+u} = y_\Omega + y'(\Omega; u) + o(\|u\|_{Lip^2}).$$

However $y_{\Omega+u}$ lives in the domain $\Omega + u$ which depends on u, but y_Ω and $y'(\Omega; u)$ live on the fixed domain Ω. Therefore this expansion cannot be satisfied in all of Ω. It can be satisfied only in the intersection, for all small enough u, of the domains $\Omega + u$.

In the following result, such an expansion is given locally, that is in any strict subdomain ω of Ω.

Theorem 4. *We assume that Ω satisfies (4) and $\partial\Omega$ is Lip^2.*

(i) *Given $u \in Lip^1(\mathbb{R}^3)^3$, there exists a unique pair $y' \in H^1(\Omega)^3$, $p' \in L^2(\Omega)$ satisfying*

$$-\nu\Delta y' + \nabla p' = 0 \quad \text{and} \quad \nabla \cdot y' = 0, \quad \text{in } \Omega, \tag{16}$$

$$y' = -u_n \frac{\partial y_\Omega}{\partial n} \quad \text{on } \partial\Omega, \quad \int_\Omega p'\, dx = -\int_{\partial\Omega} u_n p_\Omega\, ds. \tag{17}$$

Moreover, $y' = y'(\Omega; u)$ and $p' = p'(\Omega; u)$ are linear and continuous with respect to u, from $Lip^2(\mathbb{R}^3)^3$ onto $H^1(\Omega)^3$ and $L^2(\Omega)$.

(ii) *For any open set ω such that $\overline{\omega} \subset \Omega$, and for all $u \in Lip^2$ such that $\|u\|_{Lip^1} < 1$,*

$$y_{\Omega+u} = y_\Omega + y'(\Omega; u) + o(\|u\|_{Lip^2}) \quad \text{in } H^1(\omega)^3, \tag{18}$$

$$p_{\Omega+u} = p_\Omega + p'(\Omega; u) + o(\|u\|_{Lip^2}) \quad \text{in } L^2(\omega)^3. \tag{19}$$

(iii) *The variation J' defined in theorem 3 satisfies*

$$J'(\Omega; u) = \int_\Omega Ly_\Omega \cdot Ly'(\Omega; u)\, dx + \tfrac{1}{2} \int_{\partial\Omega} u_n |Ly_\Omega|^2\, ds. \tag{20}$$

(iv) *For all $u \in Lip^1$,*

$$\int_\Omega Ly_\Omega \cdot Ly'(\Omega; u)\, dx = -\int_{\partial\Omega} u_n |Ly_\Omega|^2\, ds = -2\int_{\partial\Omega} u_n |\frac{\partial y_\Omega}{\partial n}|^2\, ds. \tag{21}$$

Proof of parts i and iv of theorem 4. Here, we denote $y = y_\Omega$, $y' = y'(\Omega; u)$.

(i) To get the existence of a solution y', p' of (16) and (17), it suffices to check the condition (5) of theorem 2, which here is $\int_{\partial\Omega} u_n\, n \cdot \frac{\partial y}{\partial n}\, ds = 0$.

On each part of $\partial\Omega$, y is constant. Thus $\nabla y = n\frac{\partial y}{\partial n}$. That is $\partial_i y_j = n_i \frac{\partial y_j}{\partial n}$. Whence the required condition is satisfied, since $n \cdot \frac{\partial y}{\partial n} = \sum_i n_i \cdot \frac{\partial y_i}{\partial n} = \sum_i \partial_i y_i = 0$.

(iv) By definition, $Ly \cdot Ly' = \sum_{ij}(\partial_i y_j + \partial_j y_i)(\partial_i y'_j + \partial'_j y_i) = 2\sum_{ij}(\partial_i y_j + \partial_j y_i)\partial_i y'_j$. Thus, integrating by parts,

$$\int_\Omega Ly \cdot Ly'\, dx = -2\int_\Omega \textstyle\sum_{ij}(\partial_i\partial_i y_j + \partial_i\partial_j y_i) y'_j\, dx + 2\int_{\partial\Omega} \sum_{ij} n_i y'_j (\partial_i y_j + \partial_j y_i)\, ds. \tag{22}$$

By (1), $\sum_i \partial_i y_i = 0$ and $\sum_i \partial_i\partial_i y_j = \frac{1}{\nu}\partial_j p$. Thus the integral on Ω in the right hand side is equal to

$$-\tfrac{2}{\nu}\textstyle\int_\Omega \sum_j \partial_j p\, y'_j\, dx = \tfrac{2}{\nu}\int_\Omega \sum_j p \partial_j y'_j\, dx - \tfrac{2}{\nu}\int_{\partial\Omega} \sum_j p n_j y'_j\, dx = 0,$$

since the integral on Ω is null by $\nabla \cdot y' = 0$, and since the integral on $\partial\Omega$ is null by $\sum_j n_j y'_j = -\sum_j u_n n_j \frac{\partial y_j}{\partial n} = -\sum_j u_n \partial_j y_j = 0$.

The integral on $\partial\Omega$ in (22) is, since $n_i y'_j = -u_n n_i \frac{\partial y_j}{\partial n} = -u_n \partial_i y_j$, equal to

$$-2\int_{\partial\Omega} u_n \textstyle\sum_{ij} \partial_i y_j (\partial_i y_j + \partial_j y_i) = -\int_{\partial\Omega} u_n |Ly|^2\, ds,$$

which proves the first equality in (21).

Using again $\partial_i y_j = n_i \frac{\partial y_j}{\partial n}$ and $\sum_j n_j \frac{\partial y_j}{\partial n} = \sum_j \partial_j y_j = 0$, we get on $\partial\Omega$,

$$|Ly|^2 = \textstyle\sum_{ij}(\partial_i y_j + \partial_j y_i)^2 = \sum_{ij}(n_i \frac{\partial y_j}{\partial n} + n_j \frac{\partial y_i}{\partial n})^2 = 2\sum_j |\frac{\partial y_j}{\partial n}|^2 + |\sum_j n_j \frac{\partial y_j}{\partial n}|^2 = 2|\frac{\partial y}{\partial n}|^2,$$

which proves the second equality in (21).

The parts ii and iii of theorem 4 are proved in section 6.3.

3.2. Total variations $\dot{y}$, $\dot{p}$. To get the boundary condition (17) on y', and to get the value (20) of J', we need a uniform dependence of $y_{\Omega+u}$ on u up to the boundary $\partial\Omega + u$. This cannot be given by the expansion (18) which is necessary local. It is obtained by mapping $y_{\Omega+u}$ on the fixed domain Ω.

Denoting I the identity in $\mathbb{R}^3$, $I + u$ maps Ω onto $\Omega + u$. Thus, the function $y_{\Omega+u} \circ (I + u)$ lives in Ω. The uniform dependence is given by it's expansion in all of Ω. The following result will be proved in section 6.1.

Theorem 5. *We assume that Ω satisfies (4) and $\partial\Omega$ is Lip^2.*

Then, for all $u \in Lip^2(\mathbb{R}^3)^3$, there exists $\dot{y}(\Omega; u) \in H^2(\Omega)^3$ and $\dot{p}(\Omega; u) \in H^1(\Omega)$, which are linear and continuous with respect to u, and which satisfy, for all u such that $\|u\|_{Lip^2} < 1$,

$$y_{\Omega+u} \circ (I + u) = y_\Omega + \dot{y}(\Omega; u) + o(\|u\|_{Lip^2}) \quad \text{in } H^2(\Omega)^3,$$

$$p_{\Omega+u} \circ (I + u) = p_\Omega + \nabla\dot{p}(\Omega; u) + o(\|u\|_{Lip^2}) \quad \text{in } H^1(\Omega).$$

Remark. The existence of a total variation $\dot{y}$ is necessary for our proof of the existence of J'. However, knowing the exact value of $\dot{y}$ is not required.

To calculate the value of J', we will only use the value of the local derivative y'. In addition, a rough calculus of y', p' and J', carried on in section 4, is easy.

This is the reason why we use at the same time the two different objects $\dot{y}$ and y'. In fact, they are related by

$$\dot{y}(\mathcal{B}; u) = y'(\mathcal{B}; u) + u \cdot \nabla y_{\mathcal{B}},$$

thus equations on $\dot{y}$, $\dot{p}$ come of equations on y', p'.

4. ROUGH CALCULUS OF THE DERIVATIVES.

We check here that J' has the value announced in theorem 3, when $y_{\Omega+u}(x)$ and $p_{\Omega+u}(x)$ smoothly depend on x and u. This assumption is not satisfied, and in section 6.1 we will give a proof based on rather different ideas.

We differentiate with respect to a real parameter t, for a fixed u, since $y'(\Omega; u) = \frac{d}{dt} J(\Omega + tu)|_{t=0}$.

4.1. Differentiation of equations. From (8) we get an equation satisfied in the domain $\Omega + tu$ which varies with t. Any point x fixed in Ω, lies in $\Omega + tu$ for small enough t. Thus,

$$-\nu\Delta y_{\Omega+tu}(x) + \nabla p_{\Omega+tu}(x) = 0 \quad \forall x \in \Omega, \forall t \leq t_{x,u}. \tag{23}$$

By differentiation with respect to t, at $t = 0$, we get the first equation in (16) :

$$-\nu\Delta y'(x) + \nabla p'(x) = 0 \quad \forall x \in \Omega.$$

Similarly, $\nabla \cdot y_{\Omega+tu}(x) = 0$ yields the second equation in (16) :

$$\nabla \cdot y'(x) = 0 \quad \forall x \in \Omega.$$

4.2. Differentiation of boundary conditions. From (9) we get a boundary condition satisfied on the boundary $\partial\Omega + tu$ which varies with t. For a given x in $\partial\mathcal{B}$, the moving point $x + tu$ lies in $\partial\mathcal{B} + tu$. Thus,

$$y_{\Omega+tu}(x + tu) = 0 \quad \forall x \in \partial\mathcal{B}, \forall t \leq t_u. \tag{24}$$

By differentiation with respect to t, at $t = 0$, we get

$$y'(x) + u(x) \cdot \nabla y_\Omega(x) = 0 \quad \forall x \in \partial\mathcal{B}.$$

Since y_Ω is null on $\partial\mathcal{B}$, it's gradient at any point is proportional to the normal vector n. Thus $\nabla y_\Omega(x) = n_\Omega(x)\frac{\partial y_\Omega}{\partial n}(x)$. Whence we get the boundary condition on $\partial\mathcal{B}$ in (17) :

$$y'(x) + u(x)\cdot n_\Omega(x)\frac{\partial y_\Omega}{\partial n}(x) = 0 \quad \forall x \in \partial\mathcal{B}.$$

Similarly, the second boundary condition in (9) yields $y_{\Omega+tu}(x) = h, \forall x \in \partial\Lambda$, whence we get, on $\partial\Lambda$ and therefore on all of $\partial\Omega$, the same condition.

4.3. Differentiation of the drag. From (10), by the change of variable $I + tu$, we get, $\forall t \leq t_u$,

$$J(\Omega + tu) = \tfrac{1}{2}\int_\Omega |Ly_{\Omega+tu}|^2 \circ (I + tu)\,|\det[\partial_i(I + tu)_j]|\,dx. \tag{25}$$

Differentiating with respect to t at $t = 0$, and using $|\det[\ldots]| = \det[\ldots]$ for small t and $(\det[\ldots])'(0) = \nabla\cdot u$, we get

$$\begin{aligned} J'(\Omega; u) &= \tfrac{1}{2}\int_\Omega ((|Ly|^2)'\,dx + u\cdot\nabla(|Ly_\Omega|^2) + |Ly_\Omega|^2\nabla\cdot u)\,dx \\ &= \int_\Omega (Ly_\Omega\cdot Ly' + \tfrac{1}{2}\nabla\cdot(u|Ly_\Omega|^2))\,dx. \end{aligned}$$

This yields (20). By the formula (21), we get the value of J' stated in theorem 3. The integral condition on p in (17) is obtained by a similar calculation.

5. DIFFERENTIABILITY OF AN IMPLICIT EQUATION SOLUTION.

To prove the existence of a total variation $\dot{y}$, and of higher order total variations, we will use the following extension of the usual differentiabilty result for implicit equations solutions.

Theorem 6. *We give us*
- *an open set $\mathcal{U}$ in a Banach space U, $u_0 \in \mathcal{U}$, two reflexive Banach spaces A and B,*
- *a map $F : \mathcal{U} \times A \mapsto B$, such that $F(u;\cdot) \in \mathcal{L}(A; B)$ for all $u \in \mathcal{U}$,*
-*a function $m : \mathcal{U} \mapsto A$, and a function $f : \mathcal{U} \mapsto B$, such that*

$$F(u, m(u)) = f(u) \quad \forall u \in \mathcal{U}.$$

(i) *Assume that*

$$u \mapsto F(u;\cdot) \text{ is differentiable at } u_0 \text{ into } \mathcal{L}(A; B),\ f \text{ is differentiable at } u_0, \tag{26}$$

$$\|F(u_0; x)\|_B \geq \alpha\|x\|_A \quad \forall x \in A, \qquad \text{for some } \alpha > 0. \tag{27}$$

Then, the map $u \mapsto m(u)$ is differentiable at u_0. It's derivative $m'(u_0;\cdot)$ is the unique solution of

$$F(u_0; m'(u_0; v)) = f'(u_0; v) - \partial_u F(u_0; m(u_0); v) \quad \forall v \in U. \tag{28}$$

(ii) *In addition, assume that for some integer $k \geq 1$,*

$$u \mapsto F(u;\cdot) \text{ and } f \text{ are } k \text{ times differentiable at } u_0. \tag{29}$$

Then, the map $u \mapsto m(u)$ is k times differentiable at u_0.

The new point here is that $F(0,\cdot)$ is not necessary a one to one map, since its range is not supposed to be all of B. Whence, we have to prove the existence of a solution $m'(u_0;v)$ of (28).

Proof of part i.

Boundedness of m. By (26),

$$\|F(u;x) - F(u_0;x) - \partial_u F(u_0;x;u-u_0)\|_B \leq \|x\|_A \, o(\|u-u_0\|_U), \tag{30}$$

thus

$$\|F(u;x) - F(u_0;x)\|_B \leq \beta \|x\|_A \|u-u_0\|_U. \tag{31}$$

By equation,

$$F(u_0;m(u)) = \big(F(u_0;m(u)) - F(u;m(u))\big) + \big(f(u) - f(u_0)\big) + F(u_0;m_0), \tag{32}$$

where $m_0 = m(u_0)$. Therefore, $\alpha\|m(u)\|_A \leq \beta\|m(u)\|_A\|u-u_0\|_U + r\|u-u_0\|_U + \|F(u_0;m_0)\|_B$.

In all the sequel, we assume that $\|u-u_0\|_U \leq \frac{\alpha}{2\beta}$. Then,

$$\|m(u)\| \leq \gamma = \tfrac{2}{\alpha}\|F(u_0;m_0)\|_B + \tfrac{r}{\beta}. \tag{33}$$

Continuity of m. By (32),

$$F(u_0;m(u)-m_0) = \big(F(u_0;m(u)) - F(u;m(u))\big) + \big(f(u) - f(u_0)\big), \tag{34}$$

thus

$$\alpha\|m(u)-m_0\|_A \leq (\beta\|m(u)\|_A + r)\|u-u_0\|_U \leq (\beta\gamma + r)\|u-u_0\|_U. \tag{35}$$

Weak differentiability of m. Given a fixed $v \in U$, we denote $u_t = u_0 + tv$, $m_t = m(u_t)$, and we assume $t \leq t_v = \alpha/(2\beta\|v\|_U)$. By (35), $\frac{1}{t}(m_t - m_0)$ is bounded in A. Then, there exists $\mu \in A$ and $t_n \to 0$ such that $\frac{1}{t_n}(m_{t_n} - m_0) \to \mu$ weakly in A. By (32),

$$F(u_0;m(u)-m_0) = r(u) - \partial_u F(u_0;m_0;u-u_0) + f'(u_0;u-u_0), \tag{36}$$

where $r(u) = \Big(F(u_0;m(u)-m_0) - F(u;m(u)-m_0)\Big) - \Big(F(u;m_0) - F(u_0;m_0) - \partial_u F(u_0;m_0;u-u_0)\Big) + \Big(f(u) - f(u_0) - f'(u_0;u-u_0)\Big)$.

By (31) and (30), $\|r(u)\|_B \leq \beta\|m(u)-m_0\|_A\|u-u_0\|_U + o(\|u-u_0\|_U) = o(\|u-u_0\|_U)$.

In (36), we choose $u = u_{t_n}$, we divide by t_n, and we let $t_n \to 0$. Then $r(t_n) \to 0$, and therefore at the limit, by weak continuity of $F(u_0;\cdot)$, we get

$$F(u_0;\mu(v)) = -\partial_u F(u_0;m_0;v) + f'(u_0;v). \tag{37}$$

Differentiability of m. By (27), (37) defines a unique $\mu(v)$, which is linear and continuous with respect to v. By (36) and (37), $F(u_0; m(u) - m_0 - \mu(u - u_0)) = r(u)$. Thus,

$$\|m(u) - m_0 - \mu(u - u_0)\|_A \leq \frac{1}{\alpha} o(\|u - u_0\|_U).$$

Proof of part ii. Differentiability of m in a neighbourhood of u_0. Now, we assume that $u \mapsto F(u; \cdot)$ and f are differentiable in an open subset $\mathcal{U}_0$ of $\mathcal{U}$ containing u_0. And we restrict u to $\mathcal{V} = \mathcal{U}_0 \cap \{u \mid \|u - u_0\|_U < \frac{\alpha}{2\beta}\}$. By (27) and (31),

$$\|F(u; x)\|_B \geq \|F(u_0; x)\|_B - \|F(u; x) - F(u_0; x)\|_B \geq \frac{\alpha}{2}\|x\|_A \quad \forall x \in A.$$

Assumptions of part (i) are satisfied at any point $u \in \mathcal{V}$. Thus, $u \mapsto m(u)$ is differentiable in $\mathcal{V}$, and

$$F(u; m'(u; v)) = -\partial_u F(u; m(u); v) + f'(u; v), \quad \forall v \in U, \forall u \in \mathcal{V}. \tag{38}$$

k-differentiability in $\mathcal{V}$. Now, we assume that $F(u; \cdot)$ and f are k times differentiable in $\mathcal{U}_0$. We define $F_1 : \mathcal{V} \times \mathcal{L}(U; A) \to \mathcal{L}(U; B)$ and $f_1 : \mathcal{V} \mapsto \mathcal{L}(U; B)$ by

$$F_1(u; y)(v) = F(u; y(v)), \quad f_1(u)(v) = -\partial_u F(u; m(u); v) + f'(u; v), \qquad \forall v \in U.$$

Then, (38) yields $F_1(u; m'(u)) = f_1(u), \forall u \in \mathcal{V}$. The map F_1 satisfies (27), and F_1 and f_1 are $k-1$ times differentiable in $\mathcal{V}$. Thus, by previous step, $m'(u)$ is differentiable in $\mathcal{V}$.

Using k times the present step, we get the k times differentiability in $\mathcal{V}$.

k-differentiability at u_0. By (29), F and f are $k-1$ times differentiable in a neighbourhood $\mathcal{U}_0$ of u_0.

Then, by the previous step, $u \mapsto m(u)$ is $k-1$ times differentiable in a neighbourhood $\mathcal{V}$ of u_0, and it's derivative satisfies $F_{k-1}(u; m^{(k-1)}) = f_{k-1}(u)$ in $\mathcal{V}$.

In addition, F_{k-1} and f_{k-1} are differentiable at u_0. Then, by part (i), $m^{(k-1)}$ is differentiable at u_0. That is the k-differentiability of m at u_0.

6. PROOF OF DOMAIN VARIATION RESULTS.

Here we prove theorems 3, 4 and 5 in three steps :

- existence of an unknown total variation $\dot{y}$, that is theorem 5.
- existence of an unknown variation J' satisfying the expansion (11).
- J' has the announced value, and y' is the local variation

6.1 Existence of a total variation.

Mapped equations. By the map $I + u$, equations (8) and (9) satisfied by $y_{\Omega+u}, p_{\Omega+u}$ yield equations satisfied by $Y(u) = y_{\Omega+u} \circ (I + u)$, $P(u) = p_{\Omega+u} \circ (I + u)$.

We denote $[\partial_i(I+u)_j]$ the derivative matrix of the map $I+u$, $M(u) = {}^{tr}[\partial_i(I+u)_j]^{-1}$ the transposed inverse matrix, and $D_i(u) = \sum_{ij} M_{ij}(u)\partial j$. Using

$$(\nabla f) \circ (I + u) = D(u)(f \circ (I + u)), \tag{39}$$

we get

$$-\nu D(u) \cdot (D(u)Y(u)) + D(u)Q(u) = 0, \quad D(u) \cdot Y(u) = 0, \quad \text{in } \Omega,$$

$$Y(u) = 0 \text{ on } \partial\mathcal{B}, \quad Y(u) = h \text{ on } \partial\Lambda, \quad \int_\Omega P(u)\det[\partial_i(I+u)_j]\,dx = 0.$$

By theorem 2, $y_{\Omega+u} \in H^2(\Omega+u)^3$, $p_{\Omega+u} \in H^1(\Omega+u)$. Whence,

$$Y(u) \in H^2(\Omega)^3, P(u) \in H^1(\Omega).$$

To get homogeneous boundary condition, we consider $\tilde{h} \in \mathcal{C}^\infty(\mathbb{R}^3)^3$ such that $\nabla\cdot\tilde{h} = 0$, $\tilde{h} = 0$ on $\partial\mathcal{B}$, $\tilde{h} = h$ on $\partial\Lambda$. Setting $Z(u) = Y(u) - \tilde{h}$, we get $Z(u) \in H^2(\Omega)^3 \cap H_0^1(\Omega)^3$.

Implicit equation. We define,

$$F : Lip^2(\mathbb{R}^3)^3 \times H^2(\Omega)^3 \cap H_0^1(\Omega)^3 \times H^1(\Omega) \mapsto L^2(\Omega)^3 \times H^1(\Omega) \times \mathbb{R}$$

$$F(u; Z, P) = \Big(-\nu D(u)\cdot(D(u)Z) + D(u)P\ ;\ D(u)\cdot Z\ ;\ \int_\Omega P\det[\partial_i(I+u_j)]\,dx\Big).$$

Then, the above equations yield

$$F(u; Z(u), P(u)) = -F(u; \tilde{h}, 0). \tag{40}$$

If $F(0;\cdot)$ could be a one to one map, then, by usual differentiability properties of implicit equations solutions, we would get differentiabilty of $u \mapsto (Z(u), P(u))$. However, by condition (5) in theorem 2, the range of $F(0;\cdot)$ is not all the space. (Moreover, the range of $F(u;\cdot)$ depends on u, then we cannot restrict the range at u to the range at 0). The differentiability will come of theorem 6.

Differentiability. Now, we check assumptions of theorem 6, for equation (40).

The matrix inversion is differentiable into $Lip^1(\mathbb{R}^3)^9 = W^{1,\infty}(\mathbb{R}^3)^9$, at any point which is an invertible matrix. Then, $u \mapsto M(u)$ is differentiable from $Lip^2(\mathbb{R}^3)^3$ into $W^{1,\infty}(\mathbb{R}^3)^9$, at any u such that $\|u\|_{Lip^1} < 1$. Since F is linear with respect to (Z, P), quadratic with respect to M, and three-linear with respect to u by the determinant function, then $u \mapsto F(u,\cdot)$ is differentiable, at any such u, into $\mathcal{L}(Lip^2; L^2 \times H^1 \times \mathbb{R})$. That is the first part of assumption (26).

In addition, this yields the differentiability of $u \mapsto F(u; \tilde{h}; 0)$, which is the second part of(26).

At last, $F(0; Z, P) = (f; k; r)$ is the generalized Stokes problem stated in theorem 2, with $g = 0$. Thus, assumption (27) follows from part ii of theorem 2.

All assumptions of theorem 6 being satisfied, $u \mapsto (Z(u), P(u))$ and therefore $u \mapsto (Y(u), P(u))$ are differentiable from $Lip^2(\mathbb{R}^3)^3$ into $H^2(\Omega)^3 \times H^1(\Omega)$, at any u such that $\|u\|_{Lip^1} < 1$. This proves the existence of $\dot{y}$ and $\dot{p}$, that is theorem 5.

6.2 Existence of a total variation J'. By theorem 3.3 in [S1], the existence of $\dot{y}$ implies the existence of J' satisfying expansion (11). For sake of completness, we give the following proof.

In the right hand side of (10), we use the change of variable $I + u$. Using (39), and $\det[\ldots] > 0$ since $\|u\|_{Lip^1} < 1$, we get

$$J(\Omega + u) = \int_\Omega \textstyle\sum_{ij} |D_i(u)Y_j(u) + D_j(u)Y_i(u)|^2 \det[\partial_i(I+u)_j]\,dx. \tag{41}$$

By the differentiability of M, $u \mapsto D_i(u)Y_j(u) + D_j(u)Y_i(u)$ is differentiable from $Lip^2(\mathbb{R}^3)^3$ into $H^1(\Omega)$. Since the determinant map is multilinear, $u \mapsto \det[\partial_i(I + u)_j]$ is differentiable into $Lip^1(\mathbb{R}^3) = W^{1,\infty}(\mathbb{R}^3)$. Thus, the integrated function is differentiable with respect to u into $W^{1,1}(\Omega)$ (L^1 would be enough here), at any point such that $\|u\|_{Lip^1} < 1$. Therefore it's integral is differentiable, that is there exists J' satisfying expansion (11).

6.3 Value of the derivatives. By lemma 1.2 of [S1], the existence of total variations $\dot{y} \in H^2(\Omega)^3$ and $\dot{p} \in H^1(\Omega)$, implies the existence of local variations $y' \in H^1(\Omega)^3$ and $p' \in L^2(\Omega)$ satisfying expansions (18) and (19). By theorems 3.1 and 3.2 of [S1], equations (8) and (9) imply that y' and p' satisfy (16) and (17). This proves the part ii of theorem 4.

By theorem 3.3 of [S1], the value of J' is given by (20). This proves the part iii of theorem 4.

Due to (21), the value of J' is given by (12). This proves theorem 3.

7. APPLICATION TO DRAG MINIMIZATION FOR GIVEN VOLUME.

We are interested in the minimization of J for bodies $\mathcal{B}$ which have a given volume v, in a fixed experiment domain Λ. We assume that $\partial\Lambda$ is Lip^2, and we denote

$$\mathcal{D}_{ad} = \{\mathcal{B} \mid \overline{\mathcal{B}} \subset \Lambda, \partial\mathcal{B} \text{ is } Lip^2, \textstyle\int_{\partial\mathcal{B}} ds = v\}.$$

Theorem 7. *Assume that there exists $\mathcal{B}_0$ such that*

$$\mathcal{B}_0 \in \mathcal{D}_{ad}, \qquad J(\Lambda \setminus \overline{\mathcal{B}_0}) \leq J(\Lambda \setminus \overline{\mathcal{B}}) \quad \forall \mathcal{B} \in \mathcal{D}_{ad}.$$

Then,

$$\left|\frac{\partial y_{\mathcal{B}_0}}{\partial n}\right| \text{ is constant on } \partial\mathcal{B}_0. \tag{42}$$

Remark. This condition is satisfied by any possible $\mathcal{B}_0$. However, the existence of such a $\mathcal{B}_0$ is not proved, although it is expected.

Proof of theorem 7. Let u satisfy

$$u \in Lip^2(\mathbb{R}^3)^3, \quad \int_{\partial\mathcal{B}_0} u_n \, ds = 0, \quad u = 0 \text{ on } \partial\Lambda. \tag{43}$$

As it will be proved in a moment, there exists a path $U \in C^1(\mathbb{R}_+; Lip^2(\mathbb{R}^3)^3)$ such that,

$$\frac{dU(t)}{dt}|_{t=0} = u, \quad \int_{\mathcal{B}_0+U(t)} dx = v, \quad U(t) = 0 \text{ on } \partial\Lambda. \tag{44}$$

Then, for small t, $\mathcal{B}_0 + U(t) \in \mathcal{D}_{ad}$, thus $J(\mathcal{B}_0 + U(t)) \geq J(\mathcal{B}_0)$. By the chain rule theorem,

$$\frac{d}{dt} J(\mathcal{B}_0 + U(t))|_{t=0} = J'(\mathcal{B}_0; u) \geq 0.$$

By (12) this yields, since $u = 0$ on Λ,

$$\int_{\partial\mathcal{B}_0} u_n |\frac{\partial y_{\mathcal{B}_0}}{\partial n}|^2 \, ds \leq 0 \quad \text{for all } u \text{ such that} \quad \int_{\partial\mathcal{B}_0} u_n \, ds = 0.$$

This implies (42).

It remains to check (44). The map $u \mapsto \int_{\mathcal{B}+u} dx = \int_{\mathcal{B}} \det[\partial_i(I+u)_j] \, dx$ is multilinear, and therefore differentiable. It's derivative at 0 is $u \mapsto \int_{\mathcal{B}} \nabla \cdot u \, dx = \int_{\partial\mathcal{B}} u_n \, ds$.

Now, let u satisfy (43), let $w \in Lip^2(\mathbb{R}^3)^3$ satisfy $\int_{\partial\mathcal{B}_0} w_n \, ds \neq 0, w = 0$ on $\partial\Lambda$, and denote $V(t,c) = \int_{\mathcal{B}+(tu+cw)} dx$. Then the implicit equation

$$V(t, c(t)) = v, \quad t \geq 0,$$

has a solution $t \mapsto c(t)$ which is C^1 in a neighborhood of 0. Indeed V is C^1, and $\frac{\partial V}{\partial c}(0,0) = \int_{\partial\mathcal{B}} w_n \, ds \neq 0$. The solution satisfies, $\frac{\partial V}{\partial t}(0,0) + \frac{\partial V}{\partial c}(0,0)c'(0) = 0$. Since $\frac{\partial V}{\partial t}(0,0) = \int_{\partial\mathcal{B}} u_n = 0$, this yields $c'(0) = 0$.

Thus, $U(t) = tu + c(t)w$ satisfies $U'(0) = u$ and therefore (44).

Remark. Should we proved only directional differentiability, that is (15), we would get $J'(\Lambda \setminus \overline{\mathcal{B}_0}) \geq 0$, only for v satisfying $\mathcal{B}_0 + tv \in \mathcal{D}_{ad}, \forall t \leq t_v$. This condition yields $\int_{\mathcal{B}_0+tv} dx = \int_{\mathcal{B}_0} dx, \forall t \leq t_v$, which is satisfied only for constant v. Therefore we would obtain much less information on $\mathcal{B}_0$.

8. SECOND VARIATIONS.

We are looking for a second order expansion of J, that is

$$J(\Omega + u) = J(\Omega) + J'(\Omega; u) + \tfrac{1}{2}J''(\Omega; u, u) + o((\|u\|_{Lip^2})^2). \tag{45}$$

The following result will be proved in section 10.2.

Theorem 8. **(i)** *We assume that Ω satisfies (4) and $\partial\Omega$ is Lip^2.*

Then, there exists a bilinear continuous map $J''(\Omega; \cdot, \cdot)$ on $Lip^2(\mathbb{R}^3)^3$, such that expansion (45) holds for all $u \in Lip^2$, $\|u\|_{Lip^1} < 1$.

(ii) *We assume in addition that $\partial\Omega$ is Lip^3. Then,*

$$J''(\Omega; u, u) = \int_\Omega \left(Ly \cdot Ly'' + |Ly'|^2\right) dx$$
$$+ \int_{\partial\Omega} \left(2u_n Ly_n \cdot Ly' + \tfrac{1}{2}u_n(u \cdot \nabla)(|Ly_n|^2) + \tfrac{1}{2}(u(\nabla \cdot u) - (u \cdot \nabla)u)_n |Ly_n|^2\right) ds,$$

where unique $y'' \in H^1(\Omega)^3$ and $p'' \in L^2(\Omega)$ are defined by

$$-\nu\Delta y'' + \nabla p'' = 0 \quad \text{and} \quad \nabla \cdot y'' = 0, \quad \text{in } \Omega, \tag{46}$$

$$y'' + 2(u \cdot \nabla)y' + (u \cdot (u \cdot \nabla)\nabla)y = 0 \text{ on } \partial\Omega, \quad \int_\Omega p'' \, dx = r(p, p'; u, u), \tag{47}$$

r being a real depending linearly on all its arguments.

Moreover, $y'' = y''(\Omega; u, u)$ *and* $p'' = p''(\Omega; u, u)$ *are bilinear and continuous with respect to* u*, from* $Lip^2(\mathbb{R}^3)^3$ *into* $H^1(\Omega)^3$ *and* $L^2(\Omega)$*. For any* ω *such that* $\overline{\omega} \in \Omega$*,*

$$y_{\Omega+u} = y_\Omega + y'(\Omega; u) + y''(\Omega; u, u) + o((\|u\|_{Lip^2})^2) \quad \text{in } L^2(\omega)^3,$$

$$p_{\Omega+u} = p_\Omega + p'(\Omega; u) + p''(\Omega; u, u) + o((\|u\|_{Lip^2})^2) \quad \text{in } H^{-1}(\omega).$$

Remark. As for first variations, proofs for second variations are based on uniform dependence on $y_{\Omega+u}$ up to the varying boundary $\partial\Omega + u$. This comes of second order differentiability of $y_{\Omega+u} \circ (I+u)$. In fact, in section 10 we will get infinite differentiability order.

Remark. The second variation J'' yields a sufficient condition for local optimality. Let a domain Ω_0 in some class $\mathcal{D}_{ad}$ satisfy,

$$J'(\Omega_0; u) = 0 \quad \text{and} \quad J''(\Omega_0; u, u) > 0, \quad \forall u \in \mathcal{U}_0',$$

where $\mathcal{U}_0'$ is the tangent cone at 0 in Lip^2 to $\{u \mid \Omega_0 + u \in \mathcal{D}_{ad}\}$. Then, Ω_0 is locally minimal in $\mathcal{D}_{ad}$.

Moreover, based on J'', the velocity of gradient methods may be improved.

9. ROUGH CALCULUS OF SECOND VARIATIONS.

Equations differentiation. Twice differentiating equation (23) with respect to t, at $t = 0$, we get the first equation in (46) :

$$-\nu(\Delta y'' + \nabla p'')(x) = 0, \quad \forall x \in \Omega.$$

Similarly, we get $(\nabla \cdot y'')(x) = 0, \forall x \in \Omega$.

Boundary condition differentiation. Denoting $''$ the second order derivative with respect to t at $t = 0$, we have,

$$\big(y_{\Omega+tv} \circ (I + tv)\big)'' = y'' + 2(u \cdot \nabla)y' + (u \cdot (u \cdot \nabla)\nabla)y. \tag{48}$$

Thus, twice differentiating boundary condition (24) with respect to t, at $t = 0$, we get, $y''(x) + 2(u(x) \cdot \nabla)y'(x) + (u \cdot (u \cdot \nabla)\nabla)(y_\Omega)(x) = 0$ for all $x \in \partial\mathcal{B}$. Similarly we get the same equation on Λ, and therefore on all of Ω, that is boundary condition in (46).

Drag differentiation. Twice differentiating (25), we get

$$J''(\Omega; u, u) = \tfrac{1}{2} \int_\Omega \big(|Ly_{\Omega+tv}|^2 \circ (I + tv)\big)'' + 2\big(|Ly_{\Omega+tv}|^2 \circ (I + tv)\big)'(\nabla \cdot u)$$
$$+ |Ly|^2\big((\nabla \cdot u)^2 - \nabla u \cdot {}^{tr}\nabla u\big)\, dx,$$

since $\nabla \cdot u$ and $(\nabla \cdot u)^2 - \nabla u \cdot {}^{tr}\nabla u)$ are the first and second derivatives of the determinant function, and since $|\det[\ldots]| = \det[\ldots]$ for small t.

By (48), the right hand side is

$$\frac{1}{2}\int_\Omega \Big((|Ly|^2)'' + 2u\cdot\nabla(|Ly|^2)' + (u\cdot(u\cdot\nabla)\nabla)(|Ly|^2)$$

$$+ 2((|Ly|^2)' + u\cdot\nabla(|Ly|^2))(\nabla\cdot u) + |Ly|^2((\nabla\cdot u)^2 - \nabla u\cdot {}^{tr}\nabla u)\Big)\,dx,$$

$$= \int_\Omega \Big((Ly\cdot Ly'' + Ly'\cdot Ly') + 2\nabla\cdot(u\,Ly\cdot Ly') + \tfrac{1}{2}R\Big)\,dx,$$

where $R = \big(u\cdot(u\cdot\nabla)\nabla\big)(|Ly|^2) + u\cdot\nabla(|Ly|^2)(\nabla\cdot u) + |Ly|^2((\nabla\cdot u)^2 - \nabla u\cdot {}^{tr}\nabla u)$.

To get the announced value of J'', it remains to check, and this is left to the reader, that $R = \nabla\cdot\big(u(u\cdot\nabla)(|Ly_\Omega|^2)\big) + \nabla\cdot\big((u(\nabla\cdot u) - (u\cdot\nabla)u)|Ly_\Omega|^2\big)$.

10. INFINITE ORDER DIFFERENTIABILITY, AND SECOND VARIATIONS PROOFS.

10.1. Infinite order differentiabilty. Here, we prove that $u \mapsto J(\Omega+u)$ is indefinitely differentiable. However we do not calculate high order derivatives since, as seen for second derivatives, length of formulas increases with differentiability order.

Theorem 9. *We assume that Ω satisfies (4) and $\partial\Omega$ is Lip^2.*

Then, at any point u such that $\|u\|_{Lip^1} < 1$,

$u \mapsto J(\Omega+u)$ *is indefinitely differentiable from $Lip^2(\mathbb{R}^3)^3$ into $\mathbb{R}$,*

$u \mapsto y_{\Omega+u}\circ(I+u)$ *is indefinitely differentiable from $Lip^2(\mathbb{R}^3)^3$ into $H^2(\Omega)^3$,*

$u \mapsto p_{\Omega+u}\circ(I+u)$ *is indefinitely differentiable from $Lip^2(\mathbb{R}^3)^3$ into $H^1(\Omega)$.*

Proof. In section 6.1, we got the differentiability of the map $u \mapsto F(u;\cdot)$ defining the implicit equation (40). In fact, by a similar proof, this map is k times differentiable for any given k.

It suffices to remark that matrix inversion is k times differentiable into $W^{1,\infty}(\Omega)^9$, at any point which is an inertible matrix. Therefore, the map $u \mapsto M(u) = {}^{tr}[\partial_i(I+u)_j]^{-1}$ is k times differentiable from $Lip^2(\mathbb{R}^3)^3$ into $W^{1,\infty}(\Omega)^9$, at any u such that $\|u\|_{Lip^1} < 1$. In addition the determinant function being multilinear, $u \mapsto \det[\partial_i(I+u)_j]$ is k times differentiable into $\mathbb{R}$.

In addition this yields the k differentiability of $u \mapsto F(u;\tilde{h},0)$.

Now, we can use part ii of theorem 6, for implicit equation (40). Whence, $u \mapsto (Z(u),P(u))$ and therefore $u \mapsto (Y(u),P(u)) = (y_{\Omega+u}\circ(I+u), p_{\Omega+u}\circ(I+u))$ are k times differentiable from $Lip^2(\mathbb{R}^3)^3$ into $H^2(\Omega)^3 \times H^1(\Omega)$.

The k differentiability of $u \mapsto J(\Omega+u)$ follows by (41).

10.2. Proof of theorem 8. By theorem 9, $u \mapsto J(\Omega+u)$ is two times differentiable, which is part (i) of theorem 8.

By theorem 3.3 of [S2], the second variation f'' of any two times differentiable function $u \mapsto f(\Omega+u)$ is related to the first variation by

$$f''(\Omega;u,u) = (f')'(\Omega;u,u) - f'(\Omega;(u\cdot\nabla)u), \tag{49}$$

where $(f')'(\Omega;u,u) = \frac{d}{dt}f'(\Omega+tu;u)|_{t=0}$.

Expansions of y, p. The existence of y'' and p'' satisfying second order expansions follows from the second order differentiability of $u \mapsto (y_{\Omega+u} \circ (I+u), p_{\Omega+u} \circ (I+u))$, by using twice lemma 1.2 in [S1].

Equation differentiation. By theorem 3.1 of [S1], (16) yields $-\nu\Delta(y')' + \nabla(p')' = 0$. Thus, by (49), $-\nu\Delta y'' + \nabla p'' = 0$ in Ω, which is the first equation in (46).

Similarly, we get the second equation in (46), $\nabla \cdot y'' = 0$ in Ω.

Boundary condition differentiation. The boundary condition in (17) may be written as $y' + (u \cdot \nabla)y = 0$ on $\partial\Omega$. Therefore, by theorem 3.2 of [S2], $(y')'(\Omega; u, u) + 2(u \cdot \nabla)y'(\Omega; u) + (u \cdot \nabla)^2 y = 0$ on $\partial\Omega$. Thus, by (49), we get the boundary condition on y'' in (47).

Drag differentiation. The equality (20) may be written as

$$J'(\Omega; u) = \int_\Omega \left(Ly \cdot Ly' + \tfrac{1}{2}\nabla \cdot (u|Ly|^2)\right) dx.$$

By theorem 3.3 in [S1], we get

$$(J')'(\Omega; u, u) = \int_\Omega \Big(Ly \cdot L(y')' + Ly' \cdot Ly' + \nabla \cdot (u\, Ly \cdot Ly') + \nabla \cdot \big(u\, Ly \cdot Ly' + \tfrac{1}{2}\nabla \cdot (u|Ly|^2)\big)\Big)\, dx.$$

Therefore, using (49) and integration, we get the announced value for $J''(\Omega; u, u)$.

To get (49), the present application of theorem 3.3 in [S2] requires $u \in Lip^4(\mathbb{R}^3)^3$. Therefore, the formulas are proved for $u \in Lip^4$. All the terms of these formulas being continuous with respect to u in Lip^2, and Lip^4 being dense into Lip^2, they hold for $u \in Lip^2$.

References

[GR] V. Girault & P.A. Raviart: Finite element methods for Navier-Stokes equations. Springer-Verlag, 1986.

[L] O.A. Ladyzhenskaya: The mathematical theory of viscous incompressible fluids. Gordon & Breach, 1963.

[MS1] F. Murat & J. Simon: Quelques résultats sur le controle par un domaine géométrique. Research report, Paris 6 University, 1974.

[MS2] F. Murat & J. Simon: Sur le controle par un domaine géométrique. Research report, Paris 6 University, 1976.

[P] O. Pironneau: Optimal shape design for elliptic systems. Springer, 1983.

[S1] J. Simon: Differentiation with respect to the domain in boundary value problems. Numerical Functional Analysis & Optimization, 2, pp. 649-687, 1980.

[S2] J. Simon: Second variations for domain optimization problems. In: Control & estimation of distributed parameter systems. F. Kappel, K. Kunish & W. Scappacher eds. International Series of Numerical Mathematics, 91, pp. 361-378, Birkhauser, 1989.

The Existence of Solutions to the Infinite Dimensional Algebraic Riccati Equations with Indefinite Coefficients *

Shuping Chen

Zhejiang University, Hangzhou, China

Abstract. Necessary and sufficient conditions are established for the existence of selfadjoint solutions and positive definite solutions to the algebraic Riccati equations in Hilbert spaces with indefinite coefficients.

§1. Definitions, Notations and Formulation of Main Results.

Let X, U, V and Y be separable Hilbert spaces. Let A be the infinitesimal generator of a C_0-semigroup $e^{At}, t \geq 0$, on X with dense domain $\mathcal{D}(A)$. Let $B \in \mathcal{L}(U, X)$, $G \in \mathcal{L}(V, X)$ and $C \in \mathcal{L}(X, Y)$. Here, we use $\mathcal{L}(H_1, H_2)$ to denote the Banach space of linear bounded operators mapping from H_1 into H_2, endowed with usual operator norm, and denote $\mathcal{L}(H) = \mathcal{L}(H, H)$. The inner product in a Hilbert space H is denoted by $(\cdot, \cdot)_H$ and the induced norm is denoted by $\|\cdot\|_H$. The subscript H will be suppressed if it can be understood from the context.

In this paper, we consider the following algebraic Riccati equation

$$(ARE)_\delta : \qquad (Sx_1, Ax_2) + (Ax_1, Sx_2) - (SQSx_1, x_2) + (M_\delta x_1, x_2) = 0, \quad x_1, x_2 \in \mathcal{D}(A), \tag{1.1}$$

where $Q = C^*C$ and $M_\delta = BB^* - \delta^2 GG^*$, and $\delta \geq 0$ is a parameter. Here, T^* denotes the adjoint of the linear operator T.

It is well-known that the Riccati equations play an important role in the control theory for linear systems. In particular, $(ARE)_0$ arises from the standard linear quadratic cost control problems (L-Q) and the filtering problems of infinite dimensions [1,3], and has been studied in some details by a number of authors (see [7] for example and the references cited therein). For L-Q problems with conflicting objectives one will encounter $(ARE)_\delta$ with $\delta > 0$. Moreover, there is a close connection between $(ARE)_\delta$, $\delta > 0$, and the dual Reccati equation

$$(ARE)_\delta^* : \qquad (Px_1, Ax_2) + (Ax_1, Px_2) + (PM_\delta Px_1, x_2) - (Qx_1, x_2) = 0, \quad x_1, x_2 \in \mathcal{D}(A). \tag{1.2}$$

Whereas $(ARE)_\delta^*$ has applications to the two-person zero-sum differential games [5] and has recently been found crucial to the study of H^∞-optimal control problem via state-space approach ([2]).

The aim of the present paper is to establish conditions under which $(ARE)_\delta$, $\delta > 0$, have selfadjoint solutions and/or positove definite solutions.

---

* Project supproted by the NSF of China.

We shall use the following definitions and notational conventions throughout the paper. The generator A is called to be stable if the semigroup e^{At} is exponentially stable, i.e., there are positive constants m and ω, such that

$$\|e^{At}\| \leq me^{-\omega t}, \qquad \forall t \geq 0. \tag{1.3}$$

The pair (A, B) (resp. (C, A)) is said to be stabilizable (resp. detectable) if there exists a $K \in \mathcal{L}(X, U)$ (resp. $K \in \mathcal{L}(Y, X)$), such that $A + BK$ (resp. $A + KC$) is stable. Unless otherwise stated, we shall always assume the stabilizability of (A, B) and the detectability of (C, A). A selfadjoint operator $T \in \mathcal{L}(H)$ is said to be nonnegative, denoted by $T \geq 0$, if $(Th, h) \geq 0$, $\forall h \in H$; T is said to positive definite, denoted by $T > 0$, if $(Th, h) \geq \eta\|h\|^2$, $\forall h \in H$, for some $\eta > 0$ independent of h. In the latter case, T is boundedly invertible and $T^{-1} > 0$. We shall also denote by Γ_δ (resp. Γ_δ^*) the set of bounded selfadjoint solutions to $(ARE)_\delta$ (resp. $(ARE)_\delta^*$).

Before going further, let us review a result of Zabczyk [7].

Thereom 0. *Suppose (A, B) is stabilizable and (C, A) is detectable. Then, there exists a unique nonnegative solution $S_0 \in \Gamma_0$, and moreover,*

$$F = A - QS_0 \text{ is stable}. \tag{1.4}$$

Since the semigroup e^{Ft} and hence e^{F^*t} is stable, we can define operators $L \in \mathcal{L}(\mathcal{X})$ and $L_0 \in \mathcal{L}(\mathcal{X}, X)$ by

$$(Lf)(t) = \int_t^\infty e^{F^*(s-t)} f(s) ds, \qquad f \in \mathcal{X}, \tag{1.5}$$

and

$$L_0 f = (Lf)(0), \qquad f \in \mathcal{X}, \tag{1.6}$$

where $\mathcal{X} = L^2(0, \infty; X)$. Their respective adjoint operators $L^* \in \mathcal{L}(\mathcal{X})$ and $L_0^* \in \mathcal{L}(X, \mathcal{X})$ can be readily calculated:

$$(L^* f)(t) = \int_0^t e^{F(t-s)} f(s) ds, \qquad f \in \mathcal{X}, \tag{1.7}$$

and

$$(L_0^* x)(t) = e^{Ft} x, \qquad x \in X. \tag{1.8}$$

Our main results are now stated as follows.

Theorem 1.1 *(a) There exists an $S_\delta \in \Gamma_\delta$ with the property*

$$F_\delta = A - QS_\delta \quad \text{stable} \tag{1.9}$$

if and only if

$$I - \delta^2 G^* L^* QLG > 0. \tag{1.10}$$

Such S_δ is unique as long as it exists, and has a representation

$$S_\delta = S_0 - \delta^2 L_0 G(I - \delta^2 G^* L^* QLG)^{-1} G^* L_0^*. \tag{1.11}$$

(b) There exists a positive definite silution $S_\delta \in \Gamma_\delta$ if and only if $S_0 > 0$ and

$$I - \delta^2 G^* L^* QLG - \delta^2 G^* L_0^* S_0^{-1} L_0 G > 0. \qquad (1.12)$$

In this case, (1.11) gives the unique positive definite solution to $(ARE)_\delta$ and (1.9) is also satisfied.

§2. Preliminary Results.

In this section, we shall establish a number of lemmas which will be used to prove our main results.

Lemma 2.1. *Let S_0 be the unique nonnegative solution to $(ARE)_0$ and let $F = A - QS_0$. Then,*

(i) $S_\delta \in \mathcal{L}(X)$ is a selfadjoint solution to $(ARE)_\delta$ if and only if $\Delta - S_\delta - S_0$ is selfadjoint and satisfies

$$(\Delta x_1, Fx_2) + (Fx_1, \Delta x_2) - (Q\Delta x_1, \Delta x_2) = \delta^2 (G^* x_1, G^* x_2), \qquad x_1, x_2 \in \mathcal{D}(A). \quad (2.1)$$

(ii) If $\Gamma_\delta \neq \emptyset$, then, for all $y \in Y$ and $\omega \in I\!R$, we have

$$\|y\|^2 - \delta^2 \|G^*(i\omega I + F)^{-1} C^* y\|^2 = \|[I - C\Delta(i\omega I + F)^{-1} C^* y\|^2. \qquad (2.2)$$

Proof. (i) can be derived through a straightforward calculation. We proceed to prove (ii). Suppose there exists an $S_\delta \in \Gamma_\delta$. Then, we obtain (2.1) with $\Delta = S_\delta - S_0$. Note that (2.1) is equivalent to

$$(\Delta x_1, (i\omega I + F)x_2) + ((i\omega I + F)x_1, \Delta x_2) - (Q\Delta x_1, \Delta x_2) = \delta^2 (G^* x_1, G^* x_2), \quad (2.3)$$

and $(i\omega I + F)^{-1} \in \mathcal{L}(X)$ for all $\Omega \in I\!R$. Hence, by taking $x_1 = x_2 = (i\omega I + F)^{-1} C^* y$ in (2.3), we immediately obtain (2.2). The proof is thus completed. □

Lemma 2.2. *Let $S_\delta \in \Gamma_\delta$ and $\Delta = S_\delta - S_0$. If $F_\delta = A - QS_\delta$ is also stable, then*

$$\inf_{\omega \in I\!R} \|I - C\Delta(i\omega I + F)^{-1} C^*\|_{\mathcal{L}(Y)} > 0. \qquad (2.4)$$

Proof. If F_δ is stable, then $i\omega I + F_\delta$ is boundedly invertible for all $\omega \in I\!R$. Note that $F_\delta = F - C^* C\Delta$, one can easily verify

$$I - C\Delta(i\omega I + F)C^* = [I + C\Delta(i\omega I + F_\delta)^{-1} C^*]^{-1}. \qquad (2.5)$$

Suppose that (2.4) is violated. Then, there exist sequences $\{\omega_n\}$ and $\{y_n\}$, $\omega_n \in I\!R$, $y_n \in Y$ with $\|y_n\| = 1$, such that

$$\widetilde{y}_n = [I - C\Delta(i\omega I + F)^{-1} C^*] y_n \to 0, \quad \text{as } n \to \infty. \qquad (2.6)$$

The stability of F_δ implies

$$\alpha \equiv \sup_{\omega} \|C\Delta(i\omega I + F_\delta)^{-1} C^*\|_{\mathcal{L}(Y)} < \infty. \qquad (2.7)$$

Hence, by virtue of (2.5), we obtain

$$\begin{aligned} 1 = \|y_n\| &= \|[I + C\Delta(i\omega I + F_\delta)^{-1}C^*]\tilde{y}_n\| \\ &\leq (1+\alpha)\|\tilde{y}_n\| \to 0, \quad \text{as } n \to \infty. \end{aligned} \tag{2.8}$$

This is a contradiction and therefore (2.4) must be true. □

In what follows, we shall introduce

$$T(i\omega) = G^*(i\omega I + F)^{-1}C^*, \tag{2.9}$$

and then define

$$\|T\|_\infty = \sup_\omega [\sigma(T^*(i\omega)T(i\omega))]^{1/2}, \tag{2.10}$$

where $\sigma(\cdot)$ stands for the spectral radius of bounded linear operators. An immediate consequence of Lemma 2.1 and Lemma 2.2 then is

Corollary 2.3. *If there exists an $S_\delta \in \Gamma_\delta$ satisfying (1.9), then*

$$\|T\|_\infty < \frac{1}{\delta}. \tag{2.11}$$

The following lemma is an alternative version of the result by Yakubovich [6].

Lemma 2.4. *Let $L^* \in \mathcal{L}(X)$ be given by (1.7). Then*

$$\|G^*L^*C^*\|_{\mathcal{L}(\mathcal{Y},\mathcal{V})} = \|T\|_\infty, \tag{2.12}$$

where $\mathcal{Y} = L^2(0,\infty;Y)$ and $\mathcal{V} = L^2(0,\infty;V)$.

From Lemma 2.4 and the fact that

$$\|G^*L^*QLG\| = \|G^*L^*C^*\|^2, \tag{2.13}$$

we obtain

Corollary 2.5. *The conditions (1.10) and (2.10) are equivalent.*

We conclude this section with the following result.

Lemma 2.6. *Let $S_0 \in \Gamma_0$ be the unique nonnegative solution. Then, for any $S_\delta \in \Gamma_\delta$, we have*

$$S_\delta - S_0 \leq 0. \tag{2.14}$$

Proof. By Lemma 2.1, we see that $\Delta = S_\delta - S_0$ satisfies (2.1). For $x \in \mathcal{D}(A)$, let $\xi(t) = e^{Ft}x$. It is seen that $\xi(t) \in \mathcal{D}(A) = \mathcal{D}(F)$ for all $t \geq 0$ and $\lim_{t\to 0}\xi(t) = 0$ since F is stable. By (2.1), we can compute

$$\frac{d}{dt}(\Delta\xi(t),\xi(t)) = (Q\Delta\xi(t),\Delta\xi(t)) + \delta^2(G^*\xi(t),G^*\xi(t)). \tag{2.15}$$

Hence, it follows that

$$(\delta x, x) = -\int_0^\infty [(Q\Delta\xi(t),\Delta\xi(t)) + \delta^2(G^*\xi(t),G^*\xi(t))]dt. \tag{2.16}$$

The right-hand side of (2.16) converges because e^{Ft} is exponentially stable. Furthermore, Δ is bounded and $\mathcal{D}(A)$ is dense in X, so (2.16) is valid for all $x \in X$. Then, (2.14) follows from (2.16). □

§3. Proof of Theorem 1.1.

Proof of (a). The "only if" direction follows from Corollaries 2.3 and 2.5. We proceed to prove the "if" direction. Suppose the condition (1.10) is satisfies. Then, (1.11) makes sense and defines a bounded linear operator $S_\delta \in \mathcal{L}(X)$. By Lemma 2.1, to show $S_\delta \in \Gamma_\delta$, it suffices to show

$$\Delta = -\delta^2 L_0 G[I - \delta^2 G^* L^* QLG]^{-1} G^* L_0^* \tag{3.1}$$

satisfies (2.1). To this end, let us introduce the first order differential operator

$$D : f(t) \mapsto \frac{d}{dt} f(t). \tag{3.2}$$

Clearly, for any $T \in \mathcal{L}(H_1, H_2)$ and $f \in L^2(0, \infty; H_1)$ with $Df \in L^2(0, \infty; H_1)$, we have

$$DTf = TDf. \tag{3.3}$$

Moreover, one can easily verify the following identities:

$$G^* L_0^* Fx = DG^* L_0^* x, \qquad \forall x \in \mathcal{D}(A) = \mathcal{D}(F). \tag{3.4}$$

$$L_0 Dg = -g(0) - F^* L_0 g, \qquad \forall g \in L^2(0, \infty; \mathcal{D}(A)) \text{ with } Dg \in \mathcal{X}, \tag{3.5}$$

$$DL^* g = L_0^* g(0) + L^* Dg, \qquad \forall g \in \mathcal{X} \text{ with } Dg \in \mathcal{X}, \tag{3.6}$$

$$DLg = LDg, \qquad \forall g \in \mathcal{X} \text{ with } Dg \in \mathcal{X}. \tag{3.7}$$

For notational convenience, we shall set $E_\epsilon = I - \delta^2 G^* L^* QLG$ and henceforth. For $g \in \mathcal{V}$ with $Dg \in \mathcal{V}$, let $f = E_\delta^{-1} g$ or $g = E_\delta f$. Then, by means of (3.3)–(3.7), we can deduce

$$\begin{aligned} Dg &= Df - \delta^2 D\{G^* L^* QLGf\} \\ &= Df - \delta^2 G^* L^* QLGDf - \delta^2 G^* L_0^* QL_0 Gf. \end{aligned} \tag{3.8}$$

From (3.8), it follows that

$$DE_\delta^{-1} g = E_\delta^{-1} Dg + \delta^2 E_\delta^{-1} G^* L_0^* QL_0 GE_\delta^{-1} g. \tag{3.9}$$

Let $f = E_\delta^{-1} G^* L_0^* x$, $x \in \mathcal{D}(A)$. Then, $f \in \mathcal{V}$ since E_δ is an isomorphism on $\mathcal{V}$. With a standar argument, we can claim that $G^* L^* QLGf \in L^\infty(0, \infty; V)$. Furthermore, from

$$f(t) - \delta^2 G^* \int_0^t e^{F(t-s)} ds Q \int_s^\infty e^{F^*(\tau - s)} Gf(\tau) d\tau = G^* e^{Ft} x, \tag{3.10}$$

we can also conclude that $f \in L^\infty(0, \infty; V)$ and, in particular,

$$\overline{\lim_{t\to\infty}} \|f(t)\| < \infty, \qquad \lim_{t\to 0} f(t) = \lim_{t \to 0} G^* e^{Ft} x = G^* x. \tag{3.11}$$

With the above preparation, we can do the following verification. Let

$$\Delta = -\delta^2 L_0 G E_\delta^{-1} G^* L_0^*$$

and let $x, x_2 \in \mathcal{D}(A)$. Set $f = E_\delta^{-1} G^* L_0^* x_1$, we have

$$\begin{aligned}(\Delta F x_1, x_2) &= -\delta^2 (L_0 G E_\delta^{-1} G^* L_0^* F x_1, x_2)\\ &= -\delta^2 (L_0 G E_\delta^{-1} D G^* L_0^* x_1, x_2)\\ &= -\delta^2 (L_0 G D E_\delta^{-1} G^* L_0^* x_1, x_2) + \delta^4 (L_0 G E_\delta^{-1} G^* L_0^* Q L_0 G E_\delta^{-1} G^* L0^* x_1, x_2)\\ &= \delta^2 (G f(0), x_2) + \delta^2 (F^* L_0 G E_\delta^{-1} G^* L_0^* x_1, x_2) + (\Delta Q \Delta x_1, x_2)\\ &= \delta^2 (G^* x_1, G^* x_2) - (\Delta x_1, F x_2) + (\Delta Q \Delta x_1, x_2).\end{aligned} \tag{3.12}$$

This shows that the operator given by (3.1) does satisfy (2.1). It remains to show that the solution S_δ given by (1.11) satisfies (1.9). To prove this, it suffices to show (see [4])

$$\int_0^\infty \|e^{F_\delta t} x\|^2 dt < \infty, \qquad \forall x \in X, \tag{3.13}$$

where

$$F_\delta = A - Q S_\delta = F - C^* C \Delta(\delta), \tag{3.14}$$

with $\Delta = \Delta(\delta)$ being the operator given by (3.1). BY the well-known Parseval equality, we see that

$$\int_{-\infty}^\infty \|(i\omega I + F)^{-1} x\|^2 d\omega = \int_0^\infty \|e^{F t} x\|^2 dt < \infty, \qquad x \in X \tag{3.15}$$

since F is satble. Let us assume for the moment that $I - (i\omega I + F)^{-1} C^* C \Delta(\delta)$ is boundedly invertible for all $\omega \in I\!R$ and

$$k = \sup_\omega \|[I - (i\omega I + F)^{-1} C^* C \Delta(\delta)]^{-1}\| < \infty. \tag{3.16}$$

Then, from

$$(i\omega I + F)^{-1} = [I - (i\omega I + F)^{-1} C^* C \Delta(\delta)]^{-1} (i\omega I + F)^{-1}, \tag{3.17}$$

it follows that

$$\begin{aligned}\int_0^\infty \|e^{F t} x\|^2 dt &= \int_{-\infty}^\infty \|(i\omega I + F_\delta)^{-1} x\|^2 d\omega\\ &\le k \int_{-\infty}^\infty \|(i\omega I + F)^{-1} x\|^2 d\omega < \infty,\end{aligned}$$

as desired. We now proceed to show (3.16). If it is not true, then there exists sequences $\omega_n \in I\!R$ and $x_n \in X$, with $\|x_n\| = 1$, such that

$$[I - (i\omega_n I + F)^{-1} C^* C \Delta(\delta)] x_n \to 0, \qquad \text{as } n \to \infty. \tag{3.18}$$

Let $\widetilde{x}_n = C \Delta(\delta) x_n$. Then, from (3.18), we obtain

$$[I - C \Delta(\delta)(i\omega_n I + F)^{-1} C^*] \widetilde{x}_n \to 0, \qquad \text{as } n \to \infty. \tag{3.19}$$

By (2.2), we see that $\|I-C\Delta(\delta)(i\omega_n I+F)^{-1}C^*\|^2 \geq 1-\delta^2\|T\|_\infty > 0$, and hence (3.19) yields $C\Delta(\delta)x_n = \tilde{x}_n \to 0$, as $n \to \infty$. This fact together with (3.18) lead to $x_n \to 0$, as $n \to \infty$, a contradiction! Thus, we see that (3.16) must be true and (3.13) is proved.

Finally, let S_δ be defined by (1.11) and let $\widetilde{S}_\delta \in \Gamma_\delta$. Then, with a simple manipulation, we obtain

$$((\widetilde{S}_\delta - S_\delta)x_1, F_\delta x_2) + (F_\delta x_1, (\widetilde{S}_\delta - S_\delta)x_2) = ((\widetilde{S}_\delta - S_\delta)Q(\widetilde{S}_\delta - S_\delta)x_1, x_2). \tag{3.20}$$

Note that F_δ is stable, the same argument as employed in proving Lemma 2.6 enables us to derive

$$S_\delta \geq \widetilde{S}_\delta. \tag{3.22}$$

This shows that the selfadjoint solution S_δ with the property (1.9) is a maximal element in Γ_δ and hence is unique.

Proof of (b). With the conclusion (a) in hand, the "if" direction would be a trivial matter.Let us now turn to the "only if" direction. If there exists an $S_{\delta_0} \in \Gamma_{\delta_0}$ which is positive definite. Then, from (2.14), it follows that

$$S_0 \geq S_{\delta_0} > 0. \tag{3.23}$$

Moreover, by (2.2), we see that $1 - \delta_0^2\|T\|_\infty > 0$, which implies that (1.10) holds for all $\delta \in [0, \delta_0)$. Hence, $\Gamma_\delta \neq \emptyset$ for all $\delta \in [0, \delta_0)$ and (1.11) gives the solution S_δ that satisfies (1.9). With a similar argument as used to establish (3.23), it can be shown that

$$S_\delta \geq S_{\delta_0} > 0, \qquad \forall \delta \in [0, \delta_0). \tag{3.24}$$

Since $S_0 > 0$, we may write (1.11) into

$$S_\delta = S_0^{1/2}\{I + \delta^2 S_0^{-1/2} L_0 G[I - \delta^2 G^* L^* QLG - \delta^2 G^* L_0^* S_0^{-1} L_0 G]^{-1} G^* L_0^* S_0^{-1/2}\}^{-1} S_0^{1/2}. \tag{3.25}$$

The above is valid at least for δ sufficiently small. Then, from (3.24) and (3.25), we can derive the necessity of the condition (1.12) without much difficulty. The proof is thus completed. □

Remark. The same technique can be used to establish conditions for $(ARE)_\delta^*$ to have positive definite solutions. These results together with their applications to H^∞-optimal control problem of infinite dimensions will be given in [2].

References

[1] A. V. Balakrishnan, *Applied Functional Analysis,* Springer-Verlag, New York, 1976.

[2] S. Chen, X. Li, S. Peng and J. Yong, work in preparation.

[3] R. F. Curtain and A. J. Pritchard, *Infinite Dimensional Linear Systems Theory, Lecture Notes in Control and Information Sciences,* Vol.8, Springer-Verlag, New York, 1981.

[4] R. Datko, *Extending a theorem of A. M. Liapunov to Hilbert space, J. Math. Anal. Appl.,* 32 (1970), 610–616.

[5] A. Ichikawa, *Linear quadratic differential games in a Hilbert space*, *SIAM J. Control & Optim.*, 14 (1976), 120–136.

[6] V. A. Yakubovich, *The frequency theorem for the case in which the state space and the control space are Hilbert spaces, and its application in certain problems in the synthesis of optimal control II*, Sibrisk Mat. Z., 16 (1975), 1081–1102.

[7] J. Zabczyk, *Remarks on the algebraic Reccati equations in Hilbert space*, *Appl. Math. Optim.*, 2 (1975), 251–258.

OPTIMAL CONTROL FOR DATA ASSIMILATION IN METEOROLOGY

François-Xavier Le Dimet
Université Blaise-Pascal
63177 Aubière Cedex, France

INTRODUCTION

To carry out a numerical forecast in meteorology 2 items are needed:

- a numerical model modelizing the evolution of the atmosphere. The equations used are the general aquation of fluid dynamics plus a thermodynamic equation and an equation for water under several phases (liquid, vapor, solid). Therefore we obtain a system of 6 nonlinear partial differential equations (3 components of the wind, temperature, atmospheric pressure, humidity). A very simplified model (2-D, without thermodynamic and humidity) is given by the Shallow Water Equations:

$$\frac{\partial u}{\partial t} + u\frac{\partial u}{\partial x} + v\frac{\partial u}{\partial y} - fv + g\frac{\partial h}{\partial x} = 0$$
$$\frac{\partial v}{\partial t} + u\frac{\partial v}{\partial x} + v\frac{\partial v}{\partial y} + fu + g\frac{\partial h}{\partial y} = 0$$
$$\frac{\partial h}{\partial t} + u\frac{\partial h}{\partial x} + v\frac{\partial h}{\partial y} + h\left(\frac{\partial u}{\partial x}+\frac{\partial v}{\partial y}\right) = 0$$

In these equations u and v are the components of the wind, h is the geopotential, proportional to the height opf the atmosphere, f is the Coriolis parameter. After discretization in space we obtain a system of ordinary differential equations for which an initial condition has to be provided before a numerical integration giving the weather forecast.

-Data: the data used in meteorology are provided by the synoptic network each 12 hours, additional data are given by satellites, aircrafts and ships. From a general point of view these data arte heterogeneous in quality and density (in space and in time).

The problem is to insert these data into the model in such a way that the resulting solution of the differential equation

i) is not too far from the observation

ii) is in agreement with the general properties of the atmosphere. The dynamic system representing the atmosphere has attractors, the forecast has to be located on this attractor.

To fulfill this requirements we have proposed, see e.g. Le Dimet and Talagrand (1986), to use optimal control methods which have the advantage to transform this problem into a problem of unconstrained optimization for which standard algorithms are available.

OPTIMALITY CONDITIONS

For sake of simplicity, we will consider the problem after discretization in space. In the following X denotes the state of the atmosphere, X belongs to a space $\mathbb{X}$,

$$\frac{dX}{dt} = F(X) + B.U \qquad (1)$$

$$X(0) = V \qquad (2)$$

U is some control parameter, in a space $\mathbb{U}$, for instance representing the boundary conditions. V is the initial condition in $\mathbb{V}$, B is a linear operator from $\mathbb{U}$ to $\mathbb{X}$. We suppose that if U and V being given the differential system has a unique solution between 0 and T. The difference between a solution of the model and the observation is measured by the cost-function J defined by:

$$J(U, V) = \frac{1}{2}\int_0^T \| C.X - X_{obs} \|^2 dt$$

C is a linear mapping from $\mathbb{X}$ to $\mathbb{O}$. Therefore the problem is to determine U* and V* minimizing J. From the numerical point of view, we need to determine the gradient of J with respect to U and V before carrying out a method of unconstrained optimization.

First order optimality condition

Let H= (H_U , H_V) be a perturbation applied on the control variables U and V. For any variable Z depending on Y we define its Gateaux derivative $\hat{Z}$ in the direction H as given by:

$$\widehat{Z}(Y,H) = \lim_{\alpha \to 0} \frac{Z(Y+\alpha H) - Z(Y)}{\alpha}$$

The directional derivatives of X and J are deduced from (1) and (2):

$$\frac{d\widehat{X}}{dt} = \left[\frac{\partial F}{\partial X}\right].\widehat{X} + BH_u \quad (3)$$

$$\hat{J}(U, V, H) = \int_0^T \left(C.X - X_{obs}, C.\widehat{X}\right) dt \qquad (4)$$

$\left[\frac{\partial F}{\partial X}\right]$ is the Hessian matrix of H with respect to X.

The gradient of J is obtained by exhibiting the linear dependance of $\hat{J}$ with respect to the perturbation H. Let us introduce the adjoint variable P, take the inner product of (3), (4) with P and integrate from 0 to T. We get :

$$\int_0^T \left(\frac{d\widehat{X}}{dt}, P\right) dt = \int_0^T \left(\left[\frac{\partial F}{\partial X}\right].\widehat{X} + BH_u, P\right) dt \qquad (5)$$

After an integration by parts we see that if the adjoint system is defined as the solution of the differential equation :

$$\frac{dP}{dt} + \left[\frac{\partial F}{\partial X}\right]^t.P = C^t.(CX - X_{obs}) \qquad (6)$$

with the condition at time T:

$$P(T)=0. \qquad (7)$$

Then the components of the gradient of J with respect to U and V are:

$$Grad_U(J(U,V)) = -B^t.P$$
$$Grad_V(J(U,V)) = -P(O)$$

Second Order Optimality Conditions

The matrix C represents a linear application from the space $\mathbb{X}$ into the space of observations $\mathbb{O}$. If only few observations are given then the problem may have several solutions. For practical purposes it is important to link the uniqueness of the solution to the number of observations, this can be done by computing the hessian matrix of J with respect to the initial condition (we will suppose that B=0).

$$\frac{dX}{dt}=F(X) \qquad (10)$$

$$X(0)=U \qquad (11)$$

The second derivative of the cost-function with respect to the initial condition is obtained by deriving the mapping:

$$\nabla J : U \rightarrow -P(0)$$

Let us consider K a perturbation on the initial condition for X. The directional derivatives of X and P are obtained from (10)-(11) and (6)-(7):

$$\frac{d\hat{X}}{dt}=\left[\frac{\partial F(X)}{\partial X}\right].\hat{X} \qquad (12)$$

$$\widehat{X(0)} = K \qquad (13)$$

$$\frac{d\hat{P}}{dt}+\left[\frac{\partial^2 F}{\partial X^2}.\hat{X}\right]^t P+\left[\frac{\partial F}{\partial X}\right]^t .\hat{P}=C^t.C.\hat{X} \qquad (14)$$

$$\hat{P}(T)=0 \qquad (15)$$

Let us introduce the bidual variables Q and R, we take the inner product of (12) and (14) by Q and R respectively and integrate between 0 and T, it comes:

$$\int_0^T \left(\frac{d\hat{X}}{dt},Q\right)+\left(\frac{d\hat{P}}{dt},R\right)-\left(\left[\frac{\partial F}{\partial X}\right].\hat{X},Q\right)+\left(\left[\frac{\partial^2 F}{\partial X^2}.\hat{X}\right]^t.P+\left[\frac{\partial F}{\partial X}\right]^t.\hat{P},R\right)-\left(C^t.C.\hat{X},R\right)dt=0 \qquad (16)$$

Eq. (16) is integrated by parts, therefore if we set Q and R as solutions of the differential equations:

$$\frac{dQ}{dt} + \left[\frac{\partial F}{\partial X}\right]^t .Q - \left[\frac{\partial^2 F}{\partial X^2} .P\right]^t .Q = C^t.C.R \qquad (17)$$

$$-\frac{dR}{dt} + \left[\frac{\partial F}{\partial X}\right] . R = 0 \qquad (18)$$

with the initial condition:

$$Q(T) = 0 \quad (19)$$

Then from (16) it remains:

$$-(K, Q(0)) = (\widehat{P}(0), R(0)) \qquad (20)$$

From (11) we have:

$$\widehat{P}(0) = -(\nabla J)' . (U,K) \quad (21)$$

Let H be the Hessian matrix of J, then it verifies:

$$H.K = \nabla^2 J.K = -(\nabla J)'.(U,K) \quad (22)$$

Therefore the columns of the matrix H are equals to the N different values of the vector Q at time 0 obtained by N integrations of the differential system (17), (18), (19) with initial conditions on R:

$$R(0)=E_i, \qquad 1 \leq i \leq N \qquad (23)$$

where the E_i are the vectors of the canonical base.

Eigenvectors and eigenvalues can be computed by carrying out an iterated power method.

Numerical results.

The domain which has been considered is a rectangle of size L= 6000km by D=4400km. The discretization has been done in finite difference with a centered scheme, the parameters of the discretization in space are Δx = 300km, Δy= 220km. Inn time we have used a leap-frog scheme with a timestep of Δt = 600s.

Figure 1 shows the original initial fields of geopotential (Fig.1A) and wind (Fig.1B). The "true" fields are computed by integrating the shallowwater equations from this initial condition. "Observed" fields (Fig.3) are computed from the

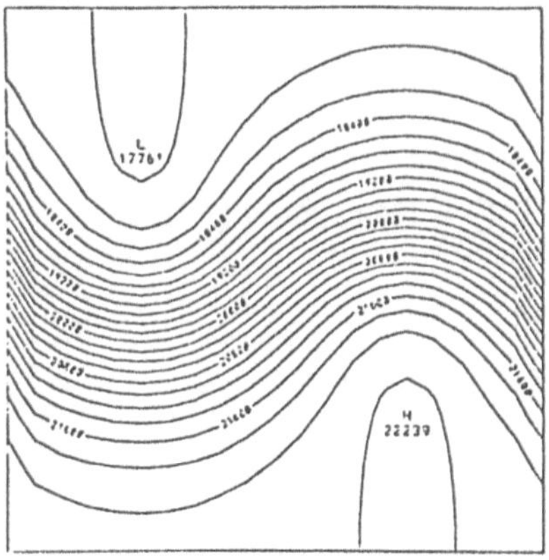

Figure 1.A

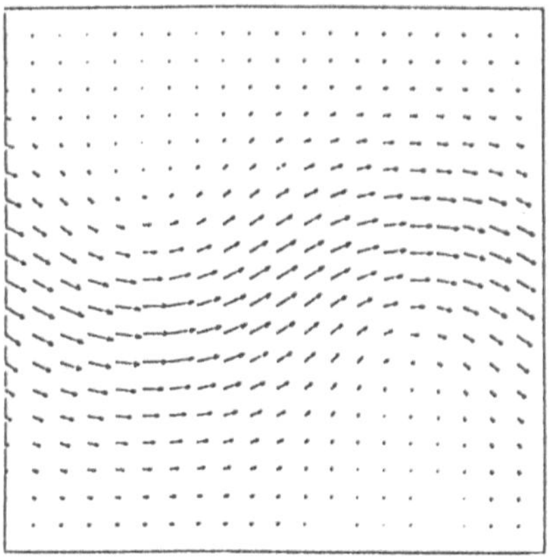

Figure 1.B

Fig1. A. Optimal Initial Condition for the geopotential
Fig1. B. Optimal Initial Condition for the wind

Figure 2.A

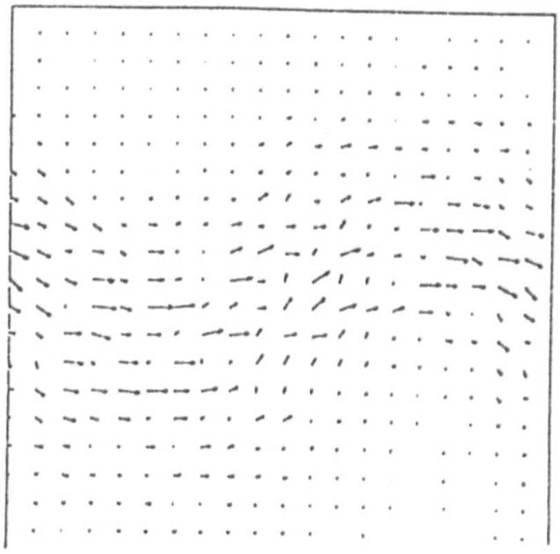

Figure 2.B

Fig 2. A. Observed Initial Condition for the geopotential
Fig 2. B. Observed Initial Condition for the wind

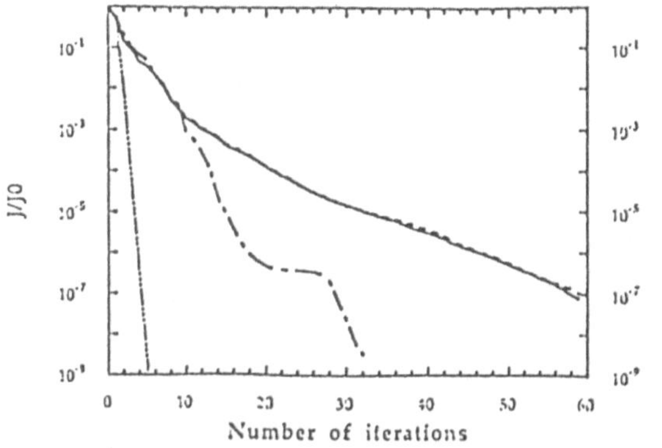

Figure 3. Evolution of the normalized cost-function

true fields by adding a random noise. The retrieved initial condition, after having performed the method, does not show any visual difference with the "true" one.

Figure 3 is the evolution of the cost-function with the number of iterations for 4 different algorithms of unconstrained optimization: QN1 and QN2 are quasi-Newton algorithms while TN1 and TN2 are truncated-Newton methods. These results show the importance of the choice of the method of unconstrained optimization on the total cost of the method. A systematic comparison of several methods of optimization applied to this problem can be found in Zou, Le Dimet, Navon, Nouailler (1990).

PHYSICAL CONSTRAINTS

From meteorological experiments it is well known that the atmosphere carries only few gravity waves in comparison of what could be expected from the governing equations . This is due to the fact that the system has an attractor and the natural evolution of the atmosphere lies on this attractor, if the initial condition ,which is provided to the model, does not belong to the attractor then the numerical solution countains gravity waves progressively damped but nevertheless preventing a short term forecast. The mathematical stucture of an attractor is very complex even for simple and low dimensional differential systems. Therefore these additional constraints may be handled into two different ways:

- by filtering gravity waves
- by using an approximation of the attractor.

Filtering of gravity waves

It is possible to filter the gravity waves by adding a constraint on the time derivative of X. In practice this can realized by using penalization-type method.

With **penalization methods** an additional term is added to the cost function which take the form:

$$J_\varepsilon(X) = J(X) + \frac{1}{\varepsilon}\left\|\frac{dX}{dt}\right\|^2$$

The effect of the penalty term is to damp the fast waves in the solution and to change the conditionning by an appropriate choice of the parameter ε. The crucial problem is the choice of ε, when ε goes to 0, then the solution will converge toward a steady state solution, there is no physical evidence on the choice of ε.

An alternative method is to use a **penalization-regularization** method. If H is a given parameter we define the functional Φ by:

$$\Phi(X) = 0 \text{ if } X < H \text{ and } \Phi(X) = (X-H)^2 \text{, if } X \geq H$$

The penalized functional will be:

$$J_\varepsilon(X) = J(X) + \frac{1}{\varepsilon} \Phi\left(\left\|\frac{dX}{dt}\right\|\right)$$

The adjoint system is derived as the solution of:

$$\frac{dP}{dt} + \left[\frac{\partial F}{\partial X}\right]^t .P = C^t.(CX-X_{obs}) + \frac{1}{\varepsilon} \frac{\partial\left(\frac{1}{\left\|\frac{dX}{dt}\right\|}\Phi'\left(\left\|\frac{dX}{dt}\right\|\right)\right)}{\partial t}$$

with the condition at time T:

$$P(T) = \frac{1}{\varepsilon} \frac{1}{\left\|\frac{dX(T)}{dt}\right\|} \Phi'\left(\left\|\frac{dX(T)}{dt}\right\|\right)$$

Then the gradient is given by:

$$\mathrm{Grad}(J(u)) = -P(0) - \frac{1}{\varepsilon} \frac{1}{\left\|\frac{dX(T)}{dt}\right\|} \Phi'\left(\left\|\frac{dX(T)}{dt}\right\|\right)$$

The advantage of such an approach is to apply the penalization only if it is necessary, furthermore the parameter H can be chosen on physical considerations such as an estimation of the velocity of gravity waves.

Let us define U_{ad} by:

$$U_{ad} = \left\{ U \in R^n, \left\|\frac{dX(U)}{dt}\right\| \leq H \right\}$$

By using standard methods it is possible to get:

Theorem 1

If F is lipschitzian and U_{ad} is nonempty then the problem of optimization has a solution U*.

Theorem 2

The penalized-regularized problem has a solution U_ε^* and $U_\varepsilon^* \to U^*$ when $\varepsilon \to 0$.

Another advantage obtained in adding some penalization term in the definition of the cost function is to control the regularity of the solution.

Approximations of the attractor

After dicretization in space, the shallow water equation can be linearized around a standard state. It is shown (see e.g. Daley (1982)) than the spectrum of the linearized operator can be splitted into two parts corresponding to two differrents waves:

- Rossby waves which the slow part of the motion
- Gravity wave are the fast components of the motion.

In a basis of eigenvectors the evolution of W=(Y,Z), the state of the atmosphere, between 0 and T verifies:

$$\frac{dY}{dt} = A_Y.Y + N_Y(Y, Z)$$

$$\frac{dZ}{dt} = A_Z.Z + N_Z(Y, Z)$$

In this expressions Y is the Rossby component of the motion belonging to $\mathbb{Y}$ (the Rossby manifold), Z is the gravity component in $\mathbb{Z}$ (the gravity manifold). A_Y and A_Z are diagonal matrices, N_Y and N_Z are the nonlinear parts of the equations.

If a noisy observation $W_{obs}= (Y_{obs}, Z_{obs})$ is given at time 0, it is not enough to use $W(0) = (Y_{obs}, 0)$ because the nonlinear term in the equation will bring back gravity waves in the solution. To prevent the development of gravity waves we can consider the problem of determining the optimal initial condition U* in Y minimizing J with:

$$J(U) = \frac{1}{2}\int_0^T \left(\|Y(t) - Y_{obs}\| + \|Z(t) - Z_{obs}\|\right) dt$$

with the constraints:

$$\frac{dY}{dt} = A_Y.Y + N_Y(Y, Z)$$

$$Y(0) = U$$

$$A_Z.Z + N_Z(Y, Z) = 0$$

The optimality condition is obtained by introducing the adjoint system. Let Q and R be the adjoint variable to Y and Z, the adjoint system is defined as the solution of the differential system:

$$\frac{dP}{dt} + A_Y.P + \left[\frac{\partial N_Y}{\partial Y}\right]^t.P - \left[\frac{\partial N_Y}{\partial Z}\right]^t.Q = Y(t) - Y_{obs}(t)$$

$$A_Z.Q + \left[\frac{\partial N_Z}{\partial Z}\right]^t.Q - \left[\frac{\partial N_Z}{\partial Y}\right]^t.P = Z(t) - Z_{obs}(t)$$

Then the gradient with repect to the initial condition on Y is given by:

$$\nabla J = -P(0)$$

It is important to point out that the dimension of the differential system is be integrated can be lower than before because the integration is carried out only on the Rossby component.

CONCLUSION

A crucial point for improving numerical weather forecast his to get efficient methods for data assimilation and which are able to be implemented on operational models of very large dimension (of the order of 10^6). Optimal control methods seems well-suited for this purpose, nevertheless a lot of studies remain to be done from the dtudies on dynamical systems modelizing the atmosphere to the numerical analysis of these problems.

ACKNOWLEDGEMENT

The numerical results presented in this paper have been made on the CRAY2 belonging to the Centre de Calcul Vectoriel pour la Recherche du CNRS.

REFERENCES

Le Dimet F.-X. and O. Talagrand (1986): Variational Algorithms for Analysis and Assimilation of Meteorological Observations: Theoretical Aspects. *Tellus, 38A, 97-110.*

Zou X, Le Dimet F.-X., Navon I.M., Nouailler A. (1990): A Comparison of Efficient Large-Scale Minimization Algorithms for Optimal Control Applications in Meteorology. *Submitted to SIAM Journal on Optimization.*

ON THE STABILITY OF OPEN POPULATION LARGE SCALE SYSTEM*

Gao Hang

(Department of Mathematics, Northeast Normal University, Changchun, jilin 130024, China)

Abstract

In this paper, We will discuss the Lyapunov's stability of open population large scale system, obtain the time-varying lower critical value and the time-varying upper critical value of absolute birth-rate, and give the sufficient conditions for the large scale system to be stable and asymptotically stable, meanwhile give the necessary condition for the large scale system to be stable. These results may provide a strict mathematics theory for population control.

In this paper, the following open population large scale system is discussed[1]:

$$L\ P(r,t)=A(r,t)P(r,t) \qquad \text{in } Q=\Omega\times(0,\infty), \qquad (1)$$

$$P(r,0)=F(r) \qquad \text{in } \Omega=(0,R), \qquad (2)$$

$$P(0,t)=\Phi(t) \qquad \text{in } (0,\infty), \qquad (3)$$

where $L=\frac{\partial}{\partial r}+\frac{\partial}{\partial t}+M(r,t)$, $M(r,t)=\mathrm{diag}(\mu_1(r,t), \ldots, \mu_n(r,t))$, $P(r,t)=(p_i(r,t))_{i=1}^{n}$, $F(r)=(f_i(r))_{i=1}^{n}$, $\Phi(t)=(\varphi_i(t))_{i=1}^{n}$ are n rows 1 column vector function, $A(r,t)=(a_{ij}(r,t))_{i,j=1}^{n}$ is a n rows n columns matrix, and the functions $p_i(r,t)$, $\mu_i(r,t)$, $f_i(r)$ and $\varphi_i(t)$ denote respectivly age density function, mortality function,

* The Project Supported by National Natural Science Foundation of China

initial age pattern, absolute birth-rate function of the ith population group at age r and time t, $a_{ij}(r,t)$ denotes the migration rate function which is migrated from the jth population group into the ith population group, R is the highest age of the human race.

We suppose that

a). $v_i(r,t)=\mu_i(r,t)-a_{ii}(r,t)$ is subjected to the condition:

$$\int_0^R v_i(r,t)dr=+\infty \text{ and } \int_0^R v_i(x,t)dx<+\infty \quad \text{if } r<R,$$

$$\operatorname*{varimin}_{Q} \quad \mu_i(r,t)=\mu_0>0, \qquad v_i(r,t)\in H^{3/2,3/2}(Q_T'),$$

where $Q_T'=\Omega'\times(0,T)$, $\Omega'=(0,R')$, $0<R'<R$ and R' may be arbitrarily close to R, T may be arbitrarily large with $T<+\infty$, and $H^{3/2,3/2}(Q_T)$ is the Sobolev space[9].

b) $a_{ij}(r,t)\in H^{3/2,3/2}(Q_T)$, $Q_T=\Omega\times(0,T)$,

$$\sum_{i=1}^{n} a_{ij}(r,t)=0, j=1,2,\ldots, n, \text{ and if } i\neq j,\ a_{ij}(r,t)\geq 0.$$

c) $f_i(r)\in H^1(\Omega)$, $f_i(r)\geq 0$, $i=1,2,\ldots,n$, $\varphi_i(t)\in H^1(0,T)$,

$\varphi_i(t)\geq 0$, $i=1,2,\ldots,n$, $T<+\infty$.

First of all, we give a lemma.

<u>Lemma 1</u>[7,8]. We assume that the assumptions a), b), and c) are fulfilled for Problem (1)—(3), and $\{f_i(r),\varphi_i(t)\}$, $i=1,2,\ldots,n$, satisfy the compatibility relations (Я. В.) [9]. Then there exists a unique function $P(r,t)$ belonging to $\prod_{i=1}^{n} H^{3/2,3/2}(Q_T)$ satisfying the problem (1)—(3). Furthermore $P(r,t)=(p_i(r,t))$ satisfies the relation:

$$p_i(r,t)=\Big\{\int_0^r\Big[\sum_{j=1}^{n}a_{ij}p_j(s,s+t-r)\Big]\exp\Big[\int_0^s v_i(x,x+t-r)dx\Big]ds$$

$$+f_i(r-t)+\varphi_i(t-r)\Big\}\exp\Big[-\int_0^r v_i(x,x+t-r)dx\Big]. \qquad (4)$$

[2-6] have discussed the stability of closed population system. In this paper, we discuss the Lyapunov's stability of the open population system (1)—(3).

Let us introduce a function and a definition.

We set

$$K_i^-(r,s,t)=\mu_i(r,t)\exp\left[-\int_s^r v_i(x,x+t-r)dx\right].$$

<u>Definition 1</u>. We define

$$\varphi_{ic}^-(t)=\int_0^R K_i^-(r,0,t)\varphi_i(t-r)dr, \qquad (5)$$

$\varphi_{ic}^-(t)$ is called the time-varying lower critical value of absolute birth-rate $\varphi_i(t)$ for the ith subsystems of population large scale system (1)—(3).

<u>Theorem 1</u>. Assume that the conditions of Lemma 1 are satisfied, if $\varphi_i(t)\leqslant\varphi_{ic}^-(t)$, $i=1,2,\dots,n$, then, the large scale system (1)—(3) is stable.

<u>Proof</u>. From Lemma 1 and [9], we know that $\|p(\cdot,t)\|_{L^1(\Omega)}\in C[0,T]$. With the aim of convenience, we write $\|p(\cdot,t)\|=\|p(\cdot,t)\|_{L^1(\Omega)}$.

In equation (1), we have that

$$\frac{\partial}{\partial r}p_i(r,t)+\frac{\partial}{\partial t}p_i(r,t)+\mu_i(r,t)p_i(r,t)=\sum_{j=1}^n a_{ij}p_j(r,t),$$

we integrate both side of this equation with respect to r in Ω, according to condition a) and $p_i(r,t)\geqslant 0$, we obtain that

$$\frac{d}{dt}\|p_i(r,t)\|=\varphi_i(t)-\int_0^R\mu_i\,p_i(r,t)dr+\sum_{j=1}^n\int_0^R a_{ij}p_j(r,t)dr.$$

From condition b), we have

$$\frac{d}{dt}\|P(r,t)\|=\frac{d}{dt}\sum_{i=1}^n\|p_i(\cdot,t)\|=\sum_{i=1}^n\left[\varphi_i(t)-\int_0^R\mu_i\,p_i(r,t)dr\right]. \qquad (6)$$

From Lemma 1 and conditions b) and c), there exists following inequality:

$$p_i(r,t)\geqslant\varphi_i(t-r)\exp\left[-\int_0^r v_i(x,x+t-r)dx\right],$$

substituting the above inequality into (6), we have

$$\frac{d}{dt}\| p_i(.,t)\| \leqslant \sum_{i=1}^{n}[\varphi_i(t)-\varphi_{ic}^-(t)] . \qquad (7)$$

From the condition $\varphi_i(t)\leqslant\varphi_{ic}^-(t)$, obtain $\frac{d}{dt}\| P(.,t)\| \leqslant 0$, hence.

$$\| P(.,t)\| \leqslant \| F\| .$$

This shows the system (1)—(3) is stable. Theorem 1 is proved.

Theorem 2. Assume that the conditions of Lemma 1 are satisfied, if $\varphi_i(t)\leqslant a\varphi_{ic}^-(t)$, $i=1,2,\ldots,n$, $0<a<1$, then, the large scale system (1)—(3) is asymptotically stable.

Proof. We first prove that

$$\lim_{t\to\infty}\int_0^R \varphi_i(t-r)dr=0,\ i=1,2,\ldots,n . \qquad (8)$$

In fact, if there exists i_o, equality (8) doesn't hold for i_o, then there exists sequence $\{t_k\}$, $t_{k+1}>t_k+R$, so that $\int_0^R \varphi_{i_o}(t_k-r)dr\geqslant C>0$. We can prove easily that $0\leqslant\varphi_i(t)\leqslant M$ for $t\in(0,\infty)$. Set $E_k\subset\subset[t_k-R,t_k]$ so that $\varphi_{i_o}(t)>\frac{C}{2R}$ for $t\in E_k$. set $[t_k-R,t_k]-E_k=H_k$, owing to

$$C\leqslant\int_0^R\varphi_{i_o}(t-r)dr=\int_{t_k-R}^{t_k}\varphi_{i_o}(s)ds=\int_{H_k}\varphi_{i_o}(s)ds+\int_{E_k}\varphi_{i_o}(s)ds$$
$$\leqslant\frac{C}{2R}.R+M\ \mathrm{mes}E_k ,$$

from this inequality, we obtain that

$$\mathrm{mes}E_k\geqslant\frac{C}{2M},\quad k=1,2,\ldots .$$

Set $S_N=\bigcup_{k=1}^{N}E_k$, we have that

$$\varphi_{i_o}(t)-\varphi_{i_oc}^-(t)=\varphi_{i_o}(t)\left[1-\frac{\varphi_{i_oc}^-(t)}{\varphi_{i_o}(t)}\right]$$
$$\leqslant\frac{C}{2R}\left[1-\frac{1}{a}\right]=-\lambda \qquad (\lambda>0)$$

for $t\in S_N$. We integrate both side of inequality (7) with respect to t on $[0,T]$ and suppose $S_N\subset[0,T]$, then, following inequality holds:

$$\| P(.,T)\| \leqslant\| F\| +\int_0^T\sum_{i=1}^{n}[\varphi_i(t)-\varphi_{ic}^-(t)]dt$$
$$\leqslant\| F\| +\int_0^T[\varphi_{i_o}(t)-\varphi_{i_oc}^-(t)]dt$$
$$\leqslant\| F\| -\int_{S_N}\lambda\ dt\leqslant\| F\| -\lambda N\frac{C}{2M} .$$

If T is sufficiently large so that N may be sufficiently large, too, then, we obtain that $\|P(.,T)\|<0$. This not ture, because $\|P(.,t)\| \geq 0$ for $t \in (0,\infty)$, hence equality (8) holds for $i=1,2,\dots,n$.

Now, we want to prove $\lim_{t\to\infty}\|P(.,t)\|=0$.

From the equality (4), we have the inequality

$$0 \leq p_i(r,t) \leq f_i(r-t)+\varphi_i(t-r)+\int_0^r \sum_{j=1}^n p_j(s,s+t-r)ds.$$

We can prove the inequality

$$\sum_{j=1}^n p_j(s,s+t-r) \leq \sum_{j=1}^n p_j(0,t-r)$$

for $s \in \Omega$ and $p_j(0,t-r)=f_j(r-t)+\varphi_j(t-r)$. In particular, if $t>R$, we have $f_j(r-t)=0$, hence

$$0 \leq p_i(r,t) \leq \varphi_i(t-r)+R\sum_{j=1}^n \varphi_j(t-r).$$

From this inequality and (8), obtain that $\lim_{t\to\infty}\|P(.,t)\|=0$, i.e. the large scale system (1)—(3) is asymptotically stable. Theorem 2 is proved.

Now, we introduce following functions:

$$a_i(r,t)=\max\{a_{ij}(r,t) \mid j\neq i,\ 1\leq j\leq n\},$$

$$K_i^+(r,s,t)=K_i^-(r,0,t)+\sum_{j=1}^n \int_s^r K_j^-(r,x,t)a_j(x,x+t-r)dx.$$

<u>Definition 2</u>. We define

$$\varphi_{ic}^+(t)=\int_0^R K_i^+(r,0,t)\varphi_i(t-r)dr, \tag{9}$$

$\varphi_{ic}^+(t)$ is called the time-varying upper critical value of absolute birth-rate $\varphi_i(t)$ for the ith subsystems of population large scale system (1)—(3).

<u>Theorem 3</u>. Assume that the conditions of Lemma 1 are satisfied, if $\varphi_i(t) \geq \alpha_1 \varphi_{ic}^+(t)$, $i=1,2,\dots,n$, $\alpha_1>1$, then, the large scale system (1)—(3) is not stable.

<u>Proof</u>. For t>R, we have that

$$\frac{d}{dt} \| P(.,t) \| = \sum_{i=1}^{n} (\varphi_i(t) - \int_0^R \mu_i(r,t) p_i(r,t)dr)$$

$$= \sum_{i=1}^{n} [\varphi_i(t) - \int_0^R \bar{K}_i(r,0,t)\varphi_i(t-r)dr -$$

$$- \int_0^R \int_0^r K_i(r,s,t) \sum_{j\neq i} a_{ij} p_j(s,s+t-r)dsdr].$$

Owing to

$$\sum_{j=1}^{n} p_j(s,s+t-r) \leqslant \sum_{j=1}^{n} \varphi_j(t-r)$$

for $s \in \Omega$ and t>R, we have that

$$\frac{d}{dt} \| P(.,t) \| \geqslant \sum_{i=1}^{n} [\varphi_i(t) - \varphi_{ic}^{+}(t)].$$

According to the condition of this theorem, we know that $\frac{d}{dt} \| p(.,t) \| > 0$, hence for t>R, $\| P(.,t) \| \geqslant \| P(.,R) \|$. From Definition 2, we have that

$$\sum_{i=1}^{n} \varphi_{ic}(t) \geqslant \sum_{i=1}^{n} \int_0^R \mu_i(r,t) p_i(r,t)dr \geqslant \mu_0 \| P(.,t) \| \geqslant C_1 > 0,$$

From this inequality, we get

$$\frac{d}{dt} \| P(.,t) \| \geqslant \sum_{i=1}^{n} [\varphi_i(t) - \varphi_{ic}^{+}(t)].$$

$$= \sum_{i=1}^{n} \varphi_i(t) \left[\frac{\sum_{i=1}^{n} \varphi_i(t)}{\sum_{i=1}^{n} \varphi_{ic}^{+}(t)} - 1 \right] \geqslant C_1(\alpha_1 - 1).$$

Integrating the both side of this inequality on [R,t], we deduce

$$\| P(.,t) \| \geqslant \| P(.,R) \| + C_1(\alpha_1 - 1)(t-R).$$

If t is sufficiently large, then, $\| p(.,t) \|$ may take arbitrarily large value. This shows: the large scale system (1)—(3) is not stable. Theorem 3 is proved.

Now, we discuss two special case.

In the first place, if n=1, system (1)—(3) degenerate into following system:

$$\frac{\partial}{\partial r} p(r,t) + \frac{\partial}{\partial t} p(r,t) + \mu(r,t)p(r,t) = 0, \qquad (10)$$

$$p(r,0) = f(r) , \qquad (11)$$

$$p(0,t) = \varphi(t) , \qquad (12)$$

Definition 3. We define

$$\varphi_c(t)=\int_0^R \mu(r,t)\varphi(t-r)\exp\left[-\int_0^r \mu(x,x+t-r)dx\right]dr, \quad (13)$$

$\varphi_c(t)$ is called time-varying cratical value of absolute birth-rate $\varphi(t)$ for the system (10)—(12).

Corolley 1. Assume that $\mu(r,t)$, $f(r)$, $\varphi(t)$ satiafy the conditions a) and c), Then, if $\varphi(t)\leq\varphi_c(t)$ the system (10)—(12) is stable; if $\varphi(t)\leq a\varphi_c(t)$, $0<a<1$, the system (10)—(12) is asymptotically stable; if $\varphi(t)\geq a\ \varphi_c(t)$, $a>1$, the system (10)—(12) is not stable.

Proof. Owing to $A(r,t)=0$, hence

$$\varphi_c(t)=\varphi_c^-(t)=\varphi_c^+(t)$$

($\varphi_c^-(t)$ and $\varphi_c^+(t)$ are given by (5) and (9) in which the index i is removed), from Theorem 1— Theorem 3, we can obtain this result of Corolley 1.

In the next place, for the system (1)—(3), if

$$\mu_1(r,t)=\ldots=\mu_n(r,t)=\mu(r,t), \quad (14)$$

we have following results.

Corolley 2. Assume that the conditions of Lemma 1 are satisfied and (14) holds. Then, there exists $\varphi_{ic}(t)$, which is called the time-varying critical value of absolute birth-rate $\varphi_i(t)$ for the ith subsystems of population large scale system (1)—(3) and given:

$$\varphi_{ic}(t)=\int_0^R \mu(r,t)\varphi_i(t-r)\exp\left[-\int_0^r \mu(x,x+t-r)dx\right]dr. \quad (15)$$

If $\varphi_i(t)\leq\varphi_{ic}(t)$, $i=1,2,\ldots,n$, the large scale system (1)—(3) is stable; if $\varphi_i(t)\leq a\varphi_{ic}(t)$, $0<a<1$, $i=1,2,\ldots,n$, the large scale system (1)—(3) is asymptotically stable; if $\varphi_i(t)\geq a_1\varphi_{ic}(t)$, $a_1>1$, $i=1,2,\ldots,n$, the large scale system (1)—(3) is not stable.

Proof. Suppose $p(r,t)=\sum_{i=1}^{n} p_i(r,t)$, $f(r)=\sum_{i=1}^{n} f_i(r)$ and $\varphi(t)=\sum_{i=1}^{n}\varphi_i(t)$, then, the fuctions $\{p(r,t), f(r), \varphi(t)\}$ satisfy problem (10)—(12). We note that

$$\| P(.,t)\| = \| p(.,t)\| \quad \text{and} \quad \| F\| = \| f\|$$

from Corolley 1, we may obtain the results of Corolley 2.

Corolley 2 shows: if (14) holds, then, population migration from one subsystems to other subsystems doesn't affect the stability of large scale system (1)—(3).

REFERENCES

[1] Song Jian, Yu Jingyuan & Li Guangyuan, Scientia Sinica, 24(1981), 3:431—444.

[2] Song Jian & Yu Jingyuan, Math. Modelling, 2(1981), 1.

[3] Feng Dexing & Zhu Guangtian, Journal of Mathematical Research and Exposition, 3(1983), 3:1—6.

[4] Song Jian & Chen Renzhao, Scientia Sinica, 26(1983), 12:1314—1325.

[5] Chen Renzhao & Gao Hang, Science in china, 33(1990), 8:909—919.

[6] Chen Renzhao, Kexue Tongbao, 30(1985), 6:711—716.

[7] Chen Renzhao & Gao Hang, Journal of Northeast Normal University, 1985, 3:1—6.

[8] Gao Hang, Journal of Northeast Normal University, 1989, 3:13—20.

[9] Lions, J.L.& Magenes, E., Non-Homogeneous Boundary Value Problems and Applications, I, II, Springer-Verlag, 1972.

TEMPERATURE CONTROL SYSTEM OF HEAT EXCHANGERS
---AN APPLICATION OF DPS THEORY

Huang Guangyuan, Nie Lei, Zhao Yaowen
Department of Mathematic, Shangdong University, Jinan,Shandong 250100, P. R. China
Wu Qibin, Yang Weiming, Liu Jian
Departiment of Automatics,Qingdao Chamical Institute of Technology, Qingdao, Shangdong 266042, P.R.China

Abstract

To the feature of long delay time,vary parameter and nonlinear ets. in the long time-delay system,using the distributed parameter theory,we give a new kind of mixed control plan,the feedforward control plus indentifying modifying feedback. Via the real time experiment on a exchanger,it is shown that it is much more superior to the popular ones based on the lumps parameter theory.

1. Introduction

Weather in the control theory or the engineering practice,the problem about the control to the system with long time-delay (more exactly,i.e. the system whose purely delay time is much more greater than its transition time) is not completely solved by now. One of the main reason for the purely time-delay is the difference between the situations of the system's input and output. In the lumps parameter theory,the delay time it is regarded as a parameter of system itself,and when is much long, there is no all-round theoretical results yet. But the outstanding of the distributed parameter theory is just considering the feather of the system's distribution,thus we can conclude the system's time-delay phenomenou by its physical parameter and special propery in solution,then the effective methods to overcome the time-delay can be found out.

As is well known,feedforward is the effective methods to overcome the time-delay. Making full use of the distributed parameter theory,we can easily achieve the compound feedforward control plan,which is obvious in phisical sense and effective to cope with disturbance meanwhile. Of course it's not enough to ensure the accurate to have the feedforward only,considering the enssetial measure and modeling error. So it's necessery to indetify the system parameter from the singnals of input and output and modify the feedforward controller. Obviously,the consequence owe to this is much more reasonable than that owe to simple direct superposition feedback. Therefore we give the general control pattern to the long time-delay system as follows:

feedback = feedforward + identification

The construction of the system control is shown as Fig.1.

The difference between the plan above and the usual adaptive control based on the lumps parameter theory is that in the plan above the feedforward controller is formed by the special distributed physical feature of the system and the parameters which need to be indentified later are all physical ones. So the calculation needed is much less.

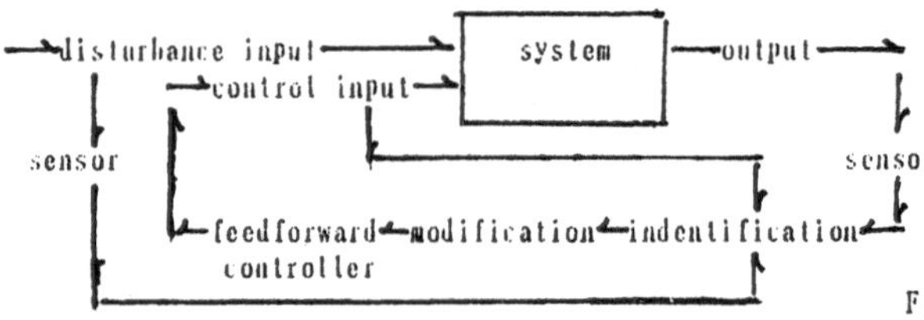

FIG.1

2. MODEL AND MATHEMATICAL ANALYSIS

The sketch of the heat exchanger is shown as Fig.2

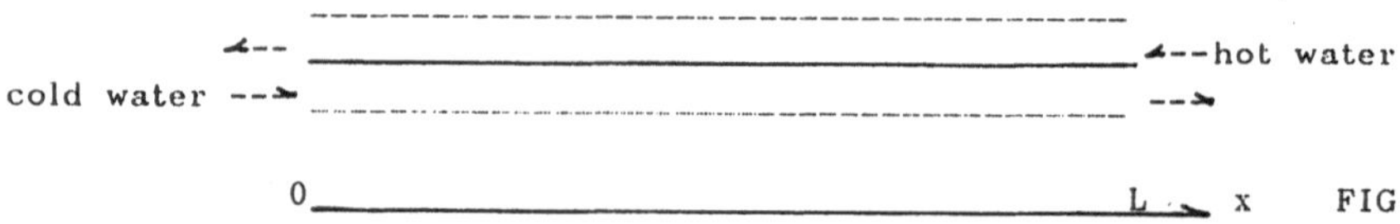

FIG.2

and its mathematical model can be obtained by physics as (1)(heat conduction omitted):

$$cpQ'_t+vcpQ'_x=a(P-Q), \quad drP'_t-udrP'_x=a(Q-P)$$

$$\text{Bound. Cond.} \quad Q(0,t)=Q_0(t), \quad P(L,t)=P_L(t) \qquad (1)$$

$$\text{Initi. Cond.} \quad Q(x,0)=Q_1(x), \quad P(x,0)=P_1(x),$$

where P and Q are the temperatures of the hot and cold waters respectively, and c,d ---specific heats, p,r---densities, v,u---velocities, a--- the heat-transfer coefficient. x---place, and t---time.

In practice, v or u is the control variable or the main disturbance and Q(L,t) is the controlled variable. Q_0 and P_L can be also disturbances. Obviously Q(L,t) and the time-delay are nonlinearly dependent on v and u.

3. COMPOUND CONTROL SYSTEM

Since all the disturbances are measurable,we obtain a scheme of feedforward control from solution to (1).the measuring signals

Q(L,t),P(L,t),Q(0,t) and P(0,t) can be used for the identification of these physical parameters. In experiments we find that the heat transfer coefficient a depends on v and u. To make the control methods more efficient, let's suppose

$$a = f(pv, ru) \qquad (2)$$

where the function f need to be defined and note $g = cp, g_1 = dr$, g(or g_1) means thermal capacitance.Then all together there are four independent parameters, i.e. g, g_1 ,a,and L. Without lossing generality we assume u is control variable. The scheme of control system is as Fig.3.

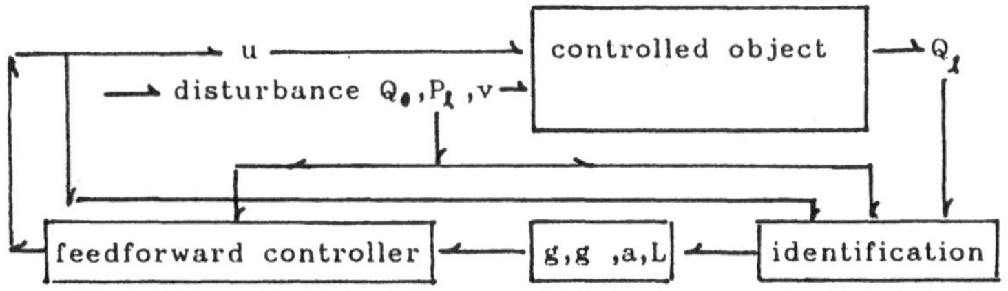

FIG.3

4. DEFFERENCE BETWEEN DPS AND LPS METHOD

There are three folds:

1). Each parameter in the model given by DPS method has clear physical sense, but in its discrete form (LPS model) all the parameters are dependent on the length of x and t, and with each other;

2). The control scheme of feedforward as above is based on the physical process, especially the relation between v,u,and Q(L,t) is nonlinear;

3). The computation time in the identification of parameters in Fig.3. is much less than that in LPS model, because all together there are 4 independent parameters to be identified here.

5. EXPERIMENS AND COMPARESON

The original control system (by LPS) :

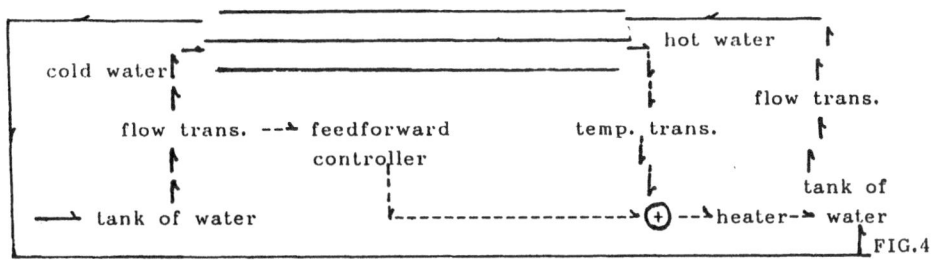

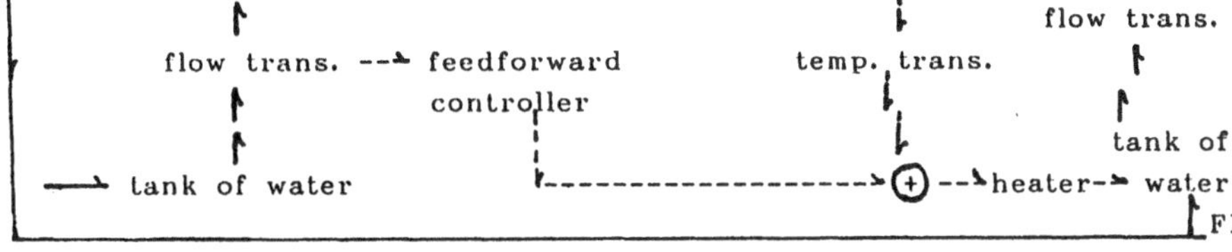

FIG.4

Only the steady situation control is considered here,and the dynamic progress control is also feasible by the principle mentioned above. In the steady situation,Q_0 ,P_l ,u,v are all constant,and the model is simplified as follows:

$$cpvQ'_x = a(P-Q) \tag{3}$$
$$druP'_x = a(P-Q)$$
$$B.C.:\ Q(0)=Q_0; \qquad P(L)=P_l \tag{4}$$

Solving the different equation above we have the steady state solution.

$$Q_l = Q_0+A(P_l-Q_0) \tag{5}$$
$$P_0 = P_l-B(P_l-Q_0) \tag{6}$$

where $A=\{1-\exp[(b_1-b)l]\}/\{1-b_1 b^{-1}\exp[(b_1-b)l]\}$

$B=\{1-\exp[(b_1-b)l]\}/\{bb_1^{-1}-\exp[(b_1-b)l]\}$

$b=a/cpv$; $b\ =a/dru$

So the relationship between the system's input and output can be obtained.

$$(Q_l-Q_0)/(P_l-Q_0) = A \tag{7}$$

1) Scheme 1. The aim is to ensure the temperature at the cool water exit to achieve the value given,using P_l as control.Its principle is shown as Fig.5.

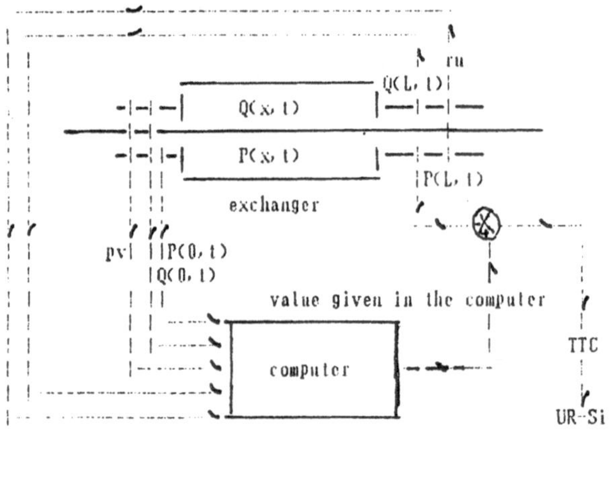

FIG.5

The feedforward control can be constructed from the formular (5),where Q_0 is the measured temperature in the cool water entrance,ru and pv are correspondingly the flowing amount of warm water and cool one,q is the temperature given.

As for the realization of identification,we can notice the relationship of steady state (5) and (6). Suppose

$$R=(Q_1-Q_0)/(P_1-Q_0); \qquad W=(P_0-P_1)/(Q_0-P_1)$$

$$X=(bL,b_1L); \qquad B=(0,\ln((R-1)/(W-1)))$$

$$A=\begin{pmatrix} R & -W \\ 1 & -1 \end{pmatrix}$$

we have

$$B=AX \tag{8}$$

Based on formular (8),we can indentify the parameter bL and b_1L,using the Least-Square method or others. consequently,the values of parameters a/c*L and a/d*L needed to be indentified in the model above are worked out ,by the follows

$$bL=a/(cpv)*L; \qquad b\ L=a/(dru)*L$$

The real time experiment is operated by an IBM-PC-XT computer with 16 D/A and 2 A/D. There are six various which can be measured,where P_0 ,P_1 ,Q_0 ,Q_1 are all transmitted into 0-10 mA electric signal by the hot-resistance temperature sensor, pv by an electric long-range rotor flowmeter and ru by a gas long- range rotor flowmeter plus gaselectric transmition.

The flowing of program is shown as Fig.6.

2) Scheme 2. Which has the same aim as Scheme 1, but using ru as control. The control principle is almost the same as shown in Fig.5. Based on formular (5)(the formular (6) can't be used,for the corresponding value of P_0 isn't known yet,and it varies much),theoretically we can give the value of ru corresponding to the given value q by solving the equation directly, but it's impossible in practice, for it's a transcendential equation concerning ru.In this case,we can get the needed value of ru by the numerical bisection seach method or the numerical iterated interpolation method.

The realization of identification in Schmem 2 is just the same as in Scheme 1. The flowing of progress is shown in Fig.7.

3) The experimential results (FIG.8 is for scheme 1 and FIG.9 for scheme 2) and the comparison with the common LPS plan. (See the form on P7)

From the comparison above,it's shown that the control plans based on DPS are much superior to that based on LPS in overcoming time-delay and deducing the transition time and so on. Besides,the accurate in steady state is satisfied and the qulity of the correspondence in the dynamic progress is high,too. The simple adaptive control can satiafy the system with all the operation situation to satisfy the ideal sequance automatically.

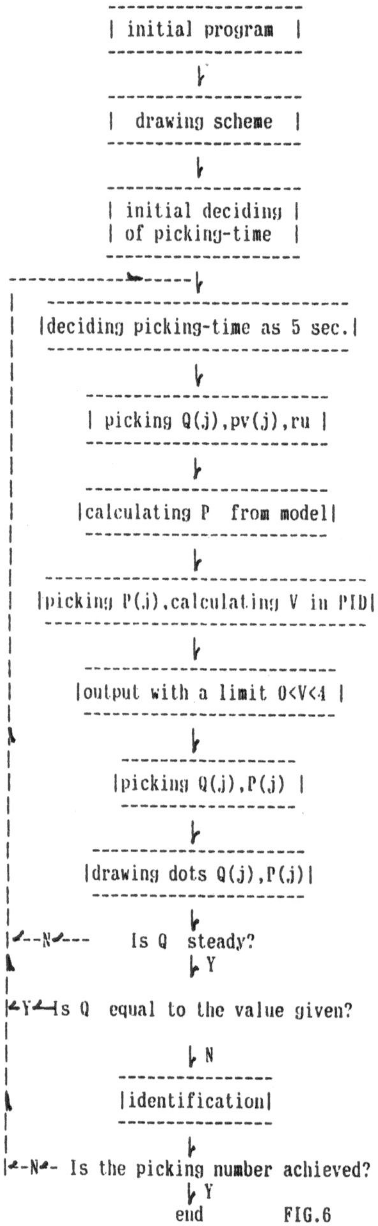
initial program
drawing scheme
initial deciding of picking-time
deciding picking-time as 5 sec.
picking Q(j),pv(j),ru
calculating P from model
picking P(j),calculating V in PID
output with a limit 0<V<4
picking Q(j),P(j)
drawing dots Q(j),P(j)
N
Is Q steady?
Y
Y
Is Q equal to the value given?
N
identification
N
Is the picking number achieved?
Y
end

FIG.6

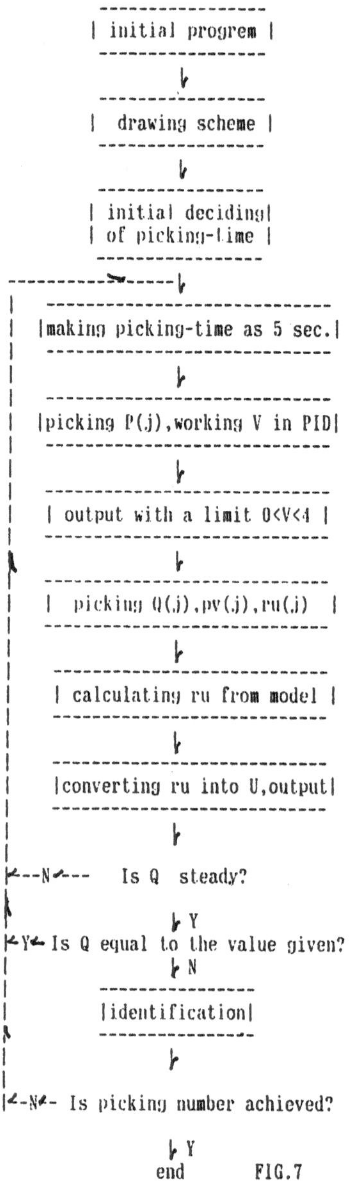
initial progrem
drawing scheme
initial deciding of picking-time
making picking-time as 5 sec.
picking P(j),working V in PID
output with a limit 0<V<4
picking Q(j),pv(j),ru(j)
calculating ru from model
converting ru into U,output
N
Is Q steady?
Y
Y
Is Q equal to the value given?
N
identification
N
Is picking number achieved?
Y
end

FIG.7

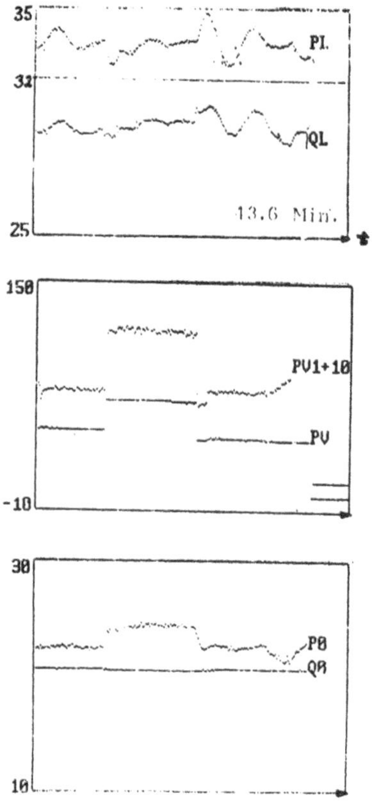

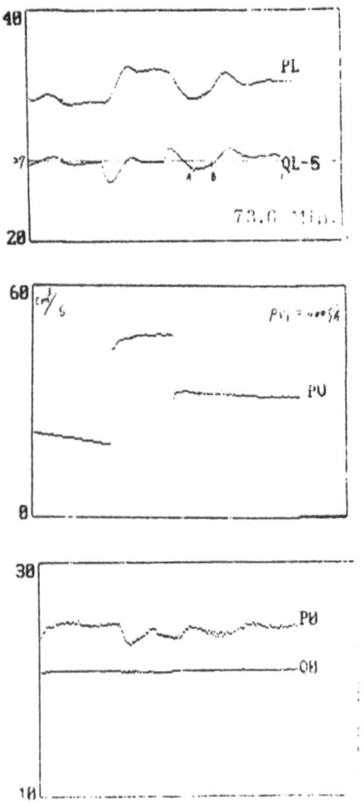

FIG.8

FIG.9

	disturbance	purely delay time	transition time	super percent	accurate of steady state
Scheme 1 (LPS)	65-137.5L/h	<20 sec.	11.75 min.	5.4 %	0.31 %
	137.5-100	<20	8.18	3.3	1.27
Scheme2(LPS)	140-200	0.18 min.	6.25	1.4	0.63
feedback with PID (one loop control) (DPS)	90-50	some 1.0	65	1.8	<1.5
	145-100	some 1.0	35	2.8	<1.5
	100-168	some 1.0	32.5	4.2	<1.5

In the other hand,we compare the Scheme 1 and 2.Shown by the comparison of their ratios of the transition time to the flowing amount varied and the ratios of the super amount to the flowing amount varied,we can definitely get a conclusion that the plan using flowing as control is much efficietion to overcome the disturbance of cool water's flowing. This conclusion give a proof of the reasonability of the mathmetical model above and on the meanwhile,it provide a reasonable control plan for the practical engineering.

reference

1.Gguang-Yuan Huang, "Application of Distributed Parameter System Methods in Exchanger" Automated Technology,1982,No.1
2.Gang-Yuan Huang, <<Theroy in Distributed Parameter System>> Published by ShanDong University
3.<<Automation in Chemical Industry>> Published by Zhejiang University

ROBUST STABILIZATION AND FINITE DIMENSIONAL CONTROLLER DESIGN ABOUT A CLASS OF DISTRIBUTED PARAMETER SYSTEMS

Hu Shun-Ju   Xu Yian-Qin

Department of computer and System Sciences , NanKai University

Abstract: In this paper, we discussed the robustness of discrete spectral systems under the perturbations of the operator A, gave the corresponding maximal robustness margin and the method of finite dimensional robust controller design . In addition, we extend major results to a larger class of systems , which satisfy the spectral decomposition assumption and the unstable part is finite dimensional .

Key words: discrete spectral systems, robustness margin , robust controller, spectral decomposition assumption

INTRODUCTION

Of late years, the robustness study of distributed parameter systems is paid attention to more and more . For different classes of systems and different classes of perturbations, people give different methods of robust controller design, for example [1],[2],[3]. But, the authors of these papers are all from frequency domain to design the robust controllers of systems for the perturbations of their transfer functions. So far , no one has studied the robustness of distributed parameter systems under parameter perturbations. However, in this paper, we are from time domain to discuss in detail the robustness of discrete spectral systems under perturbations of A, give the maximal robustness margin and design method of finite dimensional controllers. Futhermore, we extend above-mentioned major results to a larger class of systems. Compared with [1], the method of controller design in this paper has following characters : First, we have not to make finite dimensional approximation of systems, so the approximation degree has no effect on the robustness; Second, the transfer function $G+\Delta$ of the perturbed system has not to have the same number of unstable poles as the transfer function G of the original system. So, we overcome major shortcomings of the method in [1] . For discrete spectral systems, the stabilizing speed and the maximal robustness margin can be chosen sufficiently large, these can not be considered in [1] .

The paper is divided into five parts. §1, the description of discrete spectral systems and corresponding perturbations; §2 and §3 , respectively the effect of perturbations for two parts of the system ; §4, the robustness and the controller design ; §5, the extension of major results in a larger class of systems .

§1 THE DESCRIPTION OF SYSTEMS AND PERTURBATIONS

First, we give the concept of (D) operator

Definition 1: A linear operator F on Banach space X is called (D) operator, if F has a regular point at least and there exist an unconditional basis $\{\phi_n\}_{n=1}^{\infty}$ of X , a complex number sequence $\{\lambda_n\}_{n=1}^{\infty}$ and positive integer N_0 such that

$$\lim|\lambda_n| = +\infty \quad , \lambda_n \neq \lambda_m \quad , \text{for } \forall\ n\ ,m>N_0$$

$$F[\phi_1, \ \ldots\ ,\phi_{N_0}]\subseteq[\phi_1, \ \ldots\ ,\phi_{N_0}]$$

and the spectrum of F on space $[\phi_1\ , \ \ldots\ , \ \phi_{N_0}]$ are $\{\lambda_1,\lambda_2,\ldots,\lambda_{N_0}\}$, where $[\phi_1, \ \ldots\ ,\phi_{N_0}]$ is the space that $\{\phi_i\}_{i=1}^{N_0}$ span

The discrete spectral system that we will discuss is

$$\dot{x}(t) = Ax(t) + Bu(t) \qquad t\geq 0 \qquad (1)$$

where the state space is Hilbert space H , A is a (D) operator on H, its eigenvalues and generalized eigenvectors are $\{\lambda_n\}_{n=1}^{\infty}$ and $\{\phi_n\}_{n=1}^{\infty}$, without lost of generality, we assume $Re\lambda_n \geq Re\lambda_{n+1}$,n=1,2, ... Corresponding eigenvalues and generalized eigenvectors of A^* are $\{\bar{\lambda}_n\}_{n=1}^{\infty}$ and $\{\Phi_n\}_{n=1}^{\infty}$,$\{(\phi_n\ ,\Phi_n)\}$ is a biorthogonal system. Futhermore, we suppose that there is a Hilbert space V such that $H \hookrightarrow V$ is a continuous dense embedding. Input space U is Hilbert space, input operator B satisfies the condition $H_1(B,S,U;V,H)$, that is, $B\in L(U,V)$, $L(U,V)$ is the set of bounded linear operators on H, and for every $u(t)\in L_2(0,t;U)$,

$$\|\int_0^t S(t-s)Bu(s)ds\|_H \leq \varepsilon\|u\|_{L_2(0,t;U)}$$

for some constant $\varepsilon>0$, where $S(t)$ is a C_0 semigroup on H that A generates (the existence of the C_0 semigroup will be proved in the second part of the paper) .

In light of paper [4], the system (1) is well-posed under above-mentioned conditions, that is, the equation (1) on H has an unique wild solution for every $u\in L_2(0,t;U)$ and every initial values $x_0\in H$,

$$x(t)=S(t)x_0 + \int_0^t S(t-\sigma)Bu(\sigma)d\sigma$$

Let T be the perturbations for operator A , the system under perturbations T is

$$\dot{x}(t) = (A + T)x(t) + Bu(t) \qquad t\geq 0 \qquad (2)$$

Suppose the stabilizing speed of the closed loop system that we wish is not less than $|\alpha|$, where $\alpha<0$, then we can choose a nature number N such that $Re\lambda_N \geq \lambda > Re\lambda_{N+1}$. Let $H_1 = span\{\phi_i\}_{i=1}^{N}$, $H_2 = span\{\phi_i\}_{i=N+1}^{\infty}$, $A_i = A/H_i$, which is meant that A_i is the limitation of A on H_i , i=1,2 . Similarly, V can be divided into $V = V_1 \oplus V_2$, where $V_1 = H_1$, $H_2 \hookrightarrow V_2$ is a continuous dense embedding. We define B_i as $B_i u = Bu/V_i$, i=1,2, for every $u \in L_2(0,\infty;U)$. From paper [4] , we know $B_1 \in \mathcal{L}(U,V_1)$, $B_2 \in \mathcal{L}(U,V_2)$, and B_2 satisfies the condition $H_1(B_2,S_2,U;V_2,H_2)$, where $S_2(t)$ is the C_0 semigroup on H_2 that A_2 generates . Suppose perturbations T satisfy $TH_1 \subset H_1$, $TH_2 \subset H_2$, then $T = \begin{pmatrix} T_1 & 0 \\ 0 & T_2 \end{pmatrix}$, where $T_i = T/H_i$, i=1,2 . In fact , T satisfying the condition exist generally, for example, the perturbations or changes of the interior damping coefficient and the bending stiffness in elastic vibrating systems under no exterior dampings , the perturbations of every cofficient in heat conducting systems , and so on .

Thus, we divide the system (1) into the system-1 and the system-2 , divide the perturbations T for the system (1) into the perturbations T_1 for the system-1 and the perturbations T_2 for the system-2 .

Later, we let $\rho(A)$ denote the resolvent set of A, $\sigma(A)$ denote the spectrum of A .

§2 THE EFFECTS OF PERTURBATIONS FOR THE SYSTEM-2

To obtain our major results in the section , we have to introduce following lemmas.

LEMMA 1: (see[5]).Suppose that E is a linear operater on Banach space X, $F \in \beta(X)$, $\beta(X)$ is the space of bounded linear operators on X , Γ is an arbitrary compact subset of $\rho(E)$. If

$$||F|| < \min_{\zeta \in \Gamma} ||R(\zeta,E)||^{-1}$$

then $\Gamma \subset \rho(E+F)$.

LEMMA 2: (see[6]).Suppose that L is a (D) operator on Hilbert space X ,F is a linear operater on X , $\beta > 0$,they satisfy

$$||Fx|| \leq \beta ||L^{\delta}x|| \qquad \delta < 1 \text{ , for } \forall x \in D(L)$$

If the eigenvalues $\{\gamma_n\}_{n=1}^{\infty}$ of L satisfy

$$\gamma_n = cn^p[1+O(n^{-1})] \text{ , } \quad p(1-\delta) > 3/2 \text{ , where } n=1,2,3,\ldots$$

then L+F is still a (D) operator .

Now ,we can obtain our major results in this section .

THEOREM 1: If $\sup\{Re\lambda: \lambda \in \sigma(A)\} < +\infty$, then A generates a C_0 semigroup S(t) on H, A_2 generates a exponential stable semigroup $S_2(t)$ on H_2 . Futhermore , for an arbitrary constant α satisfying $Re\lambda_N \geq \alpha > \omega > Re\lambda_{N+1}$, there is some constant M>0 such that

$$\sup\{Re\lambda : \lambda \in \sigma_p(A_2+T_2)\} \leq \alpha$$

only if bounded linear perturbations T_2 on H_2 satisfy the condition

$$\|T_2\| \le (\alpha-\omega)/M$$

where ω is the exponential growth constant of $S_2(t)$. When λ_{N+1} and λ_1 are simply, we can take $\omega=Re\lambda_{N+1}$ and $\lambda_0=\sup\{Re\lambda : \lambda\in\sigma(A)\}=Re\lambda_1$.

Proof: Because A is a (D) operator, there are an unconditional basis $\{\phi_n\}_{n=1}^{\infty}$ of H, a complex sequence $\{\lambda_n\}_{n=1}^{\infty}$ and positive integer N_0 such that

$$\lim_{n=1}|\lambda_n| = +\infty \quad , \lambda_n\neq\lambda_m \quad , \text{for } n\ ,m>N_0\ ,\ n\neq m$$

$$A\phi_n = \lambda_n\phi_n \quad , \quad n>N_0$$

$$A[\phi_1,\ \ldots,\phi_{N_0}] \subseteq [\phi_1,\ \ldots,\phi_{N_0}]$$

and the spectrum of A limiting on $[\phi_1,\ \ldots,\ \phi_{N_0}]$ is $\{\lambda_i\}_{i=1}$.

We let $X_1=[\phi_1\ ,\phi_2\ ,\ \ldots\ ,\phi_{N_0}]$, $X_2=[\phi_{N+1}\ ,\phi_{N+2},\ \ldots]$.

Obviously $H=X_1\oplus X_2$, $AX_1\subseteq X_1$, $AX_2\subseteq X_2$. Operator A/X_1 is a bounded linear operator since X_1 is a finite dimensional space. So operator A/X_1 generators a C_0-semigroup. By Hille-Yosida theorem, we know there exists $M_1>0$, $\lambda_0>\sup\{Re\lambda : \lambda\in\sigma_p(A)\}$ such that

$$\|\ [(\lambda I-A)^{-1}]^k x_1\ \| \le M_1\|x_1\|/(Re\lambda-\lambda_0)^k \quad , Re\lambda>\lambda_0$$

Futhermore, through directly checking we know easily $\lambda_0=Re\lambda_1$ when λ_1 is simply.

Next, we discuss on X_2. For an arbitrary λ satisfying $Re\lambda>\lambda_0$, we know $\lambda\in\rho(A)$ by the hypotheses in theorem 1. For an arbitrary $x_2\in X_2$, we have

$$x_2=\sum_{n=N+1}^{\infty}\alpha_n\phi_n$$

$$(\lambda I-A)^{-1}x_2 = \sum_{n=N+1}^{\infty}\alpha_n\frac{1}{\lambda-\lambda_n}\phi_n$$

$$=\sum_{n=N+1}^{\infty}\alpha_n\frac{1}{Re\lambda-\lambda_0}\ \frac{Re\lambda-\lambda_0}{\lambda-\lambda_n}\ \phi_n$$

$$=\frac{1}{Re\lambda-\lambda_0}\ \sum_{n=N+1}^{\infty}\alpha_n\frac{Re\lambda-\lambda_0}{Re\lambda-\lambda_n}\ \phi_n$$

For an arbitrary given positive integer K, we may deduce by analogy :

$$[(\lambda I-A)^{-1}]^k x_2 = \frac{1}{(Re\lambda-\lambda_0)^k}\ \sum_{n=N+1}^{\infty}\alpha_n\ \frac{(Re\lambda-\lambda_0)^k}{(\lambda-\lambda_n)^k}\ \phi_n$$

Obviously ,

$$\left| \frac{(Re\lambda-\lambda_0)^k}{(\lambda-\lambda_0)^k} \right| \le 1 \quad , \text{ for all positive integer } k\ge 1$$

Let M_2 denote the unconditional basis constant with respect to basis $\{\phi_n\}_{n=1}^{\infty}$, then we have

$$\|[(\lambda I-A)^{-1}]^k x_2\| \le \frac{1}{(Re\lambda-\lambda_0)^k} \left\| \sum_{n=N+1}^{\infty} \alpha_n \frac{(Re\lambda-\lambda_0)^k}{(\lambda-\lambda_0)^k} \phi_n \right\|$$

$$\le \frac{M_2}{(Re\lambda-\lambda_0)^k} \|x\|$$

for $k=1,2, \ldots$

Again, $H=\mathcal{X}_1\oplus\mathcal{X}_2$, $(\lambda I-A)^{-1}\mathcal{X}_1\subseteq \mathcal{X}_1$, $(\lambda I-A)^{-1}\mathcal{X}_2 \subseteq \mathcal{X}_2$, $\mathcal{X}_1$ and $\mathcal{X}_2$ are closed linear subspaces of H. So, for $\forall$ $x\in H$, $\forall$ λ satisfying $Re\lambda>\lambda_0$,we have

$$x = x_1 + x_2 \qquad x_1\in\mathcal{X}_1 \;,\; x_2\in\mathcal{X}_2$$

$$R(\lambda;A)^k x = R(\lambda;A)^k x_1 + R(\lambda;A)^k x_2$$

$$\| R(\lambda;A)^k x \| \le \| R(\lambda;A)^k x_1\| + \| R(\lambda;A)^k x_2\|$$

$$\le \frac{M_1}{(Re\lambda-\lambda_0)^k} \|x_1\| + \frac{M_2}{(Re\lambda-\lambda_0)^k} \|x_2\|$$

$$\le \frac{\max(M_1,M_2)}{(Re\lambda-\lambda_0)^k} (\|x_1\| + \|x_2\|)$$

Because $\mathcal{X}_1$ and $\mathcal{X}_2$ are closed linear subspaces of H , there exists M_3 such that

$$\|x_1\| + \|x_2\| \le M_3\|x\|$$

So ,

$$\|R(\lambda;A)^k x\| \le \frac{1}{(Re\lambda-\lambda_0)^k} \max(M_1,M_2)M_3 \|x\| \quad , k=1,2, \ldots$$

So ,

$$\| R(\lambda;A)^k \| \le \frac{1}{(Re\lambda-\lambda_0)^k} \max(M_1,M_2)M_3 \quad , K=1,2, \ldots$$

By Hille-Yosida theorem, operator A generates a C_0 semigroup $S(t)$ on H . When eigenvalue λ_1 is simply, we can take $\lambda_0=Re\lambda_1$ by preceding discuss .

Similarly, A_2 generates a C_0 semigroup $S_2(t)$ on H_2 , and for $\forall$ ω satisfying $0>\omega>\sup\{Re\lambda: \lambda\in\sigma(A_2)\}=Re\lambda_{N+1}$, there exists a constant number $M>0$ such that

$$\|S_2(t)\| \le Me^{\omega t} \qquad t\ge 0$$

So $S_2(t)$ is exponentially stable. By Hille-Yosida theorem , we know for $\forall$ λ satisfying $Re\lambda>\alpha>\omega$,

$$\|R(\lambda;A_2)\| \le \frac{M}{Re\lambda-\omega} < \frac{M}{Re\lambda-\alpha}$$

Thus

$$\|T_2\| \le (\alpha-\omega)/M < \|R(\lambda;A_2)\|^{-1}$$

From lemma 1 , we know the compact subset $\{\lambda\}$ of $\rho(A_2)$ is included in $\rho(A_2+T_2)$. For λ satisfying $Re\lambda>\alpha$ is arbitrary ,

$$\sup\{Re\lambda:\ \lambda\in\sigma(A_2+T_2)\} \le \alpha$$

we complete the proof .

THEOREM 2: Suppose T_2 is a bounded linear operator on H_2 , the evigenvalues of A_2 satisfying

$$\lambda_n=cn^p[1+O(n^{-1})] \quad , \qquad n=N+1\ ,\ N+2\ ,\ \ldots$$

where c is a positive constant number , $p>3/2$. Then A_2+T_2 is a (D) operator .

Proof: Let $\delta=0$,by lemma 2 ,we know the result .

REMARK: Althogh $p>3/2$ is demanded here, yet most of practical distributed parameter systems all satisfy the condition . For example, elastic vibrating systems, heat conductive systems, and so on. So the condition is not strong .

THEOREM 3: Under the conditions of theorem 1 and theorem 2 ,the system (A_2+T_2,B_2) is well-posed on H_2 and A_2+T_2 generates an exponentially stable semigroup $\bar{S}_2(t)$, whose convergence speed can approximate α sufficiently .

Proof: By lemma 2.5 in paper [4], we know that B_2 satisfies the condition $H_1(B_2,\bar{S}_2,U;V_2,H_2)$ because T_2 is a bounded operator and B_2 satisfies the condition $H_1(B_2,S_2,U;V_2,H_2)$. So the system $(A_2+T_2\ ,B_2)$ is well-posed on H_2 .

Again, A_2+T_2 is a (D) operator by theorem 1 and theorem 2 ,and

$$\sup\{Re\lambda:\ \lambda\in\sigma(A_2+T_2)\} \le \alpha$$

So from proof of theorem 1 we know for $\forall$ $\omega_2<\alpha_2$, there exists $M_4>0$ such that

$$\|\bar{S}_2(t)\| \le M_4\, e^{\omega_2 t} \qquad \text{for all } t\ge 0$$

We take $\omega_2<0$, so A_2+T_2 generates exponentially stable semigroup . Futhermore , because ω_2 can approximate α sufficiently,we call $\bar{S}_2(t)$ α-exponentially stable . The proof is finished .

To sum up, while a α-exponentially stable discrete spectral system is perturbed by operator T_2 ,the perturbed system keeps α-exponentially stable only if T_2 satisfy $\|T_2\| \le (\alpha-\omega)/M$, where ω

is exponentially growth constant of $S_2(t)$, constant $M>0$ is determined by A_2 , α satisfying $Re\lambda_N \geq \alpha > \omega > Re\lambda_{N+1}$ can be chosen by the stabilizing speed of the closed loop system that you wish. Additionally, due to λ_n satisfying $\lambda_n = cn^p[1+O(n^{-1})]$, $n=1,2,\ldots$,under many circumstances ,

$$|Re(\lambda_n - \lambda_{n+1})| \longrightarrow +\infty \qquad \text{when } n \longrightarrow +\infty$$

In this circumstances, the maximal norm $(\alpha-\omega)/M$ of the permissible perturbations T_2 may be chosen adequate large .

§3 THE EFFECTS OF PERTURBATION FOR THE SYSTEM-2

Suppose T_1 is the perturbations for the sytem-1 , $T_1=[P_{ij}]_{N\times N}$,then T_1 can be written $T_1=\sum_{i,j=1}^{N} p_{ij}E_{ij}$ where $E_{ij}=(e_{ij})_{N\times N}$,

$$e_{kl} = \begin{cases} 1 & \text{for } k=i \text{ , } l=j \\ 0 & \text{for otherwise} \end{cases} \qquad i,j=1,2,\ \ldots\ ,N$$

Thus, the system-1 under the perturbations T_1 is

$$\dot{x}(t) = (A_1 + \sum_{i,j=1}^{N} p_{ij}E_{ij})x(t) + B_1u(t) \qquad \text{for } \forall x \in H_1$$

For finite dimensional systems with the class of parameter perturbations , their robustness have been discussed in many papers , for example [7],[8],[9],[10] . All controller design methods can be divided into two classes by parameter changes of systems: When the parameter changes being small , people design the robust controller by sensible functions, for examples [8],[9] ; When the parameter changes being big ,people discuss the robustness of systems under the condition that parameters change in some prescribed convex sets , in this circumstances , there are two kinds of design methods with the different objective in robust controller design : the first is the guaranted cost method that guaranted the cost function is upper bounded, for example method 1 and method 2 in [7]; the second only ensure the closed loop system is stable, for example method 3 in [7]. Because we consider majorly the stability of closed loop systems, we will use the design method 3 in [7] that is changed into guaranting the stabilizing speed of closed loop systems. Before we do this , we have to make the following assumptions :

Assumptions: (i) The permissible perturbations T_1 change only in a compact subset of $R^{N\times N}$, that is ,$a_{ij} \leq p_{ij} \leq b_{ij}$, $i,j=1,2,\ \ldots\ N$. Furthermore , $a_{ij} \leq 0$, $b_{ij} \geq 0$;

(ii) The stabilizing speed of the closed loop system is not less than $|\alpha|$,where $\alpha<0$. So the cost function is

$$J=\int_0^{\infty} e^{-2\alpha t}[x^T(t)Qx(t)+u^T(t)Ru(t)]dt$$

where $Q \in R^{N\times N}$, $Q \geq 0$, $R \in R^{N\times N}$, $R > 0$;

(iii) All state variables are availiable ;

(iv) $[A_1+T_1, B_1]$ forms a controllable pair , where T_1 is a arbitrary permissable perturbation operator ;

(V) $[A_1+T_1, Q^{\frac{1}{2}}]$ forms a complete observable pair , T_1 is a arbitrary permissible perturbation operator , $(Q^{\frac{1}{2}})^T(Q)^{\frac{1}{2}} = Q$, $(Q^{\frac{1}{2}})^T$ is transpose of $Q^{\frac{1}{2}}$.

Under these assumptions, we can design a robust controller by the following method because these assumptions ensure that every step of the design method is solvable .

The method is :

(1) Solve the Riccati equation :

$$S_0A_1+A_1^TS_0-S_0B_1R^{-1}B_1^TS_0 +Q_0=0 \quad , \qquad \text{where } Q_0=Q \quad ;$$

(2) Evaluate $C_0=-R^{-1}B_1^TS_0$;

(3) Evaluate $F_0=A_1+ B_1C_0$;

(4) Evaluate $T(S_0)=\delta[\varepsilon^{-1}\sum_{i,j=1}^{N}E_{ij}^TS_0E_{ij} +\varepsilon N^2S_0]$, where

$$\varepsilon=\sum_{i,j=1}^{N}||E_{ij}|| \quad , \quad \delta=\sum_{i,j=1}^{N}[\max(|a_{ij}| , |b_{ij}|)]^2 \quad ;$$

(5) Determine ρ_0 such that $(Q_d)_0= Q_0- \rho_0T(S_0)$ indefinite. Set $j=1$;

(6) Solve the equation :

$$S_jF_{j-1}+F_{j-1}^TS_j-S_jB_1R^{-1}B_1^TS_j+Q_j=0$$

where $Q_j=\mu_jQ_{j-1}+\rho_{j-1}T(S_{j-1})$, $\mu_j=||Q_j||^{-1}$;

(7) Evaluate $C_j=C_{j-1}-R^{-1}B_1^TS_j$, $F_j=A_1+B_1C_j$;

(8) Evaluate $T(S_j)=\delta[\varepsilon^{-1}\sum_{i,j=1}^{N} E_{ij}^TS_jE_{ij} + \varepsilon N^2S_j]$, where

$$\varepsilon=\sum_{i,j=1}^{N}||E_{ij}|| \quad , \quad \delta=\sum_{i,j=1}^{N}[\max(|a_{ij}| , |b_{ij}|)]^2 \quad ;$$

(9) Determine ρ_j such that $(Q_d)_j = \mu_{j-1}Q_{j-1}+(\rho_{j-1}+\rho_j)T(S_j)$ is a indefinite matrix . If $\rho_j \geq 1$, evaluate $u(t)=-e^{-\alpha t}R^{-1}B_1^TS_jx(t)$. Then $u(t)$ is the feedback controller rule. If not, set $j=j+1$, continue (6) .

ξ4 ROBUSTNESS AND CONTROLLER DESIGN OF SYSTEMS

Suppose that the stabilizing speed of the closed loop system is not less than $|\alpha|$, where $\alpha<0$ satisfies $Re\lambda_N \geq \alpha > \omega > Re\lambda_{N+1}$, then we can design the robust controller of the system (1) by the following procedure :

(1) Evaluate the eigenvalues and the normalized eigenvectors of A ;

(2) Divide the system (1) into the system-1 and the system-2 by ξ1 ;

(3) Design a robust controller K of the system-1 by ξ3, where we let the maximal robustness margin be $(\alpha-\omega)/M$,that is ,

$$\max(|a_{ij}|, |b_{ij}|)=(\alpha-\omega)/2M .$$

(4)Then for all the linear perturbation operator T_2 satisfying $\|T\|_0\le(\alpha-\omega)/M$,where $\|T\|_0=\max(\|T_1\|_1, \|T_2\|)$, $\|T_2\|$ is the operator norm of T_2 , $\|T_1\|_1$ is defined as $\|T_1\|_1=\max(|a_{ij}|, |b_{ij}|)$, K is the robust controller of the system (1) .

Thus , we give a robust controller design method when a discrete spectral system is perturbed by a bounded linear operator T in the part of operator A, and point out the maximal robustness stabilizing margin is $(\alpha-\omega)/M$. This controller make the closed loop system α-exponentially stable , where $\alpha<0$ is chosen arbitrarily. Under many circumstances, the maximal robustness margin can be chosen sufficiently big .

As compared with the controller design method in paper [1] , we overcome its two major shortcomings. Moreover, in this method, we can choose the stabilizing speed and the robustness margin , these are not consided in [1] .

ξ5 EXTENSION OF MAJOR RESULTS

Let $\sigma_\mu^+(A)=\sigma(A)\cap C_\mu^+$, where $C_\mu^+=\{\lambda\in C : Re\lambda\ge\delta\}$

$\sigma_\mu^-(A)=\sigma(A)\cap C_\mu^-$, where $C_\mu^-=\{\lambda\in C : Re\lambda<\delta\}$

Definition 1: Operator A satisfies the spectral decomposition assumption at some point μ, which is meant that $\sigma_\mu^+(A)$ is bounded and seperated from $\sigma_\mu^-(A)$ such that a rectifiable, simple, closed curve can be drawn so as to enclose an open set containing $\sigma_\mu^+(A)$ in its interior and $\sigma_\mu^-(A)$ in its exterior .

The class of systems that we consider are the systems that operator A satisfy the spectral decomposition assumption in some point $\mu<0$ and $\sigma_\mu^+(A)$ is the set containing only finite points. Futhermore, we assume that A generates a C_0 semigroup . In this circumstances, the system can be divided into the system-1 (A_1, B_1) and the system-2 (A_2, B_2) by the spectral projection P_μ , where $H_1=P_\mu H$, $H_2=(I-P_\mu)H$ are respectively the state spaces of the system-1 and the system-2, the spectral sets of A_1 and A_2 are respectively

$\sigma_{\mu}^{+}(A)$ and $\sigma_{\mu}^{-}(A)$; $B_i u = Bu/H_i$, $i=1,2$, for every $u \in L_2[0,\infty\ ;U]$. By the definition of the spectral composition assumption, we know that (A_1,B_1) is finite dimensional .

For every one of this class of systems , if A_2 generates a exponentially stable semigroup ,that is, there exists some $M>0$,$\omega<0$, such that $\|S_2(t)\| \le Me^{\omega t}$, then there is the following theorem .

THEOREM 4: For $\forall\ \alpha$ satisfying $\omega<\alpha<0$, if T_2 satisfy $\|T_2\| \le (\alpha-\omega)/M$, then $\sup\{Re\lambda\ : \lambda \in \sigma_p(A_2+T_2)\} < \alpha$.

Proof: Because $R(\zeta, A_2)x = \int_0^\infty e^{-\zeta t} S_2(t)x(t)dt$, for $\forall x \in D(A_2)$, $\forall\zeta$ satisfying $Re\zeta>\omega$, we can derive

$$\| R(\zeta, A_2) \| \le M/(Re\zeta-\omega)$$

$$\| Re(\zeta, A_2) \|^{-1} \ge (Re\zeta-\omega)/M$$

So

$$\|T_2\| \le (\alpha-\omega)/M < (Re\zeta-\omega)/M \le \|R(\zeta, A_2)\|^{-1}$$

By lemma 1, we know $\{\zeta\} \subset \rho(A_2+T_2)$, for $\forall\zeta$ satisfying $Re\zeta>\omega$. So ,

$$\sup\{Re\lambda\ : \lambda \in \sigma_p(A_2+T_2)\} < \alpha$$

we finish the proof .

From this theorem , we know the system (A_2, B_2) under the perturbations T_2 satisfying $\|T_2\| \le (\alpha-\omega)/M$ is α input-output stable ,where input-output stable is meant that the real part of eigenvalues of the system is not bigger than α . If A_2 genarates a analytic semigroup, A_2+T_2 generates still a analytic semigroup because T_2 is bounded on H_2 . If so, the system (A_2+T_2, B_2) is α-exponentially stable because analytic semigroups satisfy the spectrum determined growth assumption .

Suppose the perturbations T satisfy $TH_1 \subset H_1$, $TH_2 \subset H_2$, T can be divided into perturbations T_1 for the system-1 (A_1, B_1) and perturbations T_2 for the system-2 (A_2, B_2). From aforesaid discuss we know the system (A_2+T_2, B_2) is α input-output stable when T_2 satisfy $\|T_2\| \le (\alpha-\omega)/M$,where $\alpha>\omega$ can be chosen. So, if we design the robust controller of (A_1, B_1) with maximal robustness margin $(\alpha-\omega)/M$ by the method in ξ3 ,the controller just is the robust controller of the system (A, B) with the same maximal robustness margin . If A_2 generates a analytic semigroup , the closed loop system is α-exponentially stable. But in this section, the stabilizing speed $|\alpha|$ and the maximal robustness margin have some limits and can not be chosen arbitrarily.

REFERENCE

[1] Ruth F Curtain and Keith Glover : Robust stabilition of infinite dimensional systems by finite controllers , Systems and Control Letters 7 ,1986 .
[2] Ruth F Curtain: Robust stabilizability of normalized coprime factors: the infinite dimensional case , Report TW 291 , University of Groningen , NL ,1988
[3] Mathukumalli Vidyasagar : The graph metric for unstable plants and robustness estimates for feedback stability , IEEE Transactions on Automatic Control , vol AC-29 No5 ,1984 .
[4] T.Kato : « Perturbation theory for linear operator » , published in 1986.
[5] Lo Yu-Hu : Properties of eigenvalues of a class of discrete spectral systems, Mathmatic Journal 7,1987 . (in chinese)
[6] Ruth F .Curtain: Equivalence of input-output stability and exponential stability for infinite dimensional systems , appeared .
[7] O.I.Kosmidou and P.Bertrand: Int.J.Control 3 , 1983 .
[8] Kermin Zhou and Pramond P.Khargoneckar: SIAM J Control and Optimization 6 ,1988 .
[9] Byre and Burke: IEEE Trans. Control 21, 1976 .
[10] Kreindle E: Int J Control 8, 1968 .

THE ASYMPTOTIC REGULATOR DESIGN FOR NONLINEAR FLEXIBLE STRUCTURES WITH ARBITRARY CONSTANT DISTURBANCES

Li Chengzhi

Department of Computer science & System Science, Xiamen University, Fujian 361005, *P. R. C.*

This paper investigates the finite-dimensional asymptotic regulator design for nonlinear flexible structrues with arbitrary constant disturbances. The problem is to design a feedback controller such that resulting closed-loop nonlinear system will be stable and the controlled output will be regulated. With some assumptions, this paper presents explicit sufficient conditions for the existence of the finite dimensional asymptotic regulator which stabilizes and regulates nonlinear flexible structures.

1. INTRODUCTION.

The servomechanism problem for D. P. S. has recently been researched by some authors (for example: S. A. Pohjolainen, T. Kobayashi, U. Hiroyuki and I. Tetsuo). But they only considered the linear D. P. S. which baced on linearized models of the actual problems. In actual operation, any controller operates on the actual structure and not the linearized model. To author' s knowledge, there is no known systematic servomechanism theory for nonlinear D. P. S.. The purpose of this paper is to generalize the robust regulator theory of finite demensional nonlinear systems [3] to infinite dimensional nonlinear systems.

2. PRELIMINARIES.

The flexible structures considered here must satisfy the generalized wave equation:

$$\ddot{u}(t) + 2@\dot{u}(t) + A_0u(t) = F[u(t),\dot{u}(t),f(t)] \tag{2.1}$$

which relates the displacements $u(x,t)$ of a structure Ω from its equilibrium position due to applied force distribution $F[u(t),\dot{u}(t),f(t)]$, The operator A_0 is symmetric, time-invareant differeential operator whose domain $D(A_0)$ is dence in the Hilbert Space $H_0=L_2(\Omega)$ with the usual inner product $(.\,,.)$ and associated norm $||.||$, the operator A_0 is bounded bellow by

$$(A_0u,u) \geqslant a\,\|u\|^2 \quad , \quad a > 0 \tag{2.2}$$

and therefore has a square root $A_0^{1/2}$ and a bounded inverses operator A_0^{-1}. The damping term in (2. 1) is $2@\dot{u}(t)$, where $@>0$. The applled force distribution is seperated into the control force and the disturbance force:

$$F[u(t),\dot{u}(t),f(t)] = Fc[f(t)] + Fd[u(t),\dot{u}(t)].$$

The control force are produced by m actuators with influence forces bi in H_0

$$Fc[f(t)] = B_0 f(t) = \sum_{i=1}^{m} b_i(x) f_i(t). \tag{2. 3}$$

The disturbance forces are given by

$$Fd[u(t),\dot{u}(t)] = F_0[u(t),\dot{u}(t)] + Ew, \tag{2. 4}$$

where F_0 is uniformly Lipschitz continuous in all its arguments, i. e. , there is a constant $k>0$ such that

$$\| F_0(u_1,\dot{u}_1)-F_0(u_2,\dot{u}_2) \| \leqslant k[\ \| u_1-u_2 \|^2 + \| \dot{u}_1-\dot{u}_2 \|^2]^{1/2}$$

and $F_0(0,0)=0$. The w is an unknown constant disturbance. The E is an unknown operator beling to $L(Rq,H_0)$. Measurements are made by p sensors with output:

$$y(t) = [y_1,\cdots\cdots,y_p]^T = C_0 u(t) + D_0 u(t) \tag{2. 5}$$

where $y_i(t)=(c_i,u)+(d_i,u),i=1,\cdots,p$, and the influence functions c_i, d_i are in H_0.

When the state of (2. 1) is defined by $v(t)=[u(t),\dot{u}(t)]^T$, we have

$$\begin{cases} \dot{v}(t) = Av(t) + Bf(t) + h[v(t)] + Ew \\ y(t) = Cv(t) \end{cases} \tag{2. 6}$$

where

$$B = \begin{bmatrix} 0 \\ B_0 \end{bmatrix}, \quad h(.) = \begin{bmatrix} 0 \\ F_0(.\,,.) \end{bmatrix}, \quad C = [C_0,D_0]$$

$$A = \begin{bmatrix} 0 & I \\ -A_0 & -2@ \end{bmatrix}, \quad D(A) = D(A_0) \oplus D(A_0^{1/2})$$

and D(A) dences in the Hilbert Space $H=D(A_0^{1/2})\oplus H_0$ with the "energy norm" defined by

$$\| v \|_E^2 = \| \dot{u} \|^2 + \| A_0^{1/2} u \|^2 \tag{2. 7}$$

Lemma 2. 1: The h(v) is also uniformly Lipschits continuous with same L-constant k.

Lemma 2. 2: There exists the unique continuous "mild" solution v(t) for (2. 2) given as following:

$$v(t) = U_A(t)v(0) + \int_0^t U_A(t-s)\{Bf(s) + h[v(s)] + Ew\}ds$$

Proof: See [4. p. 183] ▌.

Now we are able to pose the following control problem.

Problem 2. 3: Find a finite dimensional control system:

$$\begin{cases} \dot{z}(t) = y(t)-\bar{y} \qquad z(0) = 0 \\ f(t) = Kz(t) \end{cases} \tag{2. 8}$$

where $K\in L(Rp,Rm)$, $\bar{y}$ is an constant reference signal in Rp, such that the resultant system

$$\begin{cases} \begin{bmatrix} \dot{v}(t) \\ \dot{z}(t) \end{bmatrix} = \begin{bmatrix} A & BK \\ C & 0 \end{bmatrix} \begin{bmatrix} v(t) \\ z(t) \end{bmatrix} + \begin{bmatrix} h[v(t)] \\ 0 \end{bmatrix} + \begin{bmatrix} Ew \\ -\bar{y} \end{bmatrix} \\ y(t) = [C,0] \begin{bmatrix} v(t) \\ z(t) \end{bmatrix} \end{cases} \tag{2. 9}$$

will behave in following way:

1) System (2.9) will be stable.

2) The output $y(t)$ will be regulated to an arbitrary constant reference $\bar{y}$, i.e., $y(t) \to \bar{y}$ as $t \to \infty$, in spite of the disturbance w.

3. Stability of The Nonlinear D. P. S.

lemma 3.1: When $@>0$, A generates a group $U_A(t)$ and there exist positive constants M and q such that

$$\| U_A(t) \|. \leqslant Me^{-qt} \qquad t \geqslant 0$$

Proof: See [1] for detail. ▌

Lemma 3.2: When $@>0$, the following equation is true

$$CA^{-1}B = -C_0A_0^{-1}B_0$$

Proof: Let

$$CA^{-1} = [G, H]$$

then

$$CA^{-1}B = HB_0$$

Becauce of

$$\begin{aligned} C &= [G, H]A \\ &= [-HA_0, G-2@H] \end{aligned}$$

we have

$$-HA_0 = C_0.$$

According to (2.2), we have

$$CA^{-1}B = -C_0A_0^{-1}B_0. \text{ ▌}$$

Theorem 3.3: When $@>0$, if rank $[C_0A_0^{-1}B_0]=p$, then there is $K \in L(Rp, Rm)$ such that

$$\bar{A} = \begin{bmatrix} A & 0 \\ C & 0 \end{bmatrix} + \begin{bmatrix} B \\ 0 \end{bmatrix} [0, K] = \begin{bmatrix} A & BK \\ C & 0 \end{bmatrix}$$

generates an exponentially stable semigroup $U_{\bar{A}}(t)$ on $\bar{H} = H \oplus Rp$, i.e., there are positive constants M_1 and g such that

$$\| U_{\bar{A}}(t) \| \leqslant M_1 e^{-gt}, \qquad t \geqslant 0.$$

Proof: See Appedix A for detail. ▌

Theorem 3.4: When $@>0$, $w=0$ and $\bar{y}=0$, if

1) rank$[C_0A_0^{-1}B_0]=p$

2) $k<g/M_1$

then the "mild" solution of system (2.9) is exponentially stable.

Proof: Considering the "mild" solution of (2.9)

$$\begin{bmatrix} v(t) \\ z(t) \end{bmatrix} = U_{\bar{A}}(t) \begin{bmatrix} v(0) \\ z(0) \end{bmatrix} + \int_0^t U_{\bar{A}}(t-s) \begin{bmatrix} h[v(s)] \\ 0 \end{bmatrix} ds,$$

we have

$$\left\| \begin{bmatrix} v(t) \\ z(t) \end{bmatrix} \right\| \leqslant M_1 e^{-gt} \left\| \begin{bmatrix} v(0) \\ z(0) \end{bmatrix} \right\| + \int_0^t M_1 k e^{-g(t-s)} \left\| \begin{bmatrix} v(s) \\ z(s) \end{bmatrix} \right\| ds$$

According to the Gronwall' s Inequality,

$$\left\| \begin{bmatrix} v(t) \\ z(t) \end{bmatrix} \right\| \leqslant M_1 \left\| \begin{bmatrix} v(0) \\ z(0) \end{bmatrix} \right\| e^{(M_1 k - g)t} = M_2 e^{-bt}$$

where

$$M_2 = M_1 \left\| \begin{bmatrix} v(0) \\ z(0) \end{bmatrix} \right\|$$

$$b = g - M_1 k > 0. \ \blacksquare$$

4. MAIN RESULT.

Considering following algebra equation:

$$\begin{bmatrix} A & BK \\ C & 0 \end{bmatrix} \begin{bmatrix} v \\ z \end{bmatrix} + \begin{bmatrix} h(v) \\ 0 \end{bmatrix} + \begin{bmatrix} Ew \\ -\bar{y} \end{bmatrix} = 0 \qquad (4.1)$$

Difinition: The $\begin{bmatrix} v \\ z \end{bmatrix}$ is called the equilibrium point if it is the solution of (4. 1).

Lemma 4. 1: There is one and only one equilibrium point if

$$\left\| \begin{bmatrix} A & BK \\ C & 0 \end{bmatrix}^{-1} \right\| < 1/k \qquad (4.2)$$

Proof: From (4. 2), we have

$$\begin{bmatrix} v \\ z \end{bmatrix} = -\begin{bmatrix} A & BK \\ C & 0 \end{bmatrix}^{-1} \begin{bmatrix} h(v) \\ 0 \end{bmatrix} - \begin{bmatrix} A & BK \\ C & 0 \end{bmatrix}^{-1} \begin{bmatrix} Ew \\ -\bar{y} \end{bmatrix} \qquad (4.3)$$

Becauce of

$$\left\| \begin{bmatrix} v_1 \\ z_1 \end{bmatrix} - \begin{bmatrix} v_2 \\ z_2 \end{bmatrix} \right\| \leqslant \left\| \begin{bmatrix} A & BK \\ C & 0 \end{bmatrix}^{-1} \right\| \left\| \begin{bmatrix} h(v^1) \\ 0 \end{bmatrix} - \begin{bmatrix} h(v_2) \\ 0 \end{bmatrix} \right\|$$

$$\leqslant k_1 \left\| \begin{bmatrix} v_1 - v_2 \\ z_1 - z_2 \end{bmatrix} \right\| \qquad (4.4)$$

where

$$k_1 = k \left\| \begin{bmatrix} A & BK \\ C & 0 \end{bmatrix}^{-1} \right\| < 1$$

According to Contraction Mapping Theorem and (4. 4), we know that the equilibrium point of (4. 1) exists and only one. ▌

Theorem 4. 3: There is a solution to Problem 2. 3, if

1) @>0.
2) $\mathrm{rank}[C_0 A_0^{-1} B_0] = p$.
3) $k < g/M_1$
4) the equilibrium point of (4. 1) exists.

Proof: Considering the system (2. 8), there K is given in theorem 3. 4. When w=0 and $\bar{y}=0$, according to theorem 3. 4, we know that the "mild" solution of resultant control system (2. 9) is exponentially stable.

According to condition 4), we have

$$\begin{bmatrix} Ew \\ -\bar{y} \end{bmatrix} = -\begin{bmatrix} A & BK \\ C & 0 \end{bmatrix}\begin{bmatrix} v \\ z \end{bmatrix} - \begin{bmatrix} h(v) \\ 0 \end{bmatrix}$$

where $\begin{bmatrix} v \\ z \end{bmatrix}$ is the equilibrium point of(4. 1).

The "mild" solution of (2. 9) is given as following:

$$\begin{bmatrix} v(t) \\ z(t) \end{bmatrix} = U_{\bar{A}}(t)\begin{bmatrix} v(0) \\ z(0) \end{bmatrix} + \int_0^t U_{\bar{A}}(t-s)\{\begin{bmatrix} h[v(s)]-h(v) \\ 0 \end{bmatrix} - \bar{A}\begin{bmatrix} v \\ z \end{bmatrix}\}ds$$

$$= U_{\bar{A}}(t)\{\begin{bmatrix} v(0) \\ z(0) \end{bmatrix} - \begin{bmatrix} v \\ z \end{bmatrix}\} + \begin{bmatrix} v \\ z \end{bmatrix} + \int_0^t U_{\bar{A}}(t-s)\begin{bmatrix} h[v(s)]-h(v) \\ 0 \end{bmatrix}ds$$

We have

$$\| \begin{bmatrix} v(t) \\ z(t) \end{bmatrix} - \begin{bmatrix} v \\ z \end{bmatrix} \| \leqslant M_1 \| \begin{bmatrix} v(0) \\ z(0) \end{bmatrix} - \begin{bmatrix} v \\ z \end{bmatrix} \| e^{-\gamma t} + \int_0^t M_1 k e^{-\gamma(t-s)} \| \begin{bmatrix} v(s)-v \\ z(s)-z \end{bmatrix} \| ds$$

Similar to the proof of theorem 3. 4, we have

$$\| \begin{bmatrix} v(t) \\ z(t) \end{bmatrix} - \begin{bmatrix} v \\ z \end{bmatrix} \| \leqslant M_3 e^{-bt}, \qquad t \geqslant 0,$$

where

$$M_3 = M_1(\| \begin{bmatrix} v(0) \\ z(0) \end{bmatrix} - \begin{bmatrix} v \\ z \end{bmatrix} \|)$$

Since, as $t \to \infty$,

$$y(t) = Cv(t) \to Cv$$

According to (4. 1), we have

$$\lim_{t\to\infty} y(t) = Cv = \bar{y} \qquad \blacksquare$$

REFFERENCES

[1] M. J. Balas, Distributed parameter control of nonlinear flexible structures with linear finite-dimensional controllers. JOURNAL OF MATHEMATICAL ANALYSIS AND APPLICATIONS 108 p528—545 (1985).

[2] S. Pohjolainen, Robust controller for systems with exponentially stable strongly continuous semigroups. JOURNAL OF MATHEMATICAL ANALYSIS AND APPLICATIONS 111 p622—636 (1985).

[3] Li Chengzhi, Robust regulator design for a class nonlinear system with unmeasurable constant disturbances. JOURNAL OF XIAMEN UNIVERSITY, to be published.

[4] A. Pazy, "Semigroups of Linear Operators and Applications to PDE". University of Maryland, College Park, 1974.

[5] R. F. Curtain and A. J. Pritchard, "Infinite Dimensional Linear Systems Theory". Springer-Verlag, Berlin/Heidelberg/ New York, 1978.

[6] T. Kato, "perturbation Theory for Linear Operators". Springer-Verlag, Berlin /Heidelberg/ New York, 1976.

APPENDIX A: Proof of Theorem 3. 3

Proof of Theorem 3. 3: According to the condition (3. 1), let $K=\varepsilon K_1$ such that

$$\sigma(-C_0A_0^{-1}B_0K_1) \subset \{\lambda \in \mathbb{C} \mid Re\lambda < 0\}, \text{ where, } \varepsilon > 0.$$

Then the operator $\bar{A}$ may be written as

$$\bar{A} = \bar{A}(\varepsilon) = A_1 + \varepsilon A_2,$$

where

$$A_1 = \begin{bmatrix} A & 0 \\ C & 0 \end{bmatrix},$$

$$A_2 = \begin{bmatrix} 0 & BK_1 \\ 0 & 0 \end{bmatrix}.$$

Obviously $\sigma(A_1)=\sigma(A) \dot{\cup} \{0\}$ and $0 \in \rho(A)$. So the spectrum of A_1 satisfies the spectrum decomposition assumption [6, P. 178]. This decomposition holds also for the perturbed operator $\bar{A}(\varepsilon)$ [6, p379] at least if

$$0 \leqslant \varepsilon < r_0 = \min_{\lambda \in \Gamma}(\left\| \begin{bmatrix} 0 & BK_1 \\ 0 & 0 \end{bmatrix} \right\| \cdot \| R(\lambda; A_1) \| (+1)^{-1},$$

where Γ is a circle of radius r^* centered at the origin, running in $U_\sigma(0)=\{\lambda \in \mathbb{C} \mid |\lambda|<\delta\} \subset \rho(A)$.

Let

$$P(\varepsilon) = \frac{-1}{2\pi i}\int_\Gamma R(\lambda; \bar{A}(\varepsilon))d\lambda,$$

then the operator $\bar{A}(\varepsilon)$ may be decomposed according to the decomposition of the original space

$$\bar{H} = H \oplus R_r = H^+(\varepsilon) \oplus H^-(\varepsilon),$$

where $H^+(\varepsilon)=P(\varepsilon)\bar{H}$, $\bar{H}(\varepsilon)=(I-P(\varepsilon))\bar{H}$, respectively. Let $A^+(\varepsilon)$and $\bar{A}(\varepsilon)$ be the restrictions of $\bar{A}(\varepsilon)$ on $H^+(\varepsilon)$ and $H^-(\varepsilon)$. Then $A^+(\varepsilon)$ is bounded and finite-dimensional, and

$$P(\varepsilon)\bar{A}(\varepsilon) = \bar{A}(\varepsilon)P(\varepsilon) = A^+(\varepsilon)P(\varepsilon)$$

$$(I - P(\varepsilon))\bar{A}(\varepsilon) \subset \bar{A}(\varepsilon)(I - P(\varepsilon)) = A^-(\varepsilon)(I - P(\varepsilon)).$$

The strongly continuous semigroup $U_{\bar{A}}(t)$ will also be decomposed according to the decomposition of the space $\bar{H}$. The parts $U_{A^+(\varepsilon)}(t)$ and $U_{\bar{A}(\varepsilon)}(t)$ are strongly continous semigroups, with infinitesimal generators $A^+(\varepsilon)$ and $A^-(\varepsilon)$ respectively [6, p212]. Clearly, the semigroup $U_{\bar{A}}(t)$ will be exponentially stable if the semigroups $U_{A^+(\varepsilon)}(t)$ and $U_{A^-(\varepsilon)}(t)$ are exponentially stable.

Stability of $U_{A^+(\varepsilon)}(t)$: Since $A^+(\varepsilon)$ is finite-dimensional, stability of $U_{A^+(\varepsilon)}(t)$ can be seen from the spectrum.

According to the Perturbation Theory [6], the eigenvalues $\lambda_i(\varepsilon)$, $i=1,\cdots,p$, of the operator $A^+(\varepsilon)$ near $\varepsilon=0$ are given as a converging Puiseux series

$$\lambda_i(\varepsilon) = \varepsilon\lambda_i + \varepsilon^{1+\frac{1}{p_j}}a_{ij} + \cdots,$$

where $p_j \geqslant 1$, and λ_i, $i=1,\cdots,p$, are the eigenvalues of operator $P(0)A_2$ on $H^+(0)$. Because of

$$P(0)A_2 = \begin{bmatrix} 0 & 0 \\ 0 & -\varepsilon C_0 A_0^{-1} B_0 K_1 \end{bmatrix},$$

we have

$$Re\lambda_i < 0 \quad , \qquad i = 1,\cdots,p.$$

Then $U_{A^+(\varepsilon)}(t)$ will be exponentially stable, for sufficiently small positive values ε.

Stability of $U_{A^-(\varepsilon)}(t)$: Since $U_{\bar{A}(0)}(t)$ is given as

$$U_{\bar{A}(0)}(t) = \begin{bmatrix} U_A(t) & 0 \\ CA^{-1}U_A(t)\text{-}CA^{-1} & I \end{bmatrix} \quad ,$$

we may compute $U_{\bar{A}(0)}(t)$ as

$$U_{\bar{A}(0)}(t)(I\text{-}P(0)\begin{bmatrix} x_0 \\ \eta_0 \end{bmatrix} = \begin{bmatrix} U_A(t)x_0 \\ CA^{-1}U_A(t)x_0 \end{bmatrix} \quad .$$

An easy computation proves that $U_{A^-(0)}(t)$ is exponentially stable on $H^-(0)$.

Let

$$U(\varepsilon) = P(\varepsilon)P(0) + (I\text{-}P(\varepsilon))(I\text{-}P(0)) \quad ,$$
$$V(\varepsilon) = [I - [P(0)\text{-}P(\varepsilon)]^2]^{-1}\{P(0)P(\varepsilon) + (I - P(0))(I\text{-}P(\varepsilon))\} \quad .$$

Becauce of

$$R(\lambda;\bar{A}(\varepsilon)) = R(\lambda;\bar{A}(0))\text{-}\varepsilon R(\lambda;\bar{A}(\varepsilon))A_2 R(\lambda;\bar{A}(0)),$$

we have

$$P(\varepsilon) = P(0) + \varepsilon\bar{P}(\varepsilon),$$

where

$$\bar{P}(\varepsilon) = \frac{1}{2\pi i}\int_\Gamma R(\lambda;\bar{A}(\varepsilon))A_2 R(\lambda;\bar{A}(0))d\lambda.$$

After an easy computation, we have

$$\| \bar{P}(\varepsilon) \| \leqslant \frac{(1\text{-}r_0)^2(2 + r_0)r^*}{ar_0^2} \quad ,$$

$$\| \bar{A}(0)\bar{P}(\varepsilon) \| \leqslant \frac{[ar_0 + r^*(1\text{-}r_0)](2 + r_0)(1\text{-}r_0)r^*}{ar_0^2} = M^* \quad ,$$

$$\| \bar{P}(\varepsilon)\bar{A}(0) \| \leqslant M^* \quad ,$$

where

$$a = \| A_2 \| \quad .$$

Thus, when $0<\varepsilon\leqslant\min\{\frac{r_0}{2},\left[\frac{2(1-r_0)^2(2+r_0)r^*}{ar_0^2}\right]^{-1}\}=\varepsilon^*$, we have

$$\begin{aligned} &\| P(0)-P(\varepsilon) \| < 1, \\ &U(\varepsilon) \in L(\overline{H}) \\ &V(\varepsilon) \in L(\overline{H}) \\ &U(\varepsilon)V(\varepsilon) = V(\varepsilon)U(\varepsilon) = I \quad , \\ &U(\varepsilon): H^-(0) \to H^-(\varepsilon) \quad , \\ &V(\varepsilon): H^-(\varepsilon) \to H^-(0) \quad , \\ &U(\varepsilon) = I + \varepsilon\overline{U}(\varepsilon) \quad , \\ &V(\varepsilon) = I + \varepsilon\overline{V}(\varepsilon) \quad , \end{aligned} \tag{5.1}$$

where $\|\overline{V}(\varepsilon)\|$, $\|\overline{U}(\varepsilon)\|$, $\|\overline{V}(\varepsilon)\cdot\overline{A}(0)\|$, $\|\overline{A}(0)\cdot\overline{U}(\varepsilon)\|$ are uniformly. bounded for $0<\varepsilon<\varepsilon^*$.

Becauce of

$$V(\varepsilon)T_{A^-(\varepsilon)}(t)U(\varepsilon) = T_{V(\varepsilon)A^-(\varepsilon)U_{(\varepsilon)}}(t),$$

on $\overline{H}(0)$, we have.

$$V(\varepsilon)A^-(\varepsilon)U(\varepsilon) = A^-(0) + \varepsilon(I - P(0))G(\varepsilon)(I - P(0))$$

where

$$\begin{aligned} G(\varepsilon) = [&A_2 + \overline{V}(\varepsilon)A_1 + A_1\overline{U}(\varepsilon) + \varepsilon\overline{V}(\varepsilon)A_2 \\ &+ \overline{V}(\varepsilon)A_1\overline{U}(\varepsilon) + \varepsilon A_2\overline{U}(\varepsilon) + \varepsilon^2\overline{V}(\varepsilon)A\overline{U}(\varepsilon)] \end{aligned}$$

Obviously there is a $G>0$, such that $\|G(\varepsilon)\|\leqslant G$ for all $0<\varepsilon<\varepsilon^*$. Thus

$$\| (I-P(0))G(\varepsilon)(I-P(0)) \| \leqslant G, \textit{ also.}$$

Since $\overline{A}(0)$ is an infinitesimal generator of an expontially stable semigroup and $G(\varepsilon)$ is uniformly bounded, $V(\varepsilon)\overline{A}(\varepsilon)U(\varepsilon)$ is an infinitesimal generator of an exponentially stable semigroup on $\overline{H}(0)$ for sufficiently small opsitive values ε.

According to (5.1), $T_{\overline{A}(\varepsilon)}(t)$ is expontially stable on $H^-(\varepsilon)$. ▌

OPTIMAL CONTROL FOR INFINITE DIMENSIONAL SYSTEMS

Xunjing Li

Department of Mathematics, Fudan University

Shanghai 200433, China

Abstract This is a survey of some of the works on optimal control theory for infinite dimensional systems carried out by the research group of Fudan University in recent years.

Key words Maximum principle, dynamic programming, optimal control, distributed parameter systems, stochastic systems.

AMS(MOS) classifications. 49B27, 93C25, 93E20.

1. Distributed Parameter Systems

Let X and Y be Banach spaces with the embedding $X \subset Y$ being dense and continuous. The dual spaces of X and Y are denoted by X^* and Y^*, respectively. Let U be a given metric space and T>0 be a constant. The admissible control set is $U_{ad}=L^\infty(0,T;U)$. Let $\Delta = \{(t,s) \in [0,T]\times[0,T] \,|\, 0\le s<t\le T\}$ and $\bar{\Delta}$ be its closure. The optimal control problem can be stated as follows:

Minimize

$$J(v(\cdot)) = \int_0^T f^0(t,x(t;v),v(t))dt$$

subject to the following

$$(1.1) \qquad x(t;v) = G(t,o)x(o;v) + \int_0^t G(t,s)f(s,x(s;v),v(s))ds \,, \qquad o \le t \le T \,,$$

$$(1.2) \qquad (x(o;v),x(o;v)) \in S \subset X \times X \,,$$

$$(1.3) \qquad v(\cdot) \in U_{ad}$$

We assume the following

This work was partially supported by the National Natural Science Fundation of China and the Chinese State Education Commission Science Fundation.

(H0) X^* is strictly convex.

(H1) The evolution operator $G: \Delta \longrightarrow L(Y,X)$ is strongly continuous in Δ and there exist constants $M>0$ and $0 \le \alpha < 1$, such that

$$\|G(t,s)\|_{\mathcal{L}(Y,X)} \le \frac{M}{(t-s)^{\alpha}} \qquad \forall\ (t,s) \in \Delta\ ,$$

Moreover, $G: \bar{\Delta} \longrightarrow \mathcal{L}(X,X)$ is also strongly continuous and

$$G(s,s) \equiv I\ , \qquad \forall\ s\in[0,T]\ ,$$

where I is the identity operator on X.

(H2) The mappings $f: [0,T]\mathrm{x}X\mathrm{x}U \longrightarrow Y$, $f^0: [0,T]\mathrm{x}X\mathrm{x}U \longrightarrow \mathbb{R}$ and their Frechet derivatives f_x and f^0_x are strongly continuous.

(H3) The set S is convex and closed in XxX.

Li and Yong [10] proved the following maximum principle.

THEOREM 1.1 *Let (H0)-(H3) hold. Let* $u(\cdot)$, $x(\cdot)=x(\cdot;u)$ *be a solution of the optimal control problem. Let* R-Q *be of finite codimension in* X, *where*

$$R = \{\xi\in X \,|\, \xi = \int_0^T G_1(T,t)\{f(t,x(t),v(t)) - f(t,x(t),u(t))\}dt,\ v(\cdot)\in U_{ad}\}$$

$$Q = \{x_1 - G_1(T,0)x_0 \,|\, (x_0,x_1)\in S\}$$

and $G_1(\cdot,\cdot)$ *satisfies*

$$(1.4)\qquad G_1(t,s)y = G(t,s)y + \int_s^T G_1(t,r)f_x(r,x(r),u(r))G_1(r,s)y\,dr \qquad \forall\ (s,t) \in \Delta\ ,\ y\in Y\ .$$

Then, there exists a $(\varphi(\cdot),\varphi^0)\neq 0$, *such that*

$$\varphi^0 \le 0\ ,$$

$$(1.5)\qquad \varphi(t) = G^*(T,t)\varphi(T) + \int_t^T G^*(s,t)f_x^*(s,x(s),u(s))\varphi(s)ds + \varphi^0 \int_t^T G^*(s,t)f_x^{0*}(s,x(s),u(s))ds, \qquad \forall\ t\in[0,T]\ .$$

$$(1.6)\qquad \langle\varphi(t),f(t,x(t),u(t))\rangle + \varphi^0 f^0(t,x(t),u(t)) = \max_{v\in U} \{\langle\varphi(t),f(t,x(t),v)\rangle + \varphi^0 f^0(t,x(t),v)\} \qquad \forall\ \text{a.e. } t\in[0,T]$$

$$\langle\varphi(0),x_0 - x(0)\rangle - \langle\varphi(T),x_1 - x(T)\rangle \le 0 \qquad \forall\ (x_0,x_1)\in S.$$

Remark 1.1 $S=\{(x_0,x_1)\}$. This is an optimal control problem with fixed end points. Then $Q=\{x_1-G_1(T,0)x_0)\}$. Hence, if R is of finite codimension in X, then the maximum principle holds. This result contains that of Fattorini [3].

Remark 1.2 $S=\{x_0\}xQ_1$, $Q=Q_1-G_1(T,0)x_0$. Hence Q is of finite codimension in X is the same as $Q_1$is so. Hence, provided Q_1 is of finite codimension in X, the maximum principle holds. This is the result of Li and Yao [9].

Remark 1.3 Optimal periodic control problem.

Assume

(H4)
$$G(t+T,s+T) = G(t,s),$$
$$f(t+T,x,v) = f(t,x,v),$$
$$f^0(t+T,x,v) = f^0(t,x,v),$$
$$S = \{(x,x)\,|\,x \in X\} .$$

THEOREM 1.2 *Let* (H0)-(H4) *hold and let Range* $(I-G_1(T,0))$ *be of finite codimension in* X. *Then the maximum principle holds for the periodic optimal control problem, i.e., there exist* $\varphi_0 \le 0$ *and* $\varphi(\cdot)$ *satisfying* (1.5)-(1.6) *with* $\varphi(0)=\varphi(T)$.

Remark 1.4 We know that if X is reflexive, then, by changing the norm to an equivalent one, we may assume that X^* is strictly convex. Also, if X is separable, then, by Day [2], may also do the above. Thus, we see that (H0) is general enough to cover almost all cases which interest us (e.g., $X=C([-r,0];\mathbb{R}^n), L^1(\mathbb{R}^n)$,etc.).

Example 1.1 Let $r>0$, $X=C([-r,0];\mathbb{R}^n)$. Then, X is separable. Thus, we may endow a new norm to X so that X^* is strictly convex. Consider the following functional differential system

$$\frac{dx(t)}{dt} = f(t,x_t,v(t)) , \tag{1.7}$$

where f: $\mathbb{R}xXx\mathbb{R}^m \to \mathbb{R}^n$ is a given map and $x_t \in X$ is defined by

$$x_t(\theta) = x(t+\theta), \qquad \forall\ \theta \in [-r,0] ,$$

whenever $x(\cdot)$ is continuous. Furthermore, let $f^0: \mathbb{R}xXx\mathbb{R}^m \to \mathbb{R}$ be given. Assume that (H2) and (H4) hold for the maps f and f^0. Now, let $G_1(\cdot,\cdot)$ be the solution operator of the variational equation

$$\frac{d\delta x(t)}{dt} = f_y(t,x_t,u(t))\delta x_t , \tag{1.8}$$

where $f_y(t,x_t,v)$ is the Frechet derivative of $f(t,x_t,v)$ in x_t. From

Hale [4], we know that $G_1(T,0)$ is compact for $T>r$. Thus, if the period $T>r$, then the Range of $I-G_1(T,0)$ be of finite codimension in X. Hence by theorem 1.2, we get the maximum principle for the optimal periodic control problem of functional differential system (1.7) without any additional condition. Here we eliminate the conditions imposed by Colonius [1] or Li and Chow [8] in proving the similar result.

2. Stochastic Systems

Let $(Q,\mathcal{F},P)$ be a probability space with filtration $\mathcal{F}^t$. Let $B(\cdot)$ be an $\mathbb{R}^d$-valued standard Wiener process. Assume

$$\mathcal{F}^t = \sigma\,\{B(s) \mid 0\le s\le t\} .$$

Let U be a given metric space. An admissible control $v(\cdot)$ is a $\mathcal{F}^t$-adapted measurable process with values in U, such that

$$\sup_{0\le t\le 1} E|v(t)|^m < \infty , \qquad \forall\ m=1,2,\cdots \quad .$$

Denote the set of all admissible controls by U_{ad} .

The optimal stochastic control problem can be stated as follows:

Minimize

$$J(v(\cdot)) = E\int_0^1 l(x(t;v),v(t))dt + Eh(x(1;v))$$

subject to the following stochastic system

$$\begin{aligned}(2.1)\qquad & dx(t;v) = g(x(t;v),v(t))dt + \sigma(x(t;v),v(t))dB(t) ,\\ & x(0;v) = x_0 ,\\ & v(\cdot) \in U_{ad} .\end{aligned}$$

Assume the mappings

$$g:\ \mathbb{R}^n \times U \longrightarrow \mathbb{R}^n , \qquad \sigma:\ \mathbb{R}^n \times U \longrightarrow \mathcal{L}(\mathbb{R}^d,\mathbb{R}^n) ,$$
$$l:\ \mathbb{R}^n \times U \longrightarrow \mathbb{R} , \qquad h:\ \mathbb{R}^n \longrightarrow \mathbb{R} .$$

and their derivatives g_x, g_{xx}, σ_x, σ_{xx}, l_x, l_{xx}, h_x, h_{xx} are continuous. Assume $g_x, g_{xx}, \sigma_x, \sigma_{xx}, l_{xx}, h_{xx}$ are bounded, g, σ, l_x, h_x are bounded by $C(1+|x|+|v|)$. The stochastic maximum principle was discussed by a lot of papers when the diffusion coefficient independs on control. Dr.Peng [11] derived a maximum principle when the diffusion coefficient depends on control. Assume $(u(\cdot), x(\cdot)=x(\cdot;u))$ is a solution of the optimal stochastic control problem. Then there exists an unique solution $(p(\cdot),K(\cdot))$ of the first order adjoint equation

$$- dp(t) = \Big[g_x^*(x(t),u(t))p(t) + \sum_{j=1}^{d} \sigma_x^{j*}(x(t),u(t))K_j(t)$$

(2.2) $$+ l_x^*(x(t),u(t))\Big] dt - K(t)dB(t)$$

$$p(1) = h_x^*(x(1)) .$$

Let $(P(\cdot),Q(\cdot))$ be the unique solution of the second order adjoint equation

$$- dP(t) = \Big[g_x^*(x(t),u(t))P(t) + P(t)g_x(x(t),u(t))$$

$$+ \sum_{j=1}^{d} \sigma_x^{j*}(x(t),u(t))P(t)\sigma_x^j(x(t),u(t))$$

$$+ \sum_{j=1}^{d} \{\sigma_x^{j*}(x(t),u(t))Q_j(t) + Q_j(t)\sigma_x^{j*}(x(t),u(t))\}$$

(2.3) $$+ l_{xx}(x(t),u(t)) + g_{xx}^*(x(t),u(t))p(t)$$

$$+ \sum_{j=1}^{d} \sigma_{xx}^{j*}(x(t),u(t))K_j(t)\Big]dt - Q(t)dB(t) .$$

$$P(1) = h_{xx}(x(1)) .$$

THEOREM 2.1 *Let* $u(\cdot)$ *and* $x(\cdot)=x(\cdot,u)$ *be an optimal solution of the optimal stochastic control problem. Let* $(p(\cdot),K(\cdot))$ *and* $(P(\cdot),Q(\cdot))$ *satisfy* (2.2) *and* (2.3), *respectively. Then the maximum principle*

$$- \frac{1}{2} \sigma^*(x(t),u(t))P(t)\sigma(x(t),u(t))$$

$$- l(x(t),u(t)) - g^*(x(t),u(t))p(t)$$

(2.4) $$- \langle K(t) - P(t)\sigma(x(t),u(t)),\sigma(x(t),u(t))\rangle$$

$$= \max_{v\in U} \{- \frac{1}{2} \sigma^*(x(t),v)P(t)\sigma(x(t),v)$$

$$- l(x(t),v) - g^*(x(t),v)p(t)$$

$$- \langle K(t) - P(t)\sigma(x(t),u(t)),\sigma(x(t),v)\rangle\} .$$

holds a.e. a.s.

Provided σ independs on v, then (2.4) reduce to

(2.5) $$- l(x(t),u(t)) - \langle p(t),g(x(t),u(t))\rangle$$

$$= \max_{v\in U} \{-l(x(t),u(t)) - \langle p(t),g(x(t),v)\rangle\} .$$

This is proved by Haussmann [5].

Dr. Hu [6] discussed the following optimal stochastic control problem:

minimize

$$J(v(\cdot)) = E\int_0^1 l(x(t),v(t))dt$$

subject to (2.1) and

$$x(1;v) \in Q$$

Assume Q is of finite codimension in $L^2(\Omega)$, then the maximum principle holds.

Let X, Y, V be Hilbert spaces. A Wiener process with values in Y is a $\mathcal{F}^t$ adapted stochastic process $B(\cdot)$, such that, for any $e\in Y$, $\langle B(\cdot),e\rangle$ is a real Wiener process in $(\Omega,\mathcal{F},F^t,P)$, with the correlation function

$$E\langle B(t_1),e\rangle\langle B(t_2),e'\rangle = (\mathrm{tr}\ Q)\min(t_1,t_2),$$

$$\forall\ e,e'\in Y, \qquad t_1,t_2\ge 0\ ,$$

where Q is a positive selfadjoint nuclear operator defined on Y, i.e., for any orthonormal basis $\{e_j\}$ of Y, we have

$$\mathrm{tr}\ Q = \sum_{j=1}^{\infty} \langle Q\,e_j,\ e_j\rangle < \infty\ .$$

Assume

$$\mathcal{F}^t = \sigma\,(B(s);\ 0\le s\le t)\ .$$

and $\{e_j\}$ be the orthonormal basis of Y such that

$$Q\,e_j = \lambda_j e_j\ ,$$

where λ_j is the eigenvalue of Q and

$$\lambda_j > 0, \quad j=1,2,\cdots, \quad \sum_{j=1}^{\infty}\lambda_j < \infty\ .$$

Let

$$g:\ [0,1]\times X\times V \longrightarrow X\ ,$$

$$\sigma:\ [0,1]\times X \longrightarrow \mathcal{L}(Y,X)\ ,$$

$$l:\ [0,1]\times X\times V \longrightarrow \mathbb{R}\ ,$$

$$h:\ X \longrightarrow \mathbb{R}\ ,$$

$$G:\ X \longrightarrow \mathbb{R}^m$$

be all continuously Gateaux differentiable, g_x, σ_x, G_x be bounded and g, σ, l_x, h_x be bounded by $c(1+\|x\|)$.

Hu and Peng [7] considered the following optimal stochastic control problem:

minimize

(2.5) $$J(v(\cdot)) = E\int_0^1 l(t,x(t;v),v(t))dt + Eh(x(1;v))$$

subject to

(2.6) $$x(t;v) = e^{tA}x_0 + \int_0^t e^{(t-s)A}\, g(s,x(s;v),v(s))ds$$
$$+ \int_0^t e^{(t-s)A}\, \sigma(s,x(s;v))dB(s) \ ,$$

$$EG(x(1;v)) = 0 \ ,$$

$$v(\cdot) \in U_{ad} \ ,$$

where A is a given infinitesimal generator of a C_0-semigroup e^{tA} and an admissible control $v(\cdot) \in U_{ad}$ is an adapted measurable process with values in $U \subset V$ such that

$$\sup_{0 \le t \le 1} E|v(t)|^2 < +\infty \ .$$

Hu and Peng [7] proved

Theorem 2.2 *Let* $u(\cdot)$ *and* $x(\cdot)=x(\cdot;u)$ *be an optimal solution of the problem* (2.5)-(2.6). *Then there exist* $\lambda \in \mathbb{R}$, $\nu \in \mathbb{R}^m$ *such that*

$$|\lambda|^2 + |\nu|^2 = 1$$

and $(p(\cdot),K(\cdot))$ *satisfying*

$$p(t) = e^{(1-t)A^*}(h_x^*(x(1))\lambda + G_x^*(x(1))\nu)$$
$$+ \int_t^1 e^{(s-t)A^*}\{g_x^*(s,x(s),u(s))p(s)$$
$$+ \sum_{j=1}^d \sigma_x^{j*}(s,x(s),u(s))K_j(s)\}ds$$
$$- \int_t^1 e^{(s-t)A^*}\, K(s)dB(s) \ ,$$

such that

$$H(t,x(t),u(t),p(t)) = \max_{v \in U} H(t,x(t),v,p(t)) \ ,$$

where

$$H(t,x,v,p) = \lambda\, l(t,x,v) + \langle p,g(t,x,v)\rangle \ .$$

3. Connection between Maximum Principle and Dynamic Programming

Given $(s,y) \in [0,1] \times \mathbb{R}^n$, assume the mappings

$$g: \quad [0,1] \times \mathbb{R}^n \times U \longrightarrow \mathbb{R}^n ,$$
$$l: \quad [0,1] \times \mathbb{R}^n \times U \longrightarrow \mathbb{R}$$
$$h: \quad \mathbb{R}^n \longrightarrow \mathbb{R}$$

and their derivatives g_x, l_x, h_x are continuous. The optimal control problem is the following:

$$\text{minimize} \qquad J(s,y;v) = \int_s^1 l(t,x(t;v),v(t))dt + h(x(1;v))$$

subject to

$$\frac{dx(t;v)}{dt} = g(t,x(t;v),v(t)) ,$$
$$(3.1) \qquad x(s;v) = y ,$$
$$v(\cdot) \in U_{ad} = L^{\infty}([s,1];U) .$$

Denote

$$V(s,y) = \inf\{J(s,y;v) \mid \text{subject to } (3.1)\} .$$

Let $u(\cdot)$ and $x(\cdot)=x(\cdot;u)$ be a solution of the optimal control problem. According to the maximum principle, there exists $\varphi(\cdot)$ satisfying

$$(3.2) \qquad \frac{d\varphi(t)}{dt} = - g_x^*(t,x(t),u(t))\varphi(t) - l_x^*(t,x(t),u(t))$$
$$\varphi(1) = h_x^*(x(1))$$

such that

$$(3.3) \quad H(t,x(t),u(t),\varphi(t)) = \max_{v\in U} H(t,x(t),v,\varphi(t)) \quad \text{a.e. } t\in[s,1] ,$$

where

$$H(t,x,v,p) = -l(t,x,v) - \langle p,g(t,x,v)\rangle .$$

The Bellman's dynamic programming says: provided $V(\cdot,\cdot)$ continuously differential, then V satifies

$$(3.4) \qquad -\frac{\partial V(t,x)}{\partial t} + \sup_{v\in U} H(t,x,V,\frac{\partial V(t,x)}{\partial x}) = 0$$
$$V(1,x) = h(x)$$

It is very know that if $V(\cdot,\cdot)$ is second order continuously differential then

$$\varphi(t) = \frac{\partial V(t,x(t))}{\partial x} .$$

But, the value function $V(\cdot,\cdot)$ may not be smooth. Dr.Zhou [12] proved

THEOREM 3.1 *Let* $u(\cdot)$ *and* $x(\cdot)=x(\cdot;u)$ *be a solution of optimal control problem. Then*

$$D_x^- V(t,x(t)) \subset \{\varphi(t)\} \subset D_x^+ V(t,x(t)) ,$$

and

$$D_{t,x}^- V(t,x(t)) \subset \{(H(t,x(t),u(t),\varphi(t)),\varphi(t))\} \subset D_{t,x}^+ V(t,x(t)) ,$$

where D^+ *and* D^- *is the superdifferential and subdifferential, respectively.*

For optimal stochastic control problem:

minimize $$J(s,y;v) = E\int_s^1 l(t,x(t;v),v(t))dt + Eh(x(1;v))$$

subject to

$$\text{(3.5)} \qquad \begin{aligned} dx(t;v) &= g(t,x(t;v),v(t))dt + \sigma(t,x(t;v),v(t))dB(t) \\ x(s;v) &= y , \\ v(\cdot) &\in U_{ad} , \end{aligned}$$

where g, σ, l, h, $B(\cdot)$, U_{ad} as §2. Dr.Zhou [13], [14] proved the following:

THEOREM 3.2 *Let* $u(\cdot)$, $x(\cdot)=x(\cdot;u)$ *be a solution of the above problem. Let* $p(\cdot)$, *and* $P(\cdot)$ *satisfy* (2.2) *and* (2.3), *respectively, and*

$$V(s,y) = \inf \{J(s,y;v) \mid \text{subject to (3.5)}\} .$$

Then

$$(p(t),P(t)) \in D_x^{2,+} V(t,x(t))$$

THEOREM 3.3 *The value function* V *is a viscosity solution of the* HJB *equation*

$$\frac{\partial V(t,x)}{\partial t} + \sup_{v\in U} G\left[t,x,v,\frac{\partial V(t,x)}{\partial x}, \frac{\partial^2 V(t,x)}{\partial x^2}\right] = 0$$

$$V(1,x) = h(x) ,$$

where

$$G(t,x,v,p,S) = -\frac{1}{2}\sigma^*(t,x,v)S\,\sigma(t,x,v) - \langle p,g(t,x,v)\rangle - l(t,x,v) .$$

THEOREM 3.4 *Let* $u(\cdot)$ *and* $x(\cdot)=x(\cdot;u)$ *be an optimal solution, then the maximum principle holds, i.e.*

$$G(t,x(t),u(t),p(t),P(t)) - \langle K(t)-P(t)\sigma(t,x(t),u(t)),\sigma(t,x(t),u(t))\rangle$$

$$= \max_{v \in U} \{G(t,x(t),v,p(t),P(t)) - \langle K(t)-P(t)\sigma(t,x(t),u(t)),\sigma(t,x(t),v)\rangle\} .$$

References

[1] F.Colonius, Optimal Periodic Control, Springer-Verlag, 1988.

[2] M.M.Day, Strict convexity and smoothness of normed spaces, Trans.Amer.Math.Soc., 78(1955), 516-528.

[3] H.O.Fattorini, A unified theory of necessary conditions for nonlinear nonconvex control systems, Appl.Math. & Optim., 15(1987), 141-185.

[4] J.K.Hale, Theory of Functional Differential Equations, Springer-Verlag, 1977.

[5] U.G.Haussmann, General necessary conditions for optimal control of stochastic systems, Math.Programming Study, 6(1976).

[6] Y.Hu, Maximum principle of optimal control for Markov processes, Acta Math. Sinica, 33(1990), 43-56.

[7] Y.Hu and S.G.Peng, Maximum principle for semilinear stochastic evolution control systems, to appear.

[8] X.J.Li and S.N.Chow, Maximum principle of optimal control for functional differential systems, J.Optim. Theory & Appl., 54(1987), 335-360.

[9] X.J.Li and Y.L.Yao, Maximum principle of distributed parameter systems with time lags, Lecture Notes in Control and Information Sciences, Springer-Verlag, No.75(1985), 410-427.

[10] X.J.Li and J.M.Yong, Necessary conditions of optimal control for distributed parameter systems, to appear in SIAM J.Contr. & Optim.

[11] S.G.Peng, A general stochastic maximum principle for optimal control problem, to appear in SIAM J.Contr. & Optim.

[12] X.Y.Zhou, Maximum principle, dynamic programming and their connection in deterministic controls, J.Optim. Theory & Appl., 65(1990), 363-373.

[13] X.Y.Zhou, The connection between the maximum principle and dynamic programming in stochastic control to appear in Stochastics and Stochastics Reports.

[14] X.Y.Zhou, Maximum principle, dynamic programming and their connection, Ph.D thesis, Fudan University, 1989.

NUMERICAL RESOLUTION OF ILL POSED PROBLEMS

R. Luce, J.P. Kernévez
Université de Technologie de Compiègne,B.P. 649 Compiègne France

Abstract

The aim of this paper is to give some numerical methods and results for the resolution of ill posed problems in linear P.D.E's.

Firstly we have studied a simple example namely the Cauchy Problem for the Laplacian Operator. To solve this problem we have applied the " Hilbert Uniqueness Method " (H.U.M.) developed by J.L. Lions [1] [2], and we have discuss the numerical problems encountered. We have then compared these results with those obtained using methods of Optimal Control : Regularization Method and Duality Method.

Secondly we have applied the same methods to an ill posed problem for a linear parabolic equation.

Key Words: Ill Posed Problems, Exact controlabillity, Optimal Control, Regularization, Duality Method.

I THE CAUCHY PROBLEM FOR THE LAPLACIAN OPERATOR

I.1 Problem formulation

Let (1) be the ill posed problem:

$$(1) \qquad \begin{cases} -\Delta u = 0 & \text{in } \Omega \\ u = 0 & \text{on } S \\ \dfrac{\partial u}{\partial \nu} = f & \text{on } S \end{cases}$$

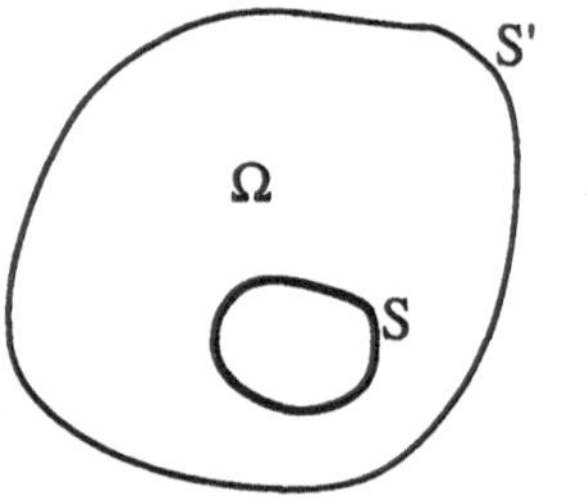

Two conditions are imposed on the same part S of the boundary. The two boundaries need not necessarily be disconnected.

The objective is to find **a function u satisfying (1) on Ω**. This example is a standard model of an ill posed problem in sense given by Tykonov in [3].

This problem can be posed as a problem of exact controlabillity:

To find u such that (1) is true is equivalent to finding a control v on S' such that u satisfies equations (2) and (3)

$$(2)\quad \begin{cases} -\Delta u = 0 & \text{in } \Omega \\ u = 0 & \text{on } S \\ u = v & \text{on } S' \end{cases} \qquad (3)\quad \frac{\partial u}{\partial \nu} = f$$

This problem of exact controlabillty does not always admit a solution, as it depends on the function f. Also The H.U.M. method permits the determination of the space G which must contain the function f must belong in order that v belong to $L^2(S')$.

I.2 Application of the H.U.M method.

We consider the following equations (4) and (5) :

$$(4)\quad \begin{cases} -\Delta \varphi = 0 & \text{in } \Omega \\ \varphi = g & \text{on } S \\ \varphi = 0 & \text{on } S' \end{cases}$$

where (4) defines a linear operator K

$$K : \mathcal{D}(S) \longrightarrow L^2(S') \qquad g \quad v = \frac{\partial \varphi}{\partial \nu}$$

$$(5)\quad \begin{cases} -\Delta \psi = 0 & \text{in } \Omega \\ \psi = 0 & \text{on } S \\ \psi = v & \text{on } S' \end{cases}$$

where (5) defines a linear operator L

$$L : L^2(S') \longrightarrow H^{-1}(S) \qquad v \quad \frac{\partial \psi}{\partial \nu}$$

Thus (4) and (5) together define a linear operator Λ, such that

$$\Lambda : \mathcal{D}(S) \longrightarrow H^{-1}(S) \qquad g \quad \frac{\partial \psi}{\partial \nu}$$

Using the frequently cited theorem of the unicity of the Cauchy problem it can be easily demonstrated that (6) defines a norm on $\mathcal{D}(S)$.

$$(6)\qquad |g| = \left\| \frac{\partial \varphi}{\partial \nu} \right\|_{L^2(S')}$$

Let $\mathcal{D}(S)$ bc endowed with the norm defined by (6). We then consider the hilbert space G be the completion of $\mathcal{D}(S)$ with respect to this norm (6)

Proposition 1: The application Λ can be extended to an isometry of G onto G'

$$\Lambda : G(S) \longrightarrow G'(S) \qquad g \quad \frac{\partial \psi}{\partial \nu}$$

Proof: The proof of this proposition is given in [4]. The demonstration is based on Hahn Banach Theorems.

Proprosition 2: K define an isometry of G onto $L^2(S')$ and L an isometry of $L^2(S')$ onto G'.

Proof : The proof is given in [4].

Propostion 3: G is a dense subspace of $H^1(S)$ and G' is included with density in $H^{-1}(S)$.

Proof : The proof is given in [4].

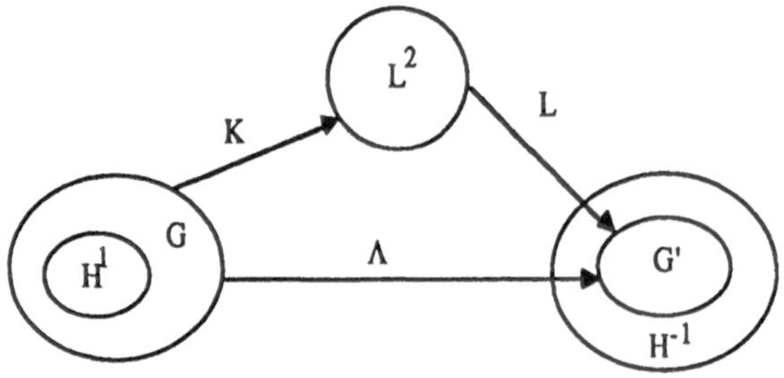

Figure n°1

The ill posed problem is now well posed. Indeed if f belongs to G' it there exists only one g in G satisfying $\Lambda g = f$, the control v is given by $v = Kg$, and the solution u is ψ.

I.3 Numerical Resolution of $\Lambda g = f$.

Proposition 4: The resolution of the linear equation $\Lambda g = f$ amounts to the minimization of J(g) with $g \in G$.

(7) $$J(g) = \frac{1}{2} < \Lambda g, g >_{G',G} - < f,g >_{G',G} = \frac{1}{2} \| \frac{\partial \varphi}{\partial v} \|^2_{L^2(S')} - < f,g >_{G',G} .$$

Proof : The proof is given in [4]

The results presented in §1.3 shows that (7) admits a minimum if f belongs to G'. But given a function f, it is difficult to determine whether or not it belongs to G' or not. Suppose that f belongs to G' so (7) admits a minimum; thus f belongs to H^{-1}. Since H^1 is included with density in G, a basis of H^1 also is a basis of G. Under these conditions 7 can be minimized using the finite element method together with a conjugate gradient method [5][4].

For a numerically attainable f the control v obtained is accurate, although the norm of the control g in $H^1(S)$ may be very large. But a perturbation δf of f, small in the norm of $H^{-1}(S)$, may be large in the norm of G' so that the corresponding δg may be large in the norms of both G(S) and $H^1(S)$. As a consequence, the control $v = \frac{\partial \varphi}{\partial v}$ is very sensitive to slight variations of f. Numerical examples which confirm these results are presented below.

The domain of the first example is a crown (fig n°2). The function f desired is represented on the figure n°5. The figures n°3 and 4 show the control g and v obtained by solving the minimization problem. The figure n°5 shows that the function numerically constructed f is correct, and agrees with the desired f.

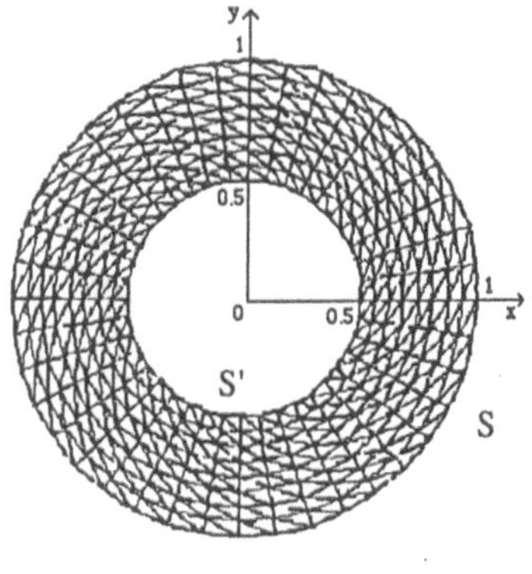

figure n°2

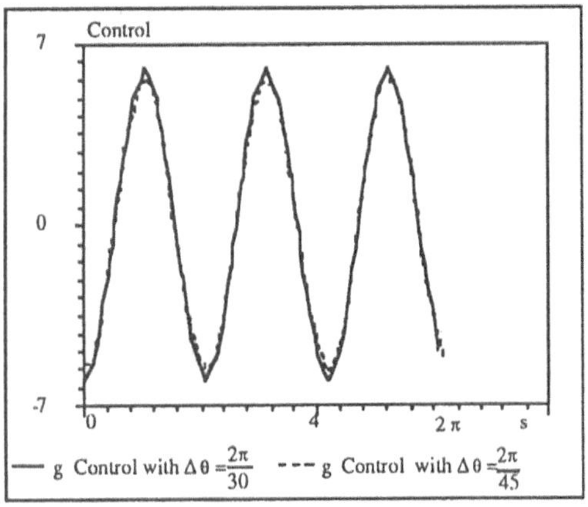

figure n°3

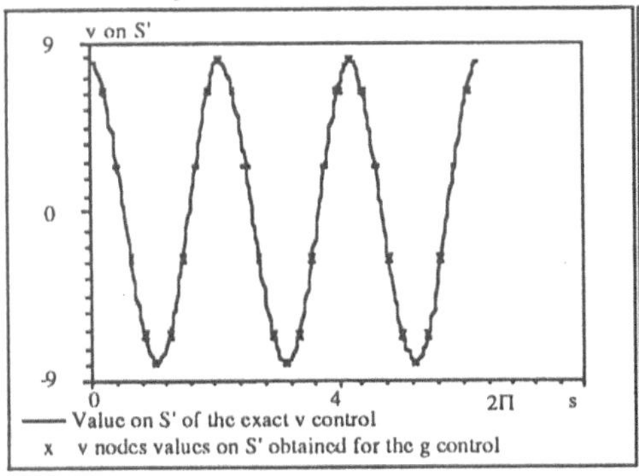

figure n°4

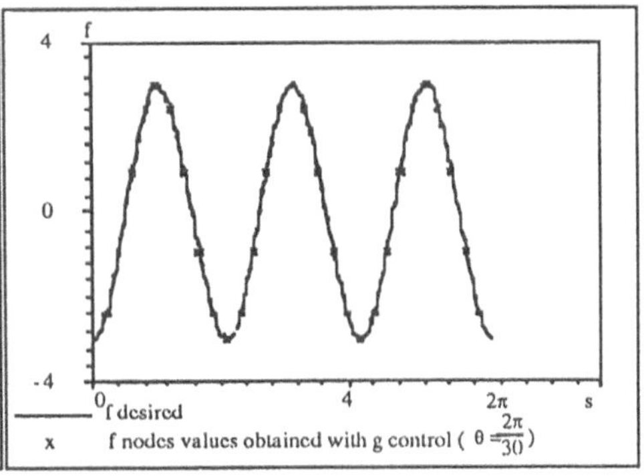

figure n°5

Thus, for a crown the results are very good, and even when the function f is pertubed, the results always are accurate.

The domain of the second example is a disk (fig n°6), the boundary S being by the part of circle comprised beetween $\frac{\pi}{2}$ and $\frac{3\pi}{2}$, the boundary S' being by the complementary part of the circle.

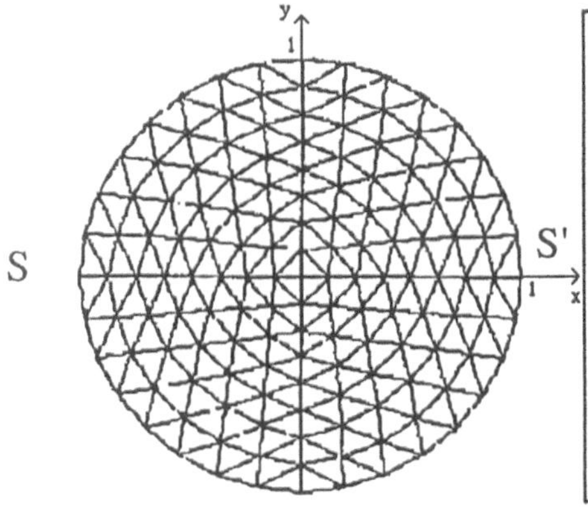

figure n°6

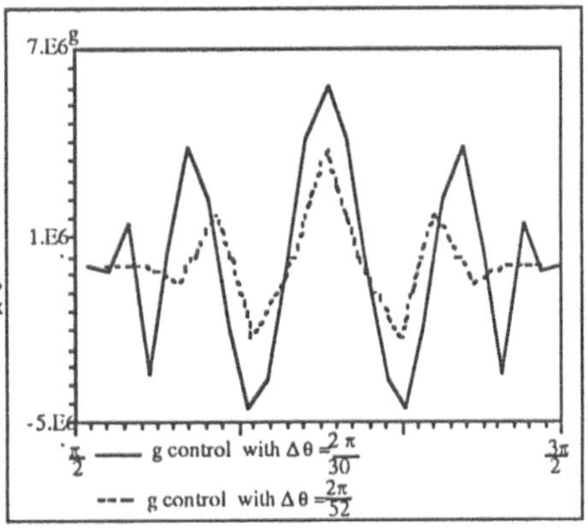

figure n°7

The figure n°7 shows the controls g obtained with different values of the step of the mesh. Notice that the control g does not converge, the amplitude of g increases with decreasing step size.

On the other hand the controls v obtained are accurate (figure n°9), and the numerically reconstructed function f is very good (figure n°8).

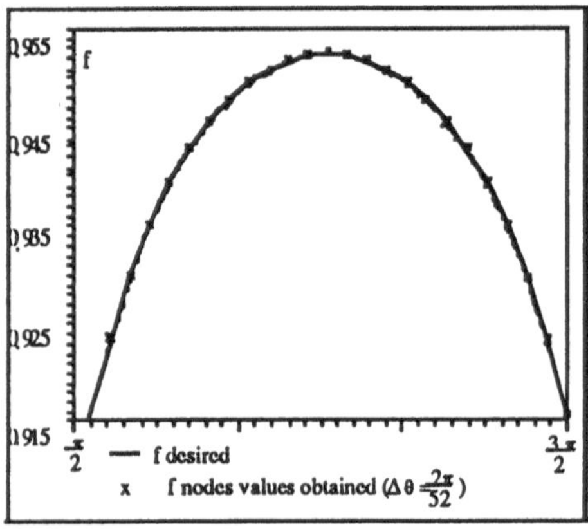

figure n°8

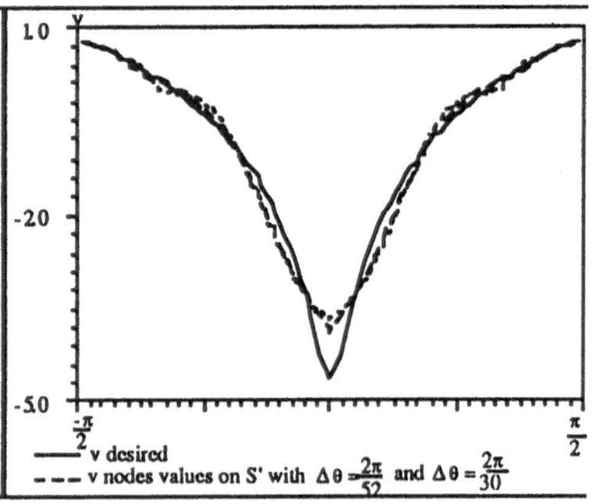

figure n°9

However when the function f is slightly pertubed according to a uniform law (fig n°10) the control v obtained is not accurate (fig n°11).

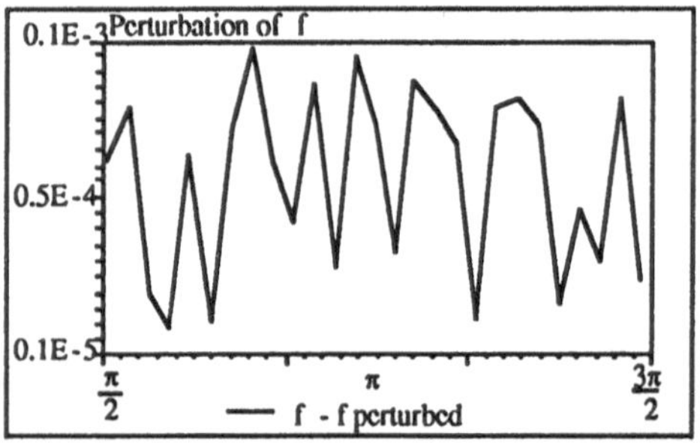

figure n°10

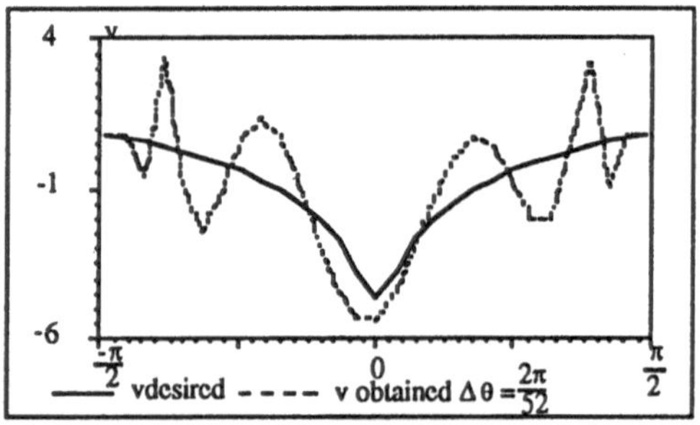

figure n°11

So the H.U.M. is not directly numerically applicable.

I.4 Regularization method.

The regularization method consists to add to the cost function the term $\frac{\varepsilon}{2}\| g \|^2_{H^1(S)}$, we define $J_\varepsilon(g)$ as

$$J_\varepsilon(g) = J(g) + \frac{\varepsilon}{2}\| g \|^2_{H^1(S)} \tag{8}$$

and we consider the minimization problem of $J_\varepsilon(g)$ for a fixed ε. For a detailed explanation see[4].

$$\min_{g \in H^1(S)} \left(J(g) + \frac{\varepsilon}{2}\| g \|^2_{H^1(S)}\right) \tag{9}$$

For a not too small ε, we find a good approximations for f, v and g. As ε tends to 0, the approximation of f always is good, while those for g and v are not (fig n°12, 13 and 14).

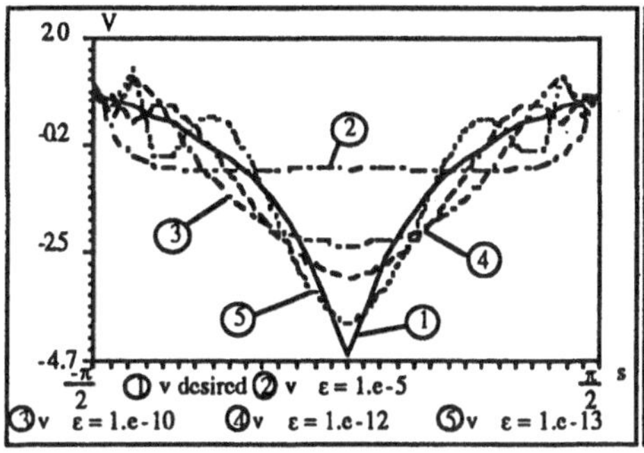

figure n°12

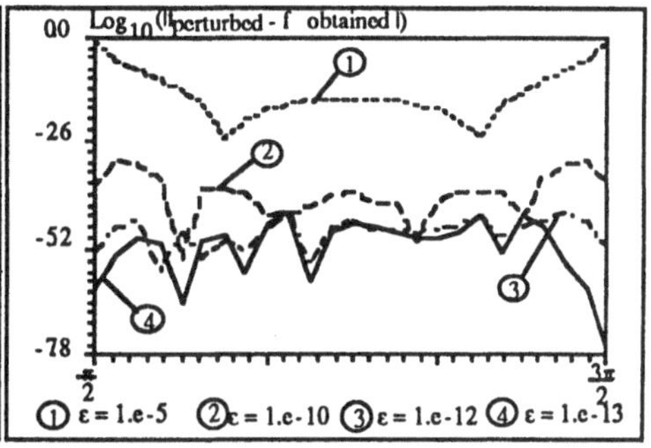

figure n°13

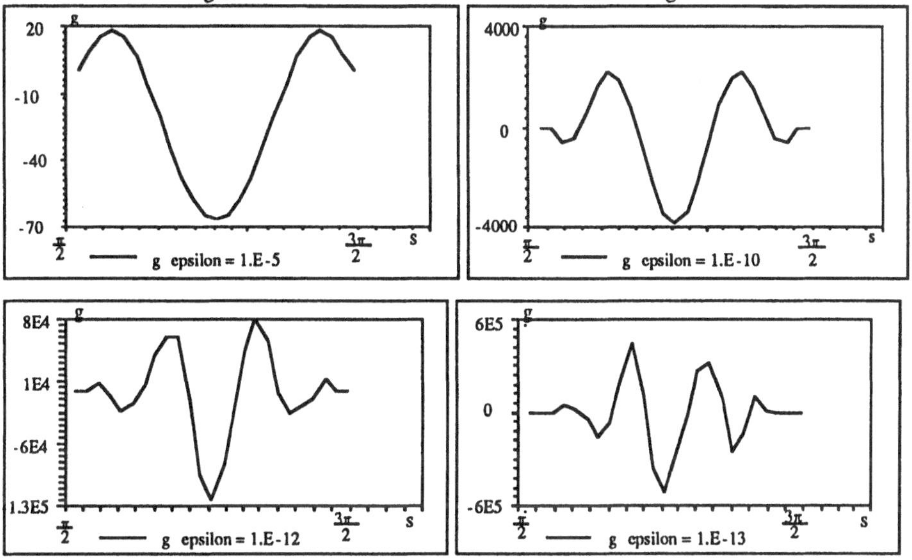

figure n°14

The regularization method gives interesting results but the difficulty lies in the choice of the parameter ε. The behaviour of the results, according to ε, depends on several parameters: the geometry of the domain, the size of the boundary S,etc. The method presented in the following section permit, in part, to resolve this problem.

I.5 Duality method[6][4].

We have

- an isometry $L : L^2(S') \mapsto G'(S)$
- $G'(S)$ dense in $H^{-1}(S)$

Let the ball $B = \left\{ z \in H^{-1}(S) \,/\, \| z - f \|_{H^{-1}(S')} \leq \beta \; (\beta > 0) \right\}$

Therefore $\forall f$ in $H^{-1}(S)$ $\exists v \in L^2(S')$ such that $Lv \in B$.

We define the $\mathcal{U}_{ad}$ the set of admissible controls $\boxed{\mathcal{U}_{ad} = \left\{ v \in L^2(S') / \, Lv \in B \right\}}$

Consider the problem (11) $\underset{v \in \mathcal{U}_{ad}}{\text{Inf}} \frac{1}{2} \| v \|^2_{L^2}$

By the Fenchel-Rockafellar theorem:

(12) $$\underset{v \in \mathcal{U}_{ad}}{\inf} \frac{1}{2} \| v \|^2_{H^1} = \underset{v \in L^2}{\text{Inf}} \{\rho(v)+\gamma(v)\} = \underset{s \in L^2}{\text{Sup}} \left\{ -\rho^*(-s)-\gamma^*(s) \right\}$$

where $\rho(v) = \frac{1}{2} \| v \|^2_{L^2}$ and $\gamma(v) = \begin{cases} 0 & \text{si } v \in \mathcal{U}_{ad} \ (Lv \in B) \\ \infty & \text{si } v \notin \mathcal{U}_{ad} \ (Lv \notin B) \end{cases}$ and

$$\rho^*(s) = \underset{v \in L^2}{\text{Sup}} \left\{ (s,v)_{L^2} - \rho(v) \right\} = \frac{1}{2} \| s \|^2_{L^2} \quad \gamma^*(s) = \underset{v \in L^2}{\text{Sup}} \left\{ (s,v)_{L^2} \gamma(v) \right\} = \underset{v \in \mathcal{U}_{ad}}{\text{Sup}} \left\{ (s,v)_{L^2} \right\}$$

After some manipulation we find

(13) $$\rho^*(-s) = \frac{1}{2} \| s \|^2_{L^2} \text{ and } \gamma^*(s) = - < g, f >_{H^1,H^{-1}} + \beta \| g \|_{H^1}$$

Where s spans a dense subset of $L^2(S')$ as g spans $H^1(S)$ and

(14) $\begin{cases} -\Delta\varphi = 0 & \text{dans } \Omega \\ \varphi = g & \text{sur } S \\ \varphi = 0 & \text{sur } S' \end{cases}$ (15) $s = \frac{\partial \varphi}{\partial v}$ sur S'

Finally the dual problem is (16) $\begin{cases} \text{Min } \rho^*(-s)+\gamma^*(s) \\ \text{subject to (2) } (g \in H^1(S)) \end{cases}$

For a not too small β, we obtain an acceptable reconstruction for v and even g. However, as β tends to 0, the results get worse.

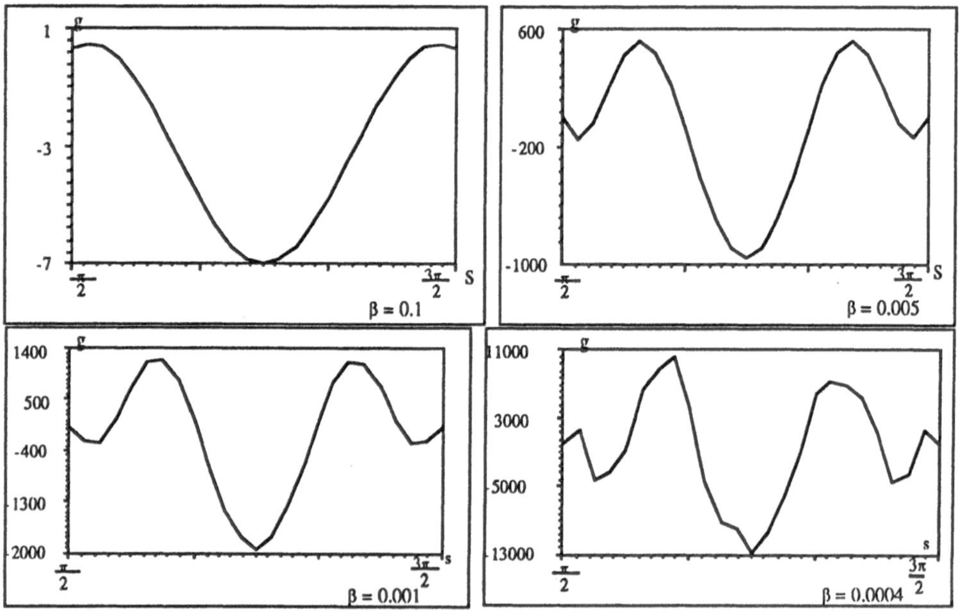

figure n°15

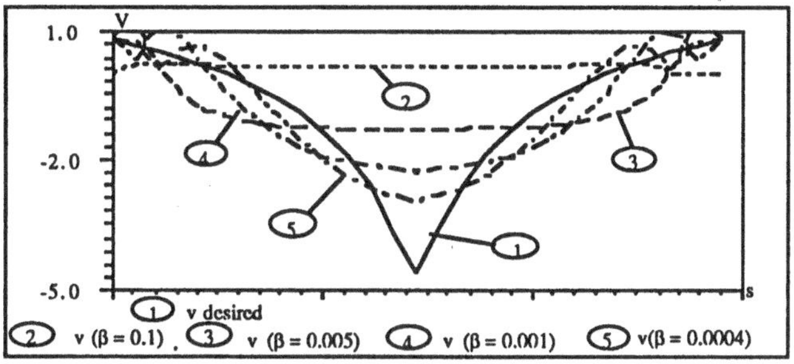

figure n°16

In fact this method is more interesting than the others because it gives a result with an error on f less than a given β, and thus β provides a measure for the accuracy of the numerically constructed f.

The methods and the results, that we have presented in this section for the solution of the Cauchy problem for the Laplacian operator, can be applied to all ill posed linear equations according to the definition of Tikonov [1].Thus, in the next section we have considered a linear parabolic equation with a control applied on a part of the domain Ω.

II A LINEAR PARABOLIC SYSTEM WITH CONTROL ON ω

II.1 Problem formulation

Let

$$(17)\quad \begin{cases} y_t - \Delta y = \chi_q v & \text{in } Q \\ \text{B.C. } y(x,t) = 0 & \text{on } \Sigma \\ \text{I.C. } y(x,0) = 0 & \end{cases}$$

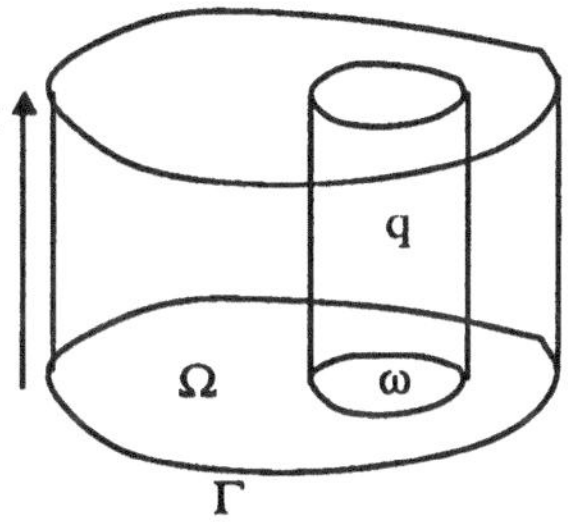

with $Q = \Omega \times]0,T[$ $\quad q = \omega \times]0,T[$ $\quad \Sigma = \Gamma \times]0,T[$ $\quad \chi_\sigma$ indicatory function of q

The objective is to find a control v such that y(T) is equal to a given function y_1. The problem, as in the preceeding examples, is to determine the space G in which y_1 must lie such that v belongs to $L^2(q)$. This problem does not have a straightforward, however, the H.U.M. method allows its resolution.

II.2 Application of the H.U.M. method

Let

$$(18)\quad \begin{cases} -\rho_t - \Delta\rho = 0 & \text{in } Q \\ \rho = 0 & \text{on } \Sigma \\ \rho(x,T) = g & \end{cases} \qquad (19)\quad \begin{cases} y_t - \Delta y = \chi_q \rho & \text{in } Q \\ y = 0 & \text{on } \Sigma \\ y(x,0) = 0 & \end{cases}$$

where (18) and (19) define a linear operator Λ

$$\boxed{\begin{array}{l} \Lambda : \mathcal{D}(\Omega) \longrightarrow L^2(\Omega) \\ \quad\quad g \qquad\qquad y(x,T) \end{array}}$$

Using the Hill - Yosida Theorem, we can show that (20) defines a norm of $\mathcal{D}(\Omega)$

$$(20) \qquad \| g \|_{\mathcal{D}(\Omega)} = \left(\int_Q (\chi_q v)^2 \, dx \, dt \right)^{\frac{1}{2}}$$

Then we consider the hilbert space G_Ω be the completion of $\mathcal{D}(\Omega)$ with respect to this norm (20). The situation is recapitulated in the figure n°17 where $L^2(\Omega)$ is include with density in G_Ω, M is a closed subspace of $L^2(q)$ and G'_Ω is include with density in $L^2(\Omega)$.

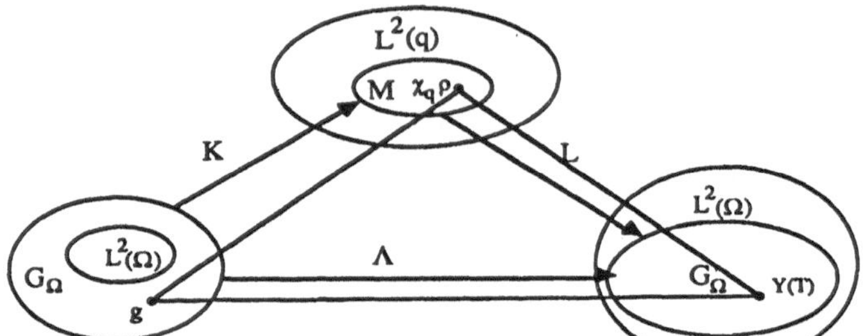

figure n°17

This problem can not be solved by solving directly $\Lambda g = y_1$ because the space G_Ω is very small [2], however it is dense in $L^2(\Omega)$ so that it is perhaps amenable to solution by the duality method.

II.3 Duality method

We consider the following ball

$$B = \left\{ y \in L^2(\Omega) \,/\, \| y - y_1 \|_{L^2} \leq \beta \right\}$$

We define $\mathcal{U}_{ad}$ as the set of admissible controls $\boxed{\mathcal{U}_{ad} = \{ v \in M \,/\, y(x,T) \in B \}}$

$$(21) \qquad \min_{v \in \mathcal{U}_{ad}} \left(J(v) = \frac{1}{2} \| v \|^2_{L^2[q]} \right)$$

This problem admits, at least, one solution if y_1 belong to $L^2(\Omega)$

We apply the duality method to this minimization problem and after manipulation we obtain:

$$\boxed{(21) \quad \Leftrightarrow \quad \operatorname*{Inf}_{g \in L^2} \left(\tilde{J}(g) = \frac{1}{2} \| s(g) \|^2_{L^2} - (g, y_1)_{L^2} + \beta \| g \|_{L^2} \right)}$$

with s defined such that (22) $\begin{cases} -\rho_t - \Delta \rho = 0 & \text{in } Q \\ \rho = 0 & \text{on } \Sigma \\ \rho(x,T) = g \end{cases}$ (23) $s = \chi_q \rho$

II.4 Numerical results[4]

For numerical simulations we consider to a trucated basis of $L^2(\Omega)$ with $\Omega =]0,1[$ and $\omega =]a,b[$ $(0 \leq a < b \leq 1)$

The control g is decomposed on the eigenfunctions of the Laplace operator $g(x) = \sum_{i=1}^{N} g_i \sin i \pi x$.

Figure n°18 shows the behaviour of g as a function of N, the number of terms in the series description of the function g, and as a function of the parameter β. This behaviour is similar to that discribed in the preceeding paragraph. Figure n°19 shows the influence of the parameter a and b on the control g. The control g tends to increase outside of the domain ω.

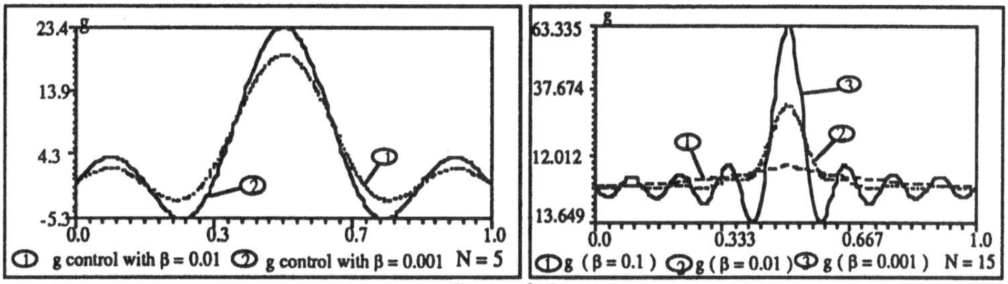

figure n°18

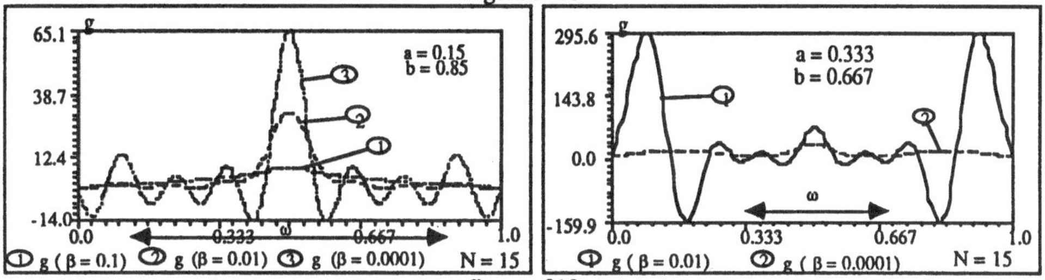

figure n°19

We demonstrated in this paper that the H.U.M. method applied to the numerical resolution of ill posed problems in PDE's does not always work, particulary when the controlabillity space G' can not be identified. In addition, with the regularization method one obtains acceptable numerical results. However, these results depend on a parameter ε, which is not easily controlled. The duality method yields a good numerical solution to these problemes depending upon a parameter β. In the limit as β tends to zero the optimality system of the H.U.M. method is recovered. In addition β gives a measure for the error in the function f.

References:

[1] J.L LIONS
Contrôlabité Exacte, Perturbation et Stabilisation de Systèmes Distribués (Tome 1)
Collection Recherches en Mathématiques Appliquées, Masson

[2] J.L. LIONS
Cours de la Sorbonne (Paris)
automne 89

[3] M. LAURENT'EV, V.G. ROMANOV, S.P. SHISTSKO
Ill Posed Problems of Mathematical Physics and Analysis
American Mathemetical Monograph

[4] R. LUCE
Study of ill posed problems
Thesis, University of tecnology of Compiegne, (To appear in december 1990)

[5] R. GLOVINSKI, C. H LI, J.L. LIONS
A Numerical Approch to The Exact Boundary Controllability of The Wave Equation, Dirichlet Controls: Description of the Numerical Methods.
Research Report UH/MD-22 January, 1980

[6] H. BREZIS
Analyse Fonctionnelle, Théorie et Applications
Collection Mathématiques Appliquées pour la Maîtrise, Masson

CONTROLLABILITY AND INDENTIFIABILITY FOR LINEAR TIME-DELAY SYSTEMS IN HILBERT SPACE

S.Nakagiri
Department of Applied Mathematics, Faculty of Engineering
Kobe University, Nada, Kobe, 657, JAPAN

1 Introduction

Let $\Omega \subset \mathbf{R}^n$ be a bounded domain with smooth boundary $\partial\Omega$, and $H = L^2(\Omega)$ be the usual L^2-space. A model control system under consideration is described by the following parabolic partial differential equation with time delays

$$(1.1) \qquad \frac{\partial u(t,x)}{\partial t} = \mathcal{A}u(t,x) + \int_{-h}^{0} a(s)\mathcal{A}u(t+s,x)ds + b(x)f(t), \quad t \geq 0, \quad x \in \Omega,$$

where $\mathcal{A}$ is an elliptic differential operator of the second order, $b \in L^2(\Omega)$ denotes a system controller, $f \in L^2_{loc}(\mathbf{R}^+;\ \mathbf{C})$ denotes a control function, $h > 0$ is a delay time and $a(s)$ is a real scalar function on $I_h \equiv [-h,\ 0]$. The boundary condition attached with (1.1) is, for simplicity, given by the Dirichlet boundary condition

$$(1.2) \qquad u|_{\partial\Omega} = 0, \qquad t \geq 0$$

and the initial data is given by

$$(1.3) \qquad u(0,x) = g^0(x), \quad u(s,x) = g^1(s,x) \quad a.e. \ s \in [-h,0), \quad x \in \Omega,$$

where $g^0 \in L^2(\Omega)$, $g^1 \in L^2(I_h;\ H_0^1(\Omega))$. This condition on the data $g = (g^0, g^1)$ is needed for the regularity of solutions for (1.1).

We are interested in the controllability problem for the system (1.1)-(1.3) both in the L^2-space and the appropriate function space over L^2 and the identifiability problem of the unknown operator $\mathcal{A}$ and the initial data $g = (g^0, g^1)$. The purpose here is to give conditions on the controller b or the initial data g such that the controllability or the identifiability holds.

Let $H = L^2(\Omega)$, $V = H_0^1(\Omega)$ and $g = (g^0, g^1) \in M_2 \equiv H \times L^2(I_h;\ V)$ and A_0 be the realization of $\mathcal{A}$ with the boundary condition (1.2) in $H = L^2(\Omega)$. Then the system (1.1)-(1.3) can be included in the following general class of abstract time-delay systems (S) in a Hilbert space H :

$$(1.4) \quad \frac{du(t)}{dt} = A_0u(t) + \gamma A_0u(t-h) + \int_{-h}^{0} a(s)A_0u(t+s)ds + \sum_{i=1}^{N} b_i f_i(t), \quad t \geq 0$$

$$(1.5) \quad u(0) = g^0, \quad u(s) = g^1(s) \ \ a.e. \ s \in [-h,0).$$

This type of equations is studied by Blasio, Kunisch and Sinestrari in [2,3] and Blasio [1] and recently by Yong and Pan [14] for quasi-linear equations. However, the system theoretical concepts such as controllability and identifiability has not been studied until now. So we investigate the problems of both H-approximate and M_2-function space controllability and spectral mode controllability for the time delay system (S), and the identifiability of an operator A_0, a constant γ, a real scalar function $a(s)$ and an initial data $g = (g^0, g^1) \in M_2$ in (S) by certain observation.

2 Linear time-delay systems and structural operator F

In this section we give exact description of the time-delay system (S). Let H and V be complex Hilbert spaces such that V is dense in H and the inclusion map $i : V \to H$ is continuous. The norms of H, V and the inner product of H are denoted by $|\cdot|$, $\|\cdot\|$ and $<\cdot,\cdot>$, respectively. By identifying the antidual of H with H we may consider $V \subset H \subset V^*$. Let $a(u,v)$ be a bounded sesquilinear form defined in $V \times V$ satisfying Gårding's inequality

$$\text{Re}\, a(u,v) \geq c_0\|u\|^2 - c_1|u|^2, \tag{2.1}$$

where $c_0 > 0$ and $c_1 \geq 0$ are real constants. Let A_0 be the operator associated with this sesquilinear form

$$< v,\ A_0 u > = -\,a(u,\ v), \qquad u,\ v \in V, \tag{2.2}$$

where $<\cdot,\cdot>$ denotes also the duality pairing between V and V^*. The operator A_0 is a bounded linear form from V into V^*. The realization of A_0 in H is also denoted by A_0. It is proved in Tanabe [11; Chap.3] that A_0 generates an analytic semigroup $e^{tA_0} = T(t)$ both in H and V^* and that $T(t) : V^* \to V$ for each $t > 0$.

Let $a(s)$ be Hölder continuous in I_h. For the brevity of notations, we introduce a Stiltjes measure η given by

$$\eta(s) = \Big(-\gamma\chi_{(-\infty,-h]}(s) - \int_s^0 a(\xi)d\xi\Big)A_0 : V \to V^*, \quad s \in I_h, \tag{2.3}$$

where $\chi_{(-\infty,-h]}$ denotes the characteristic function of $(-\infty,-h]$. Under the above conditions Tanabe [12] has constructed the fundamental solution $W(t)$ of (S) as the solution of the following integral equation with delay

$$W(t) = \begin{cases} T(t) + \int_0^t T(t-s)\int_{-h}^0 d\eta(\xi)W(\xi+s)ds, & t \geq 0 \\ O & t < 0, \end{cases} \tag{2.4}$$

where O denotes the null operator.

This fundamental solution $W(t)$ is strongly continuous both in H and V^* and $W(t) : V^* \to V$ for each $t > 0$. Then for each $t > 0$, the operator valued function $U_t(\cdot)$ given by

$$U_t(s) = \int_{-h}^0 W(t-s+\xi)d\eta(\xi) :\ V \to V, \qquad \text{a.e. } s \in I_h \tag{2.5}$$

is well defined.

Let controllers b_i, controls f_i and initial data g in the system (S) be assumed to satisfy

(2.6) $$b_i \in H, \quad f_i \in L^2_{loc}(\mathbf{R}^+; \mathbf{C}), \quad (i = 1, ..., N),$$

(2.7) $$g = (g^0, g^1) \in H \times L^2(I_h; V) \equiv M_2.$$

Under the conditions (2.6), (2.7) the (mild) solution $u(t)$ of (S) exists uniquely and is represented by

(2.8) $$u(t) = W(t)g^0 + \int_{-h}^{0} U_t(s)g^1(s)ds + \sum_{i=1}^{N} \int_0^t W(t-s)b_i f_i(s)ds, \quad t \geq 0.$$

Here we note that the function $u(t)$ in (2.8) satisfies the integrated form of (S) by $T(t)$ (cf. Nakagiri [6]).

The state space $M_2 = H \times L^2(I_h; V)$ of the system (S) is a Hilbert space and the adjoint space M_2^* of M_2 is identified with the product space $H \times L^2(I_h; V^*)$ via the duality pairing

(2.9) $$< g, f >_{M_2} = < g^0, f^0 > + \int_{-h}^{0} < g^1(s), f^1(s) > ds,$$

$$g = (g^0, g^1) \in M_2, \quad f = (f^0, f^1) \in M_2^*.$$

Now we introduce the structural operator F studied in Nakagiri [7] and Tanabe [13] on the abstract space setting.

Let $F_1 : L_2(I_h; V) \to L_2(I_h; V^*)$ be given by

(2.10) $$[F_1 g^1](s) = \int_{-h}^{s} d\eta(\xi) g^1(\xi - s) \quad \text{a.e. } s \in I_h.$$

The structural opertator $F : M_2 \to M_2^*$ is defined by $F = \begin{pmatrix} I & O \\ O & F_1 \end{pmatrix}$ i.e.,

(2.11) $$[Fg]^0 = g^0, \quad [Fg]^1 = F_1 g^1 \quad \text{for } g = (g^0, g^1) \in M_2.$$

3 Spectrum of the generator associated with (S)

Let $u(t; g)$ be the solution of (S) with $b_i = 0$ $(i = 1, ..., N)$ and the segment u_t be given by $u_t(s; g) = u(t + s; g)$, $s \in I_h$. The solution semigroup $S(t)$ associated with (S) is defined by

(3.1) $$S(t)g = (u(t; g), u_t(\cdot; g)), \quad t \geq 0, \quad g \in M_2.$$

$S(t)$ is a C_0-semigroup on M_2 and its infinitesimal generator is denoted by A. In what follows we investigate the structure of the spectrum $\sigma(A)$ of A under the following condition that

(3.2) the inclusion map $i : V \to H$ is compact.

The condition (3.2) is assumed throughout this paper. Then the resolvent $(\lambda - A_0)^{-1}$ is compact for some $\lambda \in \rho(A_0)$, so that according to the Riesz-Schauder theory the operator A_0 has a discrete spectrum

(3.3) $$\sigma(A_0) = \{\lambda_n; \ n = 1, 2, ...\}.$$

Let $m(\lambda)$ be given by

$$m(\lambda) = 1 + \gamma e^{-\lambda h} + \int_{-h}^{0} e^{\lambda s} a(s) ds \tag{3.4}$$

and define the characteristic operator $\Delta(\lambda) : V \to V^*$ by

$$\Delta(\lambda) = \lambda - m(\lambda) A_0. \tag{3.5}$$

Then the spectrum $\sigma(A)$ is completely determined by the entire function $m(\lambda)$ and $\sigma(A_0)$ (see Jeong [4]).

THEOREM 1 . *The spectrum $\sigma(A)$ is given by*

$$\begin{aligned} &\sigma(A) = \sigma_e(A) \cup \sigma_p(A), \\ &\sigma_e(A) = \{\lambda;\ m(\lambda) = 0\}, \quad \sigma_p(A) = \{\lambda;\ m(\lambda) \neq 0, \frac{\lambda}{m(\lambda)} \in \sigma(A_0)\}. \end{aligned} \tag{3.6}$$

Each nonzero point of $\sigma_e(A)$ is not an eigenvalue of A and is a cluster point of $\sigma(A)$. The point spectrum $\sigma_p(A)$ consists only of discrete eigenvalues with finite multiplicities.

If λ is in a resolvent set of A, then by Theorem 1 the inverse $\Delta(\lambda)^{-1} : V^* \to V$ exists. Further for $\mathrm{Re}\,\lambda$ sufficiently large, $\Delta(\lambda)^{-1}$ is given by the Laplace transform of $W(t)$.

LEMMA 1 . *For $\lambda \in \sigma_p(A)$,*

$$\mathrm{Ker}\,(\lambda - A) = \{(\varphi^0, e^{\lambda s}\varphi^0);\ \Delta(\lambda)\varphi^0 = 0\}. \tag{3.7}$$

For the characterization of $\mathrm{Ker}\,(\lambda - A)^l$, $l = 1, 2, \ldots$ in terms of $\Delta(\lambda)$ we refer to [4],[7]. Since each $\lambda \in \sigma_p(A)$ is an isolated eigenvalue, the order k_λ of λ as a pole of $(z - A)^{-1}$ is finite. The spectral projection P_λ and the nilpotent operator Q_λ for $\lambda \in \sigma_p(A)$ are defined respectively by

$$P_\lambda = \frac{1}{2\pi i} \int_{\Gamma_\lambda} (z - A)^{-1} dz \tag{3.8}$$

$$Q_\lambda = \frac{1}{2\pi i} \int_{\Gamma_\lambda} (z - \lambda)(z - A)^{-1} dz, \tag{3.9}$$

where Γ_λ is a small circle with center λ such that its interior and Γ_λ contains no points of $\sigma(A)$. Let $\mathcal{M}_\lambda = \mathrm{Im}\, P_\lambda$ be the generalized eigenspace corresponding to the eigenvalue λ of A. Then we have

$$Q_\lambda^{k_\lambda} = O, \quad \mathrm{Im}\, Q_\lambda \subset \mathcal{M}_\lambda \tag{3.10}$$

and the useful relation

$$\mathrm{Ker}\,(\lambda - A) = \mathcal{M}_\lambda \cap \mathrm{Ker}\, Q_\lambda \tag{3.11}$$

(cf. Suzuki and Yamamoto [10]). The following direct sum decomposition of the space M_2 (see e.g. Kato [5]) is essential in our study.

LEMMA 2 . *For $\lambda \in \sigma_p(A)$,*

$$\mathcal{M}_\lambda = \mathrm{Ker}\,(\lambda - A)^{k_\lambda}, \quad M_2 = \mathcal{M}_\lambda \oplus \mathrm{Im}\,(\lambda - A)^{k_\lambda}. \tag{3.12}$$

4 H-approximate and M_2-function space controllability

In the system (S) we define the controller $B_0 : \mathbf{C}^N \to H$ by

(4.1) $$B_0 v = \sum_{i=1}^{N} v_i b_i, \quad v = (v_1, ..., v_N) \in \mathbf{C}^N.$$

The attainable subspaces $\mathcal{R}^0$ in H and $\mathcal{R}$ in M_2 are defined by

(4.2) $$\mathcal{R}^0 = \bigcup_{t>0} \left\{ \int_0^t W(t-s) B_0 f(s) ds;\ f \in L^2([0,\ t];\ \mathbf{C}^N) \right\},$$

(4.3) $$\mathcal{R} = \bigcup_{t>0} \left\{ \int_0^t S(t-s)(B_0 f(s), 0) ds;\ f \in L^2([0,\ t];\ \mathbf{C}^N) \right\},$$

respectively.

DEFINITION 1. The system (S) is said to be H-approximately controllable (resp. M_2-function space controllable) if $Cl(\mathcal{R}^0) = H$ (resp. if $Cl(\mathcal{R}) = M_2$).

In view of (3.2), there exists a set of eigenvalues and eigenvectors $\{\mu_n,\ \psi^0_{nj};\ j = 1, 2, ..., d_n, n = 1.2....\}$ of the adoint A_0^* such that μ_n are distinct from each other and $d_n = \dim \operatorname{Ker} (\mu_n - A_0^*)$. It is well known (Kato [5]) that $\sigma(A_0^*) = \{\overline{\lambda_n};\ n = 1, 2, ...\}$.

First we give a result on the H-approximate controllability (cf. Nakagiri and Yamamoto [9]).

THEOREM 2 . *Assume that $m(0) \neq 0$ and the system of generalized eigenvectors of A_0 is complete in H. Let B_n (n=1,2,...) be $N \times d_n$ matrices given by*

(4.4) $$B_n = \left(< b_i,\ \psi^0_{nj} > ;\ i \downarrow 1, ..., N,\ j \to 1, 2, ..., d_n \right).$$

Then the following two statements are equivalent:
(i) *the control system* (S) *is H-approximately controllable;*
(ii) $\operatorname{rank} B_n = d_n$ *for each* $n \geq 1$.

For the M_2-function space controllability we require the following Lemma.

LEMMA 3 . *Assume that the system of generalized eigenvectors of A_0 is complete in H , $0 \notin \sigma(A_0)$, $\gamma \neq 0$ and $m(0) \neq 0$, then the system of generalized eigenvectors of A is complete in M_2 , that is,*

(4.5) $$Cl(\text{Span}\,\{\mathcal{M}_\lambda;\ \lambda \in \sigma_p(A)\}) = M_2.$$

THEOREM 3 . *Under the assumption in Lemma 3, the following two statements are equivalent:*
(i) *the control system* (S) *is M_2-function space controllable;*
(ii) $\operatorname{rank} B_n = d_n$ *for each* $n \geq 1$.

5 Spectral mode controllability and observability

Let U be a complex Hilbert space and $B_0 : U \to H$ be a controller which is bounded. We consider the following control system with the controller B_0 :

$$\begin{aligned} (5.1) \quad \frac{du(t)}{dt} &= A_0u(t) + \gamma A_0u(t-h) + \int_{-h}^{0} a(s)A_0u(t+s)ds + B_0f(t), \quad t \geq 0 \\ (5.2) \quad u(0) &= g^0, \quad u(s) = g^1(s) \ \ a.e.\ s \in [-h, 0), \end{aligned}$$

where $f \in L^2_{loc}(\mathbb{R}^+; U)$ denotes a control function. The observed 'transposed' system in H with initial data $(\varphi^0, \varphi^1) \in M_2$ is defined by

$$\begin{aligned} (5.3) \quad \frac{dv(t)}{dt} &= A_0^*v(t) + \gamma A_0^*v(t-h) + \int_{-h}^{0} a(s)A_0^*v(t+s)ds, \quad t \geq 0 \\ (5.4) \quad v(0) &= \varphi^0, \quad v(s) = \varphi^1(s) \ \ a.e.\ s \in [-h, 0) \\ (5.5) \quad y(t) &= B_0^*v(t) \quad t \geq 0, \end{aligned}$$

where $y(t)$ denotes the observation of transposed system and $B_0^* : H \to U$.

Let A be the infinitesimal generator of $S(t)$. Then we can imbed the system (5.1), (5.2) into the state space M_2 as the following control system (Σ) without delay :

$$\begin{aligned} (5.6) \quad \frac{d\hat{u}(t)}{dt} &= A\hat{u}(t) + Bf(t), \quad t \geq 0 \\ (5.7) \quad \hat{u}(0) &= g \in M_2, \end{aligned}$$

where $B : U \to M_2$ is given by $Bf = (B_0f, 0), \ f \in U$. The mild solution $\hat{u}(t)$ of (Σ) is given by

$$(5.8) \qquad (u(t;\ g, f), u_t(\cdot;\ g, f)) = S(t)g + \int_0^t S(t-s)Bf(s)ds, \quad t \geq 0,$$

where $u(t;\ g, f)$ is the mild solution of (5.1),(5.2).

Associated with the observed system (5.3)-(5.5) we introduce the operator $\Delta_T(\lambda)$ by

$$(5.9) \qquad \Delta_T(\lambda) = \lambda - m(\lambda)A_0^*.$$

Let us denote by $\{S_T(t)\}_{t\geq 0}$ the C_0-semigroup on M_2 corresponding to the observed system and by A_T its infinitesimal generator. Then by Theorem 1 we see that the point spectrum $\sigma_p(A_T)$ of A_T is given by

$$\sigma_p(A_T) = \{\lambda;\ m(\lambda) \neq 0, \frac{\lambda}{m(\lambda)} \in \sigma(A_0^*)\},$$

i.e., $\sigma_p(A_T)$ is the mirror image $\overline{\sigma_p(A)}$. The transposed system (Σ_T) in M_2 induced by the system (5.3)-(5.5) is given by

$$\begin{aligned} (5.10) \quad \frac{d\hat{v}(t)}{dt} &= A_T\hat{v}(t), \quad t \geq 0 \\ (5.11) \quad \hat{v}(0) &= \varphi = (\varphi^0, \varphi^1) \in M_2 \\ (5.12) \quad \hat{y}(t) &= B^*\hat{v}(t) \quad t \geq 0, \end{aligned}$$

where $B^* : M_2 \to U$ is given by $B^*g = B_0^* g^0$, $g = (g^0, g^1) \in M_2$. Then the observation $\hat{y}(t)$ of (Σ_T) is represented by

$$\hat{y}(t) = B^* S_T(t)\varphi \quad t \geq 0. \tag{5.13}$$

The attainable subspace $\mathcal{R}$ for (Σ) and the unobservable subspace $\mathcal{N}_T$ for (Σ_T) are defined by

$$\mathcal{R} = \bigcup_{t>0} \left\{ \int_0^t S(t-s)Bf(s)ds;\ f \in L^2([0,\ t];\ U) \right\}, \tag{5.14}$$

$$\mathcal{N}_T = \bigcap_{t>0} \operatorname{Ker} B^* S_T(t), \tag{5.15}$$

respectively.

DEFINITION 2. (1) The system (Σ) is said to be λ-controllable for $\lambda \in \sigma_p(A)$ if $Cl(\mathcal{R}) \supset \mathcal{M}_\lambda$.

(2) The system (Σ_T) is said to be λ-observable for $\lambda \in \sigma_p(A_T)$ if $\mathcal{N}_T \cap \mathcal{M}_\lambda^T = \{0\}$.

In Definition 2 the symbol $\mathcal{M}_\lambda^T$ denotes the generalized eigenspace corresponding to the isolated eigenvalue λ of A_T. For $\lambda \in \sigma_p(A_T)$ the symbols P_λ^T, Q_λ^T denote the spectral projection and the nilpotent operator corresponding to the eigenvalue λ of A_T, respectively. We set $Q_\lambda^0 = P_\lambda$, $(Q_\lambda^T)^0 = P_\lambda^T$ for notational convenience.

THEOREM 4 . *Let $\lambda \in \sigma_p(A)$ be given. Then the following statements* (i)-(xiii) *are equivalent:*

(i) *the system (Σ) is λ-controllable;*

(ii) *the system (Σ_T) is $\bar{\lambda}$-observable;*

(iii) $\mathcal{N}_T \subset \operatorname{Im} (\bar{\lambda} - A_T)^{k_\lambda} \subset M_2$;

(iv) $(\bigcap_{j=0}^{k_\lambda} \operatorname{Ker} B^*(Q_{\bar{\lambda}}^T)^j) \subset \operatorname{Im} (\bar{\lambda} - A_T)^{k_\lambda} \subset M_2$;

(v) $Cl(\operatorname{Span} \{Q_\lambda^j Bf\ ;\ 0 \leq j \leq k_\lambda - 1,\ f \in U\}) = \mathcal{M}_\lambda$;

(vi) $(\bigcap_{j=0}^{k_\lambda} \operatorname{Ker} B^*(Q_{\bar{\lambda}}^T)^j) \cap \mathcal{M}_{\bar{\lambda}}^T = \{0\}$;

(vii) $Cl(\operatorname{Im} (\lambda - A) + \operatorname{Im} B) = M_2$;

(viii) $Cl(\operatorname{Im} (\lambda P_\lambda - AP_\lambda) + \operatorname{Im} P_\lambda B) = \mathcal{M}_\lambda$;

(ix) $Cl(\operatorname{Im} \Delta(\lambda) + \operatorname{Im} B_0) = H$;

(x) $\operatorname{Ker} \Delta_T(\bar{\lambda}) \cap \operatorname{Ker} B_0^* = \{0\}$;

(xi) $\operatorname{Ker} (\bar{\lambda} - A_T) \cap \operatorname{Ker} B^* = \{0\}$;

(xii) $F\mathcal{M}_\lambda \subset Cl(F\mathcal{R})$;

(xiii) $\mathcal{N}_T \cap \mathcal{M}^T_{\bar\lambda} \subset \text{Ker } F^*$.

Let $\lambda \in \sigma_p(A)$ and d_λ be the dimension of the eigenspace Ker $(\lambda - A)$. Then by Lemma 1 we have

(5.16) $d_\lambda = \dim \text{Ker }(\lambda - A) = \dim \text{Ker }(\bar\lambda - A_T) = \dim \text{Ker } \Delta_T(\bar\lambda) < \infty.$

We denote the basis of Ker $\Delta_T(\bar\lambda)$ by $\{\varphi^0_{\bar\lambda 1}, \ldots, \varphi^0_{\bar\lambda d_\lambda}\}$.

For the practical case where the controller B_0 is given by (4.1), we have by applying Theorem 4, the following verifiable conditions for spectral mode controllability and observability.

THEOREM 5 . *Let* $\lambda \in \sigma_p(A)$ *be given and the controller* B_0 *be given by* (4.1). *Then the following statements* (i)-(vii) *are equivalent:*

(i) *the system* (Σ) *is* λ*-controllable;*

(ii) *the system* (Σ_T) *is* $\bar\lambda$*-observable;*

(iii) $\text{rank }\left(< b_i, \varphi^0_{\bar\lambda j} > ;\ i \downarrow 1, \ldots, N,\ j \to 1, 2, \ldots, d_\lambda\right) = d_\lambda$;

(iv) $\text{Im }(\lambda - A) + \text{Span }\{(b_i, 0);\ 1 \le i \le N\} = M_2$;

(v) $\text{Im } \Delta(\lambda) + \text{Span }\{b_i;\ 1 \le i \le N\} = H$;

(vi) $\text{Ker } \Delta_T(\bar\lambda) \cap (\{b_i;\ 1 \le i \le N\})^\perp = \{0\}$;

(vii) $\text{Ker }(\bar\lambda - A_T) \cap (\{(b_i, 0);\ 1 \le i \le N\})^\perp = \{0\}$.

6 Identifiability

Consider the following time-delay system (**RS**) with N-numbers of intial conditions:

(6.1) $$\frac{du(t)}{dt} = A_0 u(t) + \gamma A_0 u(t-h) + \int_{-h}^{0} a(s) A_0 u(t+s) ds, \quad t \ge 0$$

(6.2) $$u(0) = g^0_i, \quad u(s) = g^1_i(s) \ \ a.e. \ s \in [-h, 0) \quad (i = 1, \ldots, N).$$

Here A_0, γ, $a(s)$, $g_i = (g^0_i, g^1_i) \in M_2$ $(i = 1, \ldots, N)$ are unknown quantities to be identified.

In this section we suppose the following conditions:

(I) the operator A_0 associated with the seaquilinear form (2.2) enjoying the condition (2.1) is unknown;

(II) the real constant γ is unknown;

(III) the real scalar function $a(s)$ is unknown but is known to be Hölder continuous in I_h;

(IV) the initial conditions $g_i = (g^0_i, g^1_i) \in M_2$, $i = 1, \ldots, N$ are unknown.

Under the above conditions there exists mild solutions $u(t;\ g_i)$, $i = 1, \dots, N$ of (**RS**). By the model system $(\mathbf{RS})^m$ we understand the system (6.1), (6.2) in which A_0, γ, $a(s)$ and $g_i = (g_i^0, g_i^1)$, $i = 1, \dots, N$ are replaced by A_0^m, γ^m, $a^m(s)$ and $g_i^m = (g_i^{m0}, g_i^{m1})$, $i = 1, \dots, N$, respectively. The model states of $(\mathbf{RS})^m$ are denoted by $u(t;\ g_i^m)$, $i = 1, \dots, N$. An 'm' superscript means that the quantity is known.

DEFINITION 3. The unknown quantities A_0, γ, $a(s)$ and $g_i = (g_i^0, g_i^1)$, $i = 1, \dots, N$ in (**RS**) are said to be identifiable if

$$(6.3) \qquad A_0 = A_0^m, \quad \gamma = \gamma^m, \quad a(s) = a^m(s) \quad s \in I_h,$$

$$(6.4) \qquad g_i = g_i^m \quad in \ \ M_2 \quad (i = 1, \dots, N)$$

follows from the relations

$$(6.5) \qquad u(t;\ g_i) \ = \ u(t;\ g_i^m) \quad in \ H, \quad t \geq t_0, \quad i = 1, \dots, N$$

for some $t_0 \geq 0$.

Here we introduce the transposed model system $(\mathbf{RS})_T^m$, that is the system $(\mathbf{RS})^m$ in which A_0^m is replaced by its adjoint $(A_0^m)^*$. Let $S_T^m(t)$ be the semigroup associated with the transposed model $(\mathbf{RS})_T^m$. The infinitesimal generator of $S_T^m(t)$ is denoted by A_T^m. In view of (3.2) and Theorem 1, $\sigma_p(A_T^m)$ consists of discrete eigenvalues. Let $\lambda \in \sigma_p(A_T^m)$ and Ker $(\lambda - A_T^m)$ be the eigenspace corresponding to the eigenvalue λ of A_T^m, which can be calculated by Lemma 1. We denote the basis of Ker $(\lambda - A_T^m)$ by $\{\varphi_{\lambda 1}^m, \dots, \varphi_{\lambda d_\lambda}^m\}$, where $d_\lambda^m = \dim \operatorname{Ker} (\lambda - A_T^m)$. The structural operator associated with $(\mathbf{RS})^m$ is denoted by F^m.

THEOREM 6 . *Let A_0^m satisfy the assumption in Lemma 3. If the set of initial conditions $\{g_1^m, \dots, g_N^m\}$ satisfies*

$$(6.6) \qquad \operatorname{rank} \left(< \varphi_{\lambda j}^m, F^m g_i^m >_{M_2} ;\ i \downarrow 1, \dots, N,\ j \to 1, 2, \dots, d_\lambda \right) \ = \ d_\lambda^m \qquad for\ each \quad \lambda \in \sigma_p(A_T^m),$$

then all A_0, γ, $a(s)$ and $g_i = (g_i^0, g_i^1)$, $i = 1, \dots, N$ in (**RS**) *are identifiable.*

The above theorem improves the results in Nakagiri and Yamamoto [8] and Jeong [4]. Complete proofs of all theorems in this paper will appear elsewhere.

REFERENCES

[1] G. Di Blasio, *The linear-quadratic optimal control problem for delay differential equations*, Rend. Accad. Naz. Lincei, **71**(1981), 156-161.

[2] G. Di Blasio, K. Kunisch and E. Sinestrari, *L^2-regularity for parabolic partial integrodifferential equations with delay in the highest-order derivatives*, J. Math. Anal. Appl., **102**(1984), 38-57.

[3] G. Di Blasio, K. Kunisch and E. Sinestrari, *Stability for abstract linear functional differential equations*, Israel J. Math., **50**(1985), 231-263.

[4] J-M. Jeong, *Spectral properties of the operator associated with a retarded functional differential equations in Hilbert space*, Proc. Japan Acad., **65 A**(1989), 98-101.

[5] T. Kato, *Perturbation Theory for Linear Operators*, Second edition, Springer, Berlin-Heidelberg-New York, 1976.

[6] S. Nakagiri, *Optimal control of linear retarded systems in Banach spaces*, J. Math. Anal. Appl., **120**(1986), 169-210.

[7] S. Nakagiri, *Structural properties of functional differential equations in Banach spaces*, Osaka J. Math., **25**(1988), 353-398.

[8] S. Nakagiri and M. Yamamoto, *Identifiability of linear retarded systems in Banach spaces*, Funkcial. Ekvac., **31**(1988), 315-329.

[9] S. Nakagiri and M. Yamamoto, *Controllability and observability of linear retarded systems in Banach spaces*, Int. J. Control, **49**(1989), 1489-1504.

[10] T. Suzuki and M. Yamamoto, *Observability, controllability and feedback stabilizability for evolution equations, I*, Japan J. Appl. Math., **2**(1985), 211-228.

[11] H. Tanabe, *Equations of Evolution*, Pitman, London, 1979.

[12] H. Tanabe, *On fundamental solution of differential equation with time delay in Banach space*, Proc. Japan Acad., **64 A**(1988), 131-134.

[13] H. Tanabe, *Structural operators for linear delay-differential equations in Hilbert space*, Proc. Japan Acad., **64 A**(1988), 263-266.

[14] J. Yong and L. Pan, *Quasi-linear parabolic partial differential equations with delays in the highest order spatial derivatives*, to appear.

A Generalised Hamilton-Jacobi-Bellman Equation *

by
Shige Peng
Department of Mathematics, Shandong University
Jinan, Shandong 250100, China
and
Institute of Mathematics, Fudan University
Shanghai, 200433, China

Abstract. We interpret the following fully nonlinear second order partial differential equation

$$\begin{cases} \partial_t u + \inf_{\alpha}\{\mathcal{L}(x,\alpha)u + f(x,u,\partial_x u\sigma(x,\alpha),\alpha)\} = 0, & (x,t) \in D \times (0,T), \\ \text{for } (x,t) \in D \times [0,T]; \quad u(x,T) = g(x). \end{cases}$$

as the value function of certain optimal controlled diffusion problem. Where $A \in \mathbb{R}^k$ is control domain. $\mathcal{L}(x,\alpha)$ is a second order elliptic partial differential operator parametrized by the control variable $\alpha \in A \subset \mathbb{R}^k$. A particular case of this equation is when $f = f(x,\alpha)$. In this case, the equation is the well known Hamilton Jacobi Bellman equation.

The problem is formulated as follows: The state equation of the control problem is as classical one. The cost function is described by a solution of certain backward stochastic differential equation.

§1. Introduction

It was known that a solution of a second order linear parabolic (or elliptic) equation can be formulated as a functional of a solution of some stochastic differential equation. This kind of interpretation has been found it's important applications both in theory of partial differential equations and that of stochastic differential equations, such as large derivation, optimal control theory, martingale problem, variational and quasi variational inequality, e.c.t.

A natural and interesting problem is to obtain a similar interpretation for a system of parabolic (or elliptic) partial differential equation. A recent result in backward stochastic differential equation (see Pardoux and Peng [7]) produces new insights to this direction. In [11], we interpreted a systems of second order quasilinear parabolic partial differential equation as a solution of a backward stochastic differential equation . This backwrad equation is associated with some classical Ito's forward stochastic differential equation. When the backward equation is linear and one dimensional, the corresponding system of equation becomes a linear one. So the classical probabilistic interpretation can be regarded as a special case of our formulation.

The same idea can be also applied to generalize Hamilton-Jacobi-Bellman (HJB) equation.

* partially supported by the Chinese National Natural Science Fundation.

Let D be a domain in $\mathbb{R}^n$, Let A be a nonempty set in $\mathbb{R}^k$. Let

$$b(x,\alpha): \mathbb{R}^n \times \mathbb{R}^k \to \mathbb{R}^n,$$
$$\sigma(x,\alpha): \mathbb{R}^n \times \mathbb{R}^k \to \mathcal{L}(\mathbb{R}^d; \mathbb{R}^n)$$

Consider the following stochstic control problem parametrized by the initial data $x \in \mathbb{R}^n$

$$\text{(1.1)} \qquad \begin{cases} dy(s) = b(y(s),\alpha(s))ds + \sigma(y(s),\alpha(s))dW(s), \\ y(0) = x \in \mathbb{R}^n, \end{cases}$$

where $W(s); s \geq 0$ is an d-dimensional standard Wiener process. $\alpha(s), s \geq 0$ is an $\mathcal{F}_s^W$-adapted process taking values in A, called control process. $y(\cdot)$ is called trajectory corresponding to $\alpha(\cdot)$. If b and σ satisfy some suitable condition, then this system is well defined. With this system, we can introduce the following backward stochastic differential equation: For any given $t \in [0,T)$, we look for an adapted pair $(p(\cdot), q(\cdot))$ that solves uniquely

$$\text{(1.2)} \qquad \begin{cases} p(s) = g(y(T-t)) + \int_s^{T-t} f(p(r), q(r), y(r), \alpha(r))dr - \int_s^{T-t} q(r)dW(r), \\ \qquad s \in [0, T-t]. \end{cases}$$

where $f(x,u,q,\alpha)$ is a given real function defined on $\mathbb{R}^n \times \mathbb{R} \times \mathbb{R}^d \times A$, and g is a given real function defined on $\mathbb{R}^n$. The existence and uniqueness for the above equation is obtained by Pardoux and Peng [7]. Then we can define the so called cost function as follows

$$J_{x,t}(\alpha(.)) = Ep(0) = p(0).$$

The value function of this optimal control problem is defined by

$$u(x,t) = \inf_{\alpha(.)} J_{x,t}(\alpha(.)).$$

We will show that this value function can be charecterized by the following generalized HJB equation

$$\begin{cases} \partial_t u + \inf_{\alpha}\{\mathcal{L}(x,v)u + f(x,u,\partial_x u\sigma(x,\alpha),\alpha)\} = 0, \quad (x,t) \in D \times (0,T), \\ u(x,t) = \Psi(x), \text{ for } (x,t) \in dD \times [0,T]; \quad u(x,T) = \Psi(x). \end{cases}$$

where $\mathcal{L}(x,v)$ is a second order elliptic differential oprator parametrized by $\alpha \in A$. When p depends only on (x,α), the above equation becomes a classical HJB equation.

This kind of problems can be applied to financial problem, where the term f is called utility function (C.Ma, private communication.).

§2. Backward SDE and Systems of Parabolic PDE

2.1. Backward Stochastic Differential Equation

We begin with presenting a recent result of adapted solution of backward stochastic differential equation. (see Pardoux and Peng [7]).

Let $(\Omega, \mathcal{F}, P)$ be a probability space equipped with filtration $\mathcal{F}_t$. Let $\{W(t), \tau \geq 0\}$ be a d-dimensional standard Wiener prosess in this space. We assume

$$\mathcal{F}_t = \sigma\{W_s; 0 \leq s \leq t\}.$$

We denote by $\mathcal{M}^2(0,T;\mathbb{R}^n)$ or $\mathcal{M}^2(\mathbb{R}^n)$ the set adapted proceeses such that

$$E\int_0^T |x(t)|^2 dt \leq \infty, \quad \forall y(.) \in \mathcal{M}^2(\mathbb{R}^n)$$

Let the following functions be given.

$$f(p,q,t,\omega) : \mathbb{R}^m \times \mathcal{L}(\mathbb{R}^d;\mathbb{R}^m) \times [0,T] \times \Omega \to \mathbb{R}^m,$$
$$Q(\omega) : \Omega \to \mathbb{R}^m,$$

We assume

$(H2.1.1)$ for each $(p,q) \in \mathbb{R}^m \times \mathcal{L}(\mathbb{R}^d;\mathbb{R}^m)$, $f(p,q,\cdot) \in \mathcal{M}^2(\mathbb{R}^m)$

$(H2.1.2)$ $\left\{\begin{array}{l}\text{for each } (t,\omega) \in [0,T] \times \Omega,\ f(p,q,t,\omega) \text{ is continuously} \\ \text{differentiable with respect to } (p.q), \text{ their derivatives} \\ f_p \text{ and } f_q \text{ are bounded.}\end{array}\right.$

$(H2.1.3)$ $Q(\omega)$ is $\mathcal{F}_T$ measurable, and $E|Q|^2 < \infty$

Consider the following backward stochastic differential equation

$$(2.1.1) \qquad p(t) = Q + \int_t^T f(p(s),q(s),s)ds - \int_t^T q(s)dW(s).$$

Our problem is to look for a pair of adapted $\mathbb{R}^m \times\ \mathcal{L}(\mathbb{R}^d;\mathbb{R}^m)$ valued processes $(p(s),q(s))$ which solves equation (2.1.1). We have

Proposition 2.1.1. We assume (H2.1.1)-(H2.1.3), . Then, there exists an unique pair $(p(\cdot),q(\cdot))$ in $\mathcal{M}^2(0,T;\mathbb{R}^m) \times \mathcal{M}^2(0,T;\mathcal{L}(\mathbb{R}^d;\mathbb{R}^m))$ which solves equation (2.1.1). We have

$$E\sup_t |p(t)|^2 < \infty.$$

If we assume further more

$$K = \sup_\omega\{|Q(\omega)|^2 + \int_0^T |f(0,0,t)|^2 dt\} < \infty,$$

then we have

$$(2.1.2) \qquad \sup_\omega |p(t,\omega)|^2 \leq Ke^{C(T-t)}.$$

The proof for the existence and uniqueness can be found in Pardoux and Peng [7]. The proof for the boundness can be found in Peng [9], or [10].

We can also consider the following type of backward equation

$$p(t\wedge\tau)=Q+\int_{t\wedge\tau}^{\tau}f(p(s),q(s),s)ds-\int_{t\wedge\tau}^{\tau}q(s)dW(s). \tag{2.1.3}$$

Indeed, we have

Proposition 2.1.2. We assume (H2.1.1), (H2.1.2) and (H2.1.4). Then there exists a pair $(p(\cdot),q(\cdot))$ in $\mathcal{M}^2(0,T;I\!R^m)\times\mathcal{M}^2(0,T;\mathcal{L}(I\!R^d;I\!R^m))$ which solves equation (2.1.3). Such solution is unique in the following sence: if both $(p^1(\cdot),q^1(\cdot))$, $(p^2(\cdot),q^2(\cdot))$ solve (2.1.2), then,

$$(p^1(s\wedge\tau),q^1(s\wedge\tau))=(p^2(s\wedge\tau),q^2(s\wedge\tau)),\quad ,\forall s\in[0,\tau].$$

The proof can be found in Peng [11].

2.2. Probabilistic formulation for System of parabolic PDE

In this subsection, we will formulate a system of quasilinear parabolic partial differential equations as a solution of certain backward stochastic differential equation, of type (2.2), associated with some forward (classical) stochastic differential equation. We first introduce the forward equation.

Let

$$b(x):I\!R^m\to I\!R^m,$$
$$\sigma(x):I\!R^m\to\mathcal{L}(I\!R^d;I\!R^m).$$

Let D a domain in $I\!R^m$ with boundary $\partial D=S$. We denote $Q=D\times(0,T)$. assume

$$(H2.2.1)\qquad\begin{cases}\text{(i) } \sigma(x) \text{ is of class } C^2(\bar{Q}),\\ \quad\ b(x) \text{ is of class } C^1(\bar{Q});\\ \text{(ii) } S \text{ is a manifold of } C^3;\\ \text{(iii) } \sum a_{ij}(x)\xi_i\xi_j\geq\beta|\xi|^2,\quad\forall x\in\bar{Q}.\end{cases}$$

where $\beta>0$ is a constant and $a_{ij}=\frac{1}{2}[\sigma\sigma^*]_{ij}$. For any given $(x,t)\in\bar{Q}$, consider the following forward equation defined on $[0,T]$

$$\begin{cases}dy(s)=b(y(s))ds+\sigma(y(s))dW(s),\\ y(0)=x.\end{cases}\tag{2.2.1}$$

For any given $t\in[0,T]$, we define the follwing stopping time

$$\tau=\tau_{x,t}=\inf\{s\in[0,T-t];(y(s),s)\notin D\times[0,T-t)\}.$$

From (H2.2.1), the diffusion process $y(\cdot)$ and related stopping time τ are well defined.

Then, we consider the associated backward stochastic equation: defined on $s \in [0,\tau]$

$$p(s\wedge\tau) = \Psi(y(\tau)) + \int_{s\wedge\tau}^{\tau} f(p(r),q(r),y(r)\quad)dr + \int_{s\wedge\tau}^{\tau} q(r)dW(r). \tag{2.2.2}$$

Where

$$f(p,q,x) : I\!R^n \times \mathcal{L}(I\!R^d \times I\!R^n) \times I\!R^m \to I\!R^n,$$
$$\Psi(x) I\!R^m \to I\!R^n.$$

We assume

$$(H2.2.2) \qquad \begin{cases} \text{(i) } f(p,q,x) \text{ is continuously differentiable} \\ \quad \text{in } I\!R^n \times \mathcal{L}(I\!R^d \times I\!R^n) \times \bar{Q}, \text{ the derivatives are bounded;} \\ \text{(ii) } \Psi(x) \text{ is of } C^3. \end{cases}$$

We can now define $\mathbf{u} : \bar{Q} \to I\!R^n$ by

$$\mathbf{u}(x,t) = E^{x,t}p(0).$$

We will show that the function $\mathbf{u}(x,t)$ solves the following system of parabolic PDE

$$\begin{cases} \mathbf{u}_t + \mathcal{L}_{(x)}\mathbf{u}(x,t) + f(\mathbf{u}(x,t), \mathbf{u}_x(x,t)\sigma(x), x) = 0, \quad (x,t) \in Q, \\ \mathbf{u}(x,T) = \Psi(x), \quad \forall x \in D, \\ \mathbf{u}(x,t) = \Psi(x), \quad \forall x \in S,\ t \in [0,T). \end{cases} \tag{2.2.3}$$

Here we denote, for $\mathbf{u}(x) = (u^1(x), ..., u^n(x))$,

$$u_t = \partial_t u, \quad \mathbf{u}_t = \begin{pmatrix} u_t^n \\ \vdots \\ u_t^n \end{pmatrix}, \quad u_x = \partial_x u, \quad \mathbf{u}_x = \begin{pmatrix} u_{x_1}^1 & \cdots & u_{x_m}^1 \\ \vdots & \ddots & \vdots \\ u_{x_1}^n & \cdots & u_{x_m}^n \end{pmatrix},$$

where

$$\mathcal{L}_{(x)} u = \sum_{i,j} a_{ij}(x)\partial_{x_i x_j} u + \sum_i b_i(x)\partial_{x_i} u, \quad \mathcal{L}\mathbf{u} = \begin{pmatrix} \mathcal{L}u^1 \\ \vdots \\ \mathcal{L}u^n \end{pmatrix}.$$

We need the following lemma

Proposition 2.2.1. We assume (H2.2.1) and (H2.2.2), we assume also the following compatibility condition

$$[\mathcal{L}_{(x,T)}\Psi(x) + f(\Psi(x), \Psi_x(x)\sigma(x,T), x, T)]_{x\in S} = 0.$$

If D is bounded, then, (2.2.3) has an unique solution in $C^{2,1}$.

The proof of this lemma can be found in [4], (Th.7.1, Ch.VII).

With this lemma, we can assert

Theorem 2.2.2. We assume the same conditions in lemma 2.2.1. Then, for any given (x, t), the solution of the system of parabolic equation (2.2.3) has the following interpretation

$$u(x,t) = E^{x,t}p(0) = p(0),$$

where $p(t)$ is determined uniquely by (2.2.1), (2.2.2).

Remark. Particularly, when $p(\cdot)$ (of (2.2.2)) is valued in $\mathbb{R}$ (n=1), (2.2.3) becomes a parabolic equation. Even in this case, it is still a nontrivial extension of the classical probabilistic interpretation. In fact, the classical case can be described by setting

$$f = f_0(x) + c(x)p.$$

§3. Optimal Stochastic Control: Dynamic Programming

In this section, we introduce a generalized form of optimal control system where the cost function is determined by certain backward stochastic differential equation discussed in §2. We will show that the principle of dynamic programming, known as Bellman's principle, can also apply in this situation.

We introduce the set of admissible control in usual sence: Let A be a compact set of $\mathbb{R}^k$. An admissible control is a collection of

(i) a probability space $(\Omega, \mathcal{F}, P)$ equipped with a filtrition $\mathcal{F}_t$;

(ii) a d-dimensional standard Wiener process $(W(t); t \geq 0)$, such that $\mathcal{F}_t = \{W(s); 0 \leq s \leq t\}$;

(iii) a progressively measurable process $\alpha(t); t \geq 0$ taking its value in A.

We denote the set of admissible controls by $\mathcal{A}$. For given admissible control and initial data $x \in \mathbb{R}^n$, we can consider the following stochastic control problem

$$\text{(3.1)} \qquad \begin{cases} dy(s) = b(y(s), \alpha(s))ds + \sigma(y(s), \alpha(s))dW(s), \\ y(0) = x \end{cases}$$

where $b(x, \alpha)$, $\sigma(x, \alpha)$ are respectively $\mathbb{R}^n-$ valued and $\mathcal{L}(\mathbb{R}^d; \mathbb{R}^n)$ valued functions defined on $\mathbb{R}^n \times \mathbb{R}^k$. We assume that

$$(H3.1) \qquad \begin{cases} b \text{ and } \sigma \text{ are continuous in } (x, \alpha), \text{ and continuously} \\ \text{differentiable in } x, \text{ their derivatives } b_x, \sigma_x \text{ are bounded.} \end{cases}$$

Obvioursly, the solution of $y(\cdot)$, called the trajectory corresponding to the control $\alpha(\cdot)$, is well defined and

$$\text{(3.2)} \qquad E|y^{x,\alpha(.)}(s)|^2 \leq C|x|^2,$$

where C is a constant independent of x, $\alpha(.)$. We can define a stopping time by

$$\tau = \tau_{x,t} = \inf_s\{(y(s), s) \notin D \times [0, t)\}.$$

We now introduce the following backward stochastic differential equation: let $f(p,q,x,\alpha)$ be a real function defined on $I\!R\times (I\!R^d)^*\times I\!R^n\times I\!R^k$ ($(I\!R^d)^*$ denotes the adjoint space of $I\!R^d$). For any given continuous function $g(x): I\!R^n \to I\!R$ satisfying $|g(x)| \le C(1+|x|)$, we consider

$$(3.3)\qquad \begin{cases} p(s\wedge\tau) = g(y(\tau)) + \int_{s\wedge\tau}^{\tau} f(p(r),q(r),y(r),\alpha(r))dr - \int_{s\wedge\tau}^{\tau} q(r)dW(r), \\ \qquad s\in[0,\tau]. \end{cases}$$

We assume

$$(H3.2)\qquad \begin{cases} f \text{ is continuous in } (p,q,x,\alpha) \text{ and continuously differentiable} \\ \text{in } (p,q,x), \text{ its derivatives } f_p,\ f_q,\ f_x \text{ are bounded.} \end{cases}$$

According to proposition 2.3, (3.1) is well defined. Since $p(s)$ is $\mathcal{F}_s$-adapted and $W(0)=0$, thus $p(0)=Ep(0)$. We can introduce the following generalized cost function

$$p(0) = J(x,t;g(\cdot),\alpha(\cdot));\quad (x,t)\in D\times(0,T).$$

From proposition 2.1, for given $g(\cdot)$, $J(x,t;g(\cdot),\alpha(\cdot))$ is uniformly bounded. Thus we can define

$$V(x,t;g(\cdot)) = \inf_{\alpha(\cdot)\in\mathcal{A}} J(x,t;g(\cdot),\alpha(\cdot)).$$

We have the following regularity result about V. To simplify, we set $D=I\!R^n$. In this case, $\tau = t$, thus

$$(3.4)\qquad \begin{cases} p(s) = g(y(t)) + \int_s^t f(p(r),q(r),y(r),\alpha(r))dr - \int_s^t q(r)dW(r), \\ \qquad s\in[0,t]. \end{cases}$$

Lemma 3.1. We suppose (H3.1) and (H3.2). We have

(i) Let $g(x)$ be uniformly Lipschitzian. Then we have

$$(3.5)\qquad |V(z,t;g(\cdot)) - V(x,t;g(\cdot))| \le C_g|z-x| \quad \forall z,x\in I\!R^n\ \forall t.$$

where C_g is a constant depending only on the Lipschitian constant of $g(\cdot)$.

(ii) Let $g_1(x)$ be continuous and bounded, Then

$$(3.6)\qquad |V(x,t;g(\cdot)) - V(x,t;g(\cdot)+g_1(\cdot))| \le C\sup_z |g_1(z)|.$$

(iii) If $g_1(x)$ is continuous, nonnegative and bounded, then

$$J(x,t;g(\cdot)+g_1(\cdot),\alpha(\cdot)) \ge J(x,t;g(\cdot),\alpha(\cdot)),$$
$$V(x,t;g(\cdot)+g_1(\cdot)) \ge V(x,t;g(\cdot)).$$

Now we can generalize the well known dynamic programming principle.

We assume

$(H3.3)$ $\qquad g(x)$ is uniformly Lipschitzian.

We have the so called Bellman's principle (in form of Nisio [6]). We still set $D = I\!R^n$.

Theorem 3.2. Let (H3.1)-(H3.3) hold. Then we have

$$V(x,t+h;g(\cdot)) = V(x,t;V(\cdot,h,g)); \qquad \forall x, \quad \forall t+h \le T. \tag{3.9}$$

§4. Hamilton-Jacobi-Bellman equation

In this section, we fix $g = g(x)$ and set

$$u(x,t) = V(x,T-t;g(\cdot)), \quad t \in [0,T].$$

Under certain conditions, this function is the solution of the following fully nonlinear partial differential equation

$$\partial_t u + \inf_{\alpha \in A}\{\mathcal{L}(x,\alpha)u + f(u, u_x\sigma(x,\alpha), x, \alpha)\} = 0, \tag{4.1}$$

$$u(x,T) = g(x), \tag{4.2}$$

$$u(x,t)|_{x\in\partial D} = g(x), \tag{4.3}$$

where

$$\mathcal{L}(x,\alpha)u = \sum_{i,j=1}^{n} a_{ij}(x,\alpha)\partial_{x_i x_j} u + \sum_{i=1}^{n} b_i(x,\alpha)\partial_{x_i} u,$$
$$a_{ij}(x,\alpha) = \frac{1}{2}[\sigma(x,\alpha)\sigma^*(x,\alpha)]_{ij},$$
$$u_x = (u_{x_1}, ..., u_{x_n}).$$

To simplify, we set $D = I\!R^n$. We have

Theorem 4.1. Let $D = I\!R^n$ and let us assume (H2.1)-(H2.4). Let $\tilde{u} \in C_b^{2,1}(I\!R^n)$ be a solution of (4.1), (4.2). Then $u \equiv \tilde{u}$.

Theorem 4.2. Let $D = I\!R^n$ and let us assume (H2.1)-(H2.4). Let $u \in C_b^{2,1}(I\!R^n)$. Then u is a solotion of HJB equation (4.1), (4.2).

If we assume

$(H4.1)$
- (i) (not loss of generality) let $b \equiv 0$;
- (ii) $a_{ij}(x,\alpha)$, $i,j, = 1,...,n$, $f(p,q,x,\alpha)$ are continuous in (x,α), for each $\alpha \in A$, they are twice continuously differentiable with respect to (p,q,x). All those derivatives are bounded;
- (iii) $\lambda_1|\xi|^2 \le \sum_{ij} a_{ij}(x,v)\xi_i\xi_j \le \lambda_2|\xi|^2$ $\lambda_2 \ge \lambda_1 > 0$ are constants;
- (iv) $g(x)$ is $C^3(I\!R^n)$.

Then, the following proposition obtained by Krylov [3] provides the existence (and uniqueness) result of HJB equation (4.1) for non degenerate case. For the domain $D \subset I\!R^n$, we assume

$$(H4.2) \qquad \begin{cases} D = \{x; \psi(x) > 0\} \text{ be a bounded domain such that } \psi \text{ is in } C^3(I\!R^n) \\ \text{and } |\psi_x| \geq 1 \text{ on } \partial D. \end{cases}$$

We have

Proposition 4.3. Let (H4.1), (H4.2) hold, Then The HJB equation (4.1) with the boundary conditions (4.2), (4.3) has unique solution $u \in C^{2,1}(D \times (0,T))$. The norm of u in $C^{2,1}(D \times (0,T))$ is bounded.

Detail treatment of the above problems can be found in our forthcoming papers [11], [13].

References

[1] R.Bellman, *Dynamic Programming*, Princeton Univ. Press, N.J. 1957.

[2] W.H. Fleming and R.W. Rishel, *Deterministic and Stoshastic Optimal Control,* Springer-Verlag, 1975.

[3] Krylov, *Nonlinear Elliptic and Parabolic Equations of Second Order,* D. Reidel Publishing Company, Dordrecht, 1987.

[4] O. Ladyzenskaja, V. Slonnikov, N. Uralceva, *Linear and Quasi-linear Equations of Parabolic Type*, Translations of Maths Monograghs, Vol.23 AMS, Providence Rhode Island, 1968.

[5] P.L.Lions, *Optimal Control of Diffusion Processes and Hamilton Jacobi Equations,* Part 1, 2, 3, Commu. in PDE, 1983.

[6] M.Nisio, *Some Remarks on Stochastic Optimal Controls,* Proc. 3rd U.S.S.R.-Japan Symp. Prob. Theory, Lecture Notes in Maths 550, Springer, Berlin, 1976.

[7] E. Pardoux and S. Peng, *Adapted Solution of Backward Stochastic Equation,* to appear in Systems and Control Letters.

[8] S. Peng, *A General Stochastic Maximum Principle for Optimal Control Problems,* SIAM J. Cont. Vol.28 No.4, 1990.

[9] S.Peng, *Backward Stochastic Differential Equation and It's Application in Optimal Control,* to appear.

[10] S. Peng, *Stochastic Hamilton-Jacobi-Bellman Equations,* to appear.

[11] S. Peng, *Probabilistic Interpretation for Systems of Parabolic Partial Differential Equations,* submitted.

[12] S. Peng, *Maximum Principle for Stochastic Optimal Control with Nonconvex Control Domain,* to appear in Proceedings of 9th International Conference on Analysis and Optimization of Systems, Antibes, 1990.

[13] S.Peng, *A Generalized Dynamic Programming Principle* , to appear.

DYNAMICS AND CONTROL OF BENDING AND TORSIONAL VIBRATIONS OF FLEXIBLE BEAMS

Yoshiyuki Sakawa and Zheng Hua Luo

Department of Control Engineering, Osaka University
Toyonaka, Osaka 560, Japan

1. INTRODUCTION

Modeling and control of flexible beams has received a great deal of attention in recent years. This problem has arisen in the area of space and industrial robots with lightweight and flexible arms as well as in the area of flexible space structures. In this paper, we consider a flexible beam having a rigid tip body at the free end. It is assumed that the cross section of the beam is geometrically symmetric and the shear center axis coincides with the centroidal axis. We considered the coupled bending and torsional vibrations of the flexible beam, when the mass center of the tip body does not lie on the shear center axis of the flexible beam [1]. Here, we consider the flexible beam having a rigid tip body, of which the mass center lies on the centroidal axis of the beam. In this case, it will be shown that the decoupled bending and torsional vibrations occur, and the motion is governed by decoupled partial differential equations and boundary conditions.

We first derive decoupled partial differential equations with decoupled boundary conditions as a mathematical model of the vibrations. Then each set of equations will be rewritten together as an evolution equation in a properly defined Hilbert space. We design a finite-dimensional dynamic compensator using sensor outputs to construct a feedback control system on the basis of the previous results [2].

2. MATHEMATICAL MODELING OF A BEAM WITH A TIP BODY

We consider a slender flexible beam which is rotated in a horizontal plane by a motor M_1 which may suppress the bending vibration. It is also rotated around a centroidal axis of the beam by another motor M_2 which may suppress the torsional vibration. The beam of length L, having uniform mass density ρ per unit length, uniform flexural rigidity EI, and uniform torsional rigidity GJ, is clamped on a vertical shaft of the motor M_1 at one end and has a tip body rigidly attached at the free end, as shown in Fig. 1. Let X_0, Y_0, Z_0 designate an inertial Cartesian coordinate axes, where X_0 and Y_0 axes span a horizontal plane, and Z_0 axis is taken so that it coincides with the vertical rotation shaft of the motor M_1. Let $X_1, Y_1, Z_1 (Z_1 = Z_0)$ denote a coordinate axes rotating with the motor M_1, and let $X_2, Y_2, Z_2 (X_2 = X_1)$ denote a coordinate axes rotating with the motor M_2. Let $\theta(t)$ be the angle of rotation of the motor M_1, and let $\varphi_m(t)$ be the angle of rotation of the second motor M_2.

Let Q denote the mass center of the rigid tip body. It lies on the centroidal axis of the beam. Let c be the distance between the beam's tip point and the point Q. It is assumed that c is small. We take another coordinate axes X_3, Y_3, Z_3 attached to the tip body, where X_3 is the beam's tip tangent. Since the tip body is a rigid body, it is characterized by mass m, and two moments of inertia J_E and J_0 with respect to the lines QX_3 and QZ_3, respectively.

Let $y(t,x)$ and $\varphi(t,x)$ denote the transverse displacement and the angle of twist of the beam, respectively, at position x $(0 < x < L)$ and at time t. Both y and φ are assumed to be small. For the transverse vibration we use the Euler-Bernoulli model with internal viscous damping of the Voigt type [3]

$$\frac{\partial^2 y(t,x)}{\partial t^2} + 2\delta_1 \frac{EI}{\rho}\frac{\partial^5 y(t,x)}{\partial t \partial x^4} + \frac{EI}{\rho}\frac{\partial^4 y(t,x)}{\partial x^4} = -x\ddot{\theta}(t), \tag{1}$$

where $\delta_1 > 0$ is a small damping constant of the beam material. Since the beam is clamped at $x = 0$, we obtain

$$y(t,0) = 0, \qquad y'(t,0) = 0. \tag{2}$$

The torsional vibration is governed by [4]

$$\frac{\partial^2 \varphi(t,x)}{\partial t^2} - 2\delta_2 \frac{GJ}{\rho\kappa^2}\frac{\partial^3 \varphi(t,x)}{\partial t \partial x^2} - \frac{GJ}{\rho\kappa^2}\frac{\partial^2 \varphi(t,x)}{\partial x^2} = -\ddot{\varphi}_m(t), \tag{3}$$

where $\delta_2 > 0$ is a small damping constant for the torsional vibration, $\rho\kappa^2$ is the mass polar moment of inertia per unit length of the beam. Since the beam is clamped at $x = 0$, we obtain

$$\varphi(t, 0) = 0. \tag{4}$$

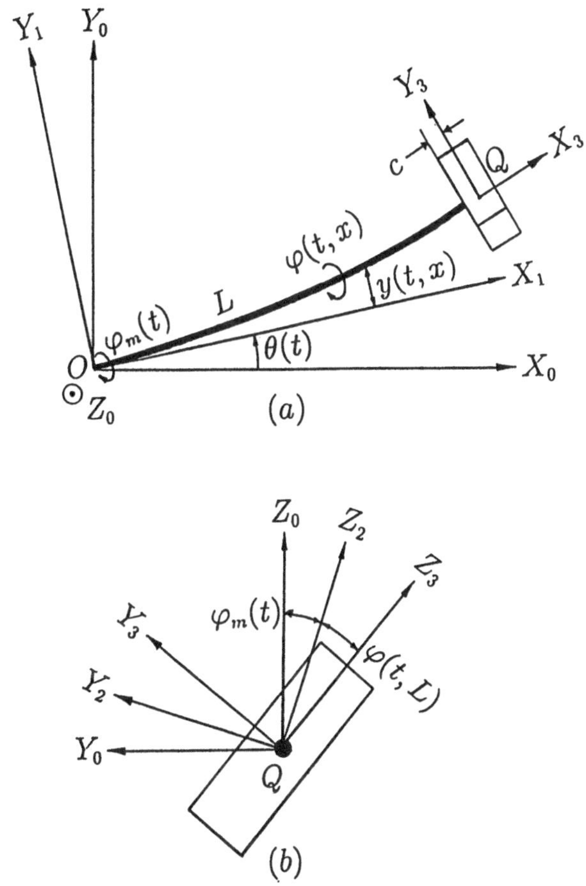

Fig. 1. Bending and torsion of a flexible beam with a rigid tip body.

In the same way as in [1], using the Lagrangian method, we obtain the equations of motion of the tip body as follows:

$$\begin{aligned} m[\ddot{y}(t, L) + c\ddot{y}'(t, L)] - EI[y'''(t, L) + 2\delta_1 \dot{y}'''(t, L)] \\ = -m(L + c)\ddot{\theta}(t), \end{aligned} \tag{5}$$

$$mc[\ddot{y}(t,L)+c\ddot{y}'(t,L)]+J_0\ddot{y}'(t,L)+EI[y''(t,L)+2\delta_1\dot{y}''(t,L)]$$
$$=-[J_0+mc(L+c)]\ddot{\theta}(t), \tag{6}$$
$$J_E\ddot{\varphi}(t,L)+GJ[\varphi'(t,L)+2\delta_2\dot{\varphi}'(t,L)]=-J_E\ddot{\varphi}_m(t), \tag{7}$$

where a dot denotes the time derivative, and a prime denotes the derivative with respect to the spatial variable x. Equations (5)-(7) give the decoupled boundary conditions for the bending and the torsional vibrations.

The equations of motion of the rotation motors can be respectively written as

$$J_1\ddot{\theta}(t)+\gamma_1\dot{\theta}(t)=T_1(t)+EIy''(t,0), \tag{8}$$

$$J_2\ddot{\varphi}_m(t)+\gamma_2\dot{\varphi}_m(t)=T_2(t)+GJ\varphi'(t,0), \tag{9}$$

where J_1 and J_2 are the moments of inertia of each motor, γ_1 and γ_2 are the viscous friction coefficients, and $T_1(t)$ and $T_2(t)$ are the torques developed by each motor. It is clear that the bending moment $EIy''(t,0)$ works on the motor shaft of M_1 as a reaction of the bending motion, and the torsional moment $GJ\varphi'(t,0)$ works on the motor shaft of M_2 as a reaction of the torsional motion.

3. EVOLUTION EQUATIONS AND CONTROL SYSTEM

We treat the partial differential equation (1) together with the boundary conditions (2), (5), and (6) for the bending vibration in the form of an abstract second-order evolution equation in an appropriate Hilbert space $H_1=L_2(0,L)\times R^2$ with the inner product

$$< (u_1,u_2,u_3),(v_1,v_2,v_3) >_{H_1}$$
$$=\rho\int_0^L u_1(x)v_1(x)dx+[u_2\ u_3]\begin{bmatrix} m & mc \\ mc & J_0+mc^2\end{bmatrix}\begin{bmatrix} v_2 \\ v_3\end{bmatrix}, \tag{10}$$

where the first component of an element of H_1 corresponds to the bending distribution $y(t,\cdot)$ and the remaining components correspond to $y(t,L)$ and $y'(t,L)$, respectively.

Define the operator Λ_1 on H_1 into H_1 by

$$\Lambda_1 u=(u_1(\cdot),\ mu_2+mcu_3,\ mcu_2+(J_0+mc^2)u_3)^T, \tag{11}$$

Furthermore, let us define the operator Σ_1 by

$$\Sigma_1 u=\Big(\frac{EI}{\rho}u_1^{(4)}(\cdot),\ -EIu_1'''(L),\ EIu_1''(L)\Big)^T, \tag{12}$$

$$D(\Sigma_1) = \{u = (u_1, u_2, u_3)^T | u_1(\cdot) \in H^4(0, L), u_2 = u_1(L), \\ u_3 = u_1'(L), u_1(0) = 0, u_1'(0) = 0\}, \tag{13}$$

where $D(\Sigma_1)$ denotes domain of Σ_1. Also let us define an element $\Omega_1 \in H_1$ by

$$\Omega_1 = -(x,\ m(L+c),\ J_0 + mc(L+c))^T. \tag{14}$$

Then it is easily seen that, by using the operators Λ_1 and Σ_1, the differential equations (1) together with the boundary conditions (2), (5) and (6) can be written as

$$\Lambda_1 \ddot{u}(t) + 2\delta_1 \Sigma_1 \dot{u}(t) + \Sigma_1 u(t) = \Omega_1 \ddot{\theta}(t), \tag{15}$$

where $u(t) = (u_1(t), u_2(t), u_3(t))^T \in D(\Sigma_1)$. If we define an operator A_1 on H_1 by $A_1 u = \Lambda_1^{-1} \Sigma_1 u$, $D(A_1) = D(\Sigma_1)$, (15) can be written as

$$\ddot{u}(t) + 2\delta_1 A_1 \dot{u}(t) + A_1 u(t) = \Lambda_1^{-1} \Omega_1 f(t), \tag{16}$$

where $f(t) = \ddot{\theta}(t)$.

It can be proved in the same manner as in [1] that the operator A_1 is selfadjoint and positive definite on H_1, and the inverse operator A_1^{-1} exists and is compact. Consequently, the operator A_1 has the eigenvalues $\{\lambda_n\}$ and the corresponding eigenfunctions $\{\psi_n(x)\}$ satisfying

$$A_1 \psi_n(x) = \lambda_n \psi_n(x). \tag{17}$$

Let $\psi_n(\cdot) = (y_n(\cdot), y_n(L), y_n'(L))^T$. Then (17) implies

$$\left.\begin{aligned} &(EI/\rho) y_n^{(4)}(x) = \lambda_n y_n(x), \\ &-EI y_n'''(L) = m\lambda_n [y_n(L) + c y_n'(L)], \\ &EI y_n''(L) = \lambda_n [mc y_n(L) + (J_0 + mc^2) y_n'(L). \end{aligned}\right\} \tag{18}$$

From (18) the eigenvalues λ_n and the eigenfunctions $\psi_n(x)$ can be calculated easily. The eigenvalues are given by

$$\lambda_n = (EI/\rho)(\beta_n/L)^4,$$

where β_n are the positive solutions of

$$\begin{aligned} &1 + \cos\beta \cosh\beta - (m/\rho)(\beta/L)(\sin\beta \cosh\beta - \sinh\beta \cos\beta) \\ &- 2\left(\frac{mc}{\rho}\right)\left(\frac{\beta}{L}\right)^2 \sin\beta \sinh\beta - \left(\frac{J_0 + mc^2}{\rho}\right)\left(\frac{\beta}{L}\right)^3 (\sin\beta \cosh\beta + \sinh\beta \cos\beta) \\ &+ (mJ_0/\rho^2)(\beta/L)^4 (1 - \cos\beta \cosh\beta) = 0. \end{aligned} \tag{19}$$

It is easily seen that

$$\beta_n \to (\pi/2) + n\pi \qquad (n \to \infty).$$

Similarly, the partial differential equation (3) together with the boundary conditions (4) and (7) for the torsional vibration can be written in the form of an evolution equation in an appropriate Hilbert space $H_2 = L_2(0, L) \times R$ as

$$\ddot{v}(t) + 2\delta_2 A_2 \dot{v}(t) + A_2 v(t) = \Lambda_2^{-1}\Omega_2 g(t), \tag{20}$$

where $g(t) = \ddot{\varphi}_m(t)$. The inner product of H_2 is defined by

$$< (u_1, u_2), (v_1, v_2) >_{H_2} = \rho\kappa^2 \int_0^L u_1(x)v_1(x)dx + J_E u_2 v_2,$$

and the first component of an element of H_2 corresponds to the torsional distribution $\varphi(t, \cdot)$ and the second component of H_2 corresponds to $\varphi(t, L)$. The operator A_2 has the same property as A_1, and has the eigenvalues $\{\mu_n\}$ and the corresponding eigenfunctions $\{\tau_n(x)\}$.

By cementing several strain gage foils at the root of the beam, it is possible to measure the bending moment and the torsional moment at $x = 0$, i.e., $y''(t, 0)$ and $\varphi'(t, 0)$. Therefore, the output of each system can be defined by $z_1(t) = y''(t, 0)$ and $z_2(t) = \varphi'(t, 0)$, respectively.

The solution of (16) can be expressed in terms of the eigenfunctions $\psi_n(\cdot) = (y_n(\cdot), y_n(L), y_n'(L))^T$ as

$$u(t, \cdot) = \sum_{n=1}^{\infty} \frac{1}{\lambda_n} u_n(t)\psi_n(\cdot), \tag{21}$$

where $u_n(t)$ are the solution of

$$\ddot{u}_n(t) + 2\delta_1 \lambda_n \dot{u}_n(t) + \lambda_n u_n(t) = -EIy_n''(0)f(t). \tag{22}$$

In the same way, the solution of (20) can be expressed as

$$v(t, \cdot) = \sum_{n=1}^{\infty} \frac{1}{\mu_n} v_n(t)\tau_n(\cdot), \tag{23}$$

where $v_n(t)$ are the solution of

$$\ddot{v}_n(t) + 2\delta_2 \mu_n \dot{v}_n(t) + \mu_n v_n(t) = -GJ\varphi_n'(0)g(t). \tag{24}$$

In (22) and (24), $f(t) = \ddot{\theta}(t)$ and $g(t) = \ddot{\varphi}_m(t)$ are regarded as the control inputs for each distributed parameter system. By defining variables $\tilde{u}_n(t)$ by $\tilde{u}_n(t) = -1/\omega_n[\dot{u}_n(t) + \delta_1\lambda_n u_n(t)]$, where $\omega_n = \sqrt{\lambda_n - \delta_1^2\lambda_n^2}$, let us consider two finite-dimensional modal equations

$$\dot{x}_1(t) = F_1 x_1(t) + B_1 f(t), \tag{25}$$

$$\dot{\tilde{x}}_1(t) = \widetilde{F}_1 \tilde{x}_1(t) + \widetilde{B}_1 f(t), \tag{26}$$

where $x_1(t) = (u_1(t), \tilde{u}_1(t), \cdots, u_l(t), \tilde{u}_l(t))^T$ is a $2l$-dimensional state vector corresponding to the first l modes of (16), and $\tilde{x}_1(t) = (u_{l+1}(t), \tilde{u}_{l+1}(t), \cdots, u_N(t), \tilde{u}_N(t))^T$ is a $2(N-l)$-dimensional state vector corresponding to the subsequent $(N-l)$ modes of (16). The matrices in (25) and (26) are defined by

$$F_1 = \text{block diag}(M_1, \cdots, M_l), \qquad B_1 = [b_1^T, \cdots, b_l^T]^T,$$

$$\widetilde{F}_1 = \text{block diag}(M_{l+1}, \cdots, M_N), \quad \widetilde{B}_1 = [b_{l+1}^T, \cdots, b_N^T]^T,$$

$$M_n = \begin{bmatrix} -\delta_1\lambda_n & -\omega_n \\ \omega_n & -\delta_1\lambda_n \end{bmatrix}, \qquad b_n = \begin{bmatrix} 0 & \dfrac{EIy_n''(0)}{\omega_n} \end{bmatrix}^T.$$

The output can be correspondingly approximated as

$$z_1(t) = y''(t,0) \simeq \sum_{n=1}^{N} \frac{1}{\lambda_n} u_n(t) y_n''(0) = C_1 x_1(t) + \widetilde{C}_1 \tilde{x}_1(t), \tag{27}$$

where $C_1 = [c_1, \cdots, c_l]$, $\widetilde{C}_1 = [c_{l+1}, \cdots, c_N]$, $c_n = (y_n''(0)/\lambda_n, 0)$. If $y_n''(0) \neq 0$, $n = 1, \cdots, l$, then $b_n \neq 0$ and $c_n \neq 0$, $n = 1, \cdots, l$, and the linear system (F_1, B_1, C_1) is controllable and observable [2].

By constructing two kinds of observers [2]

$$\begin{aligned} \dot{w}_1(t) &= (F_1 - G_1C_1)w_1(t) + G_1[z_1(t) - \widetilde{C}_1\tilde{w}_1(t)] + B_1 f(t), \\ \dot{\tilde{w}}_1(t) &= \widetilde{F}_1\tilde{w}_1(t) + \widetilde{B}_1 f(t), \end{aligned} \tag{28}$$

the state vector $x_1(t)$ can be estimated with an arbitrarily high convergence rate. In (28), the term $\widetilde{C}_1\tilde{w}_1(t)$ is introduced to avoid observation spillover.

The following results was proved in [2]: Given an arbitrary damping constant σ such that $\delta_1\lambda_1 < \sigma < (1/2\delta_1)$, if l is chosen corresponding to σ and N is large and $b_n \neq 0, n = 1, \cdots, l$, then a finite-dimensional controller $f(t) = K_1 w_1(t)$, where $w_1(t)$ is the solution of (28) with an arbitrary initial condition, can be constructed in such a way that the solution of the evolution equation (16) satisfies

$$||u(t)|| \leq \text{const}.e^{-\sigma t}||u(0)||.$$

This means that the original infinite-dimensional system is exponentially stabilizable with the exponent σ by using the finite-dimensional observer-based compensator.

Let $r_1(t)$ be a given reference input function for the rotation angle $\theta(t)$, and let

$$\tilde{\theta}(t) = \theta(t) - r_1(t), \qquad \tilde{\omega}_1(t) = \dot{\theta}(t) - \dot{r}_1(t). \tag{29}$$

Combining (25) with (29) yields

$$\dot{x}(t) = \mathcal{A}_1 x(t) + \mathcal{B}_1 f(t) + \mathcal{D}_1 \ddot{r}_1(t), \tag{30}$$

where $x(t) = (x_1(t)^T, \tilde{\theta}(t), \tilde{\omega}_1(t))^T$, and

$$\mathcal{A}_1 = \begin{bmatrix} F_1 & 0 & 0 \\ 0 & 0 & 1 \\ 0 & 0 & 0 \end{bmatrix}, \quad \mathcal{B}_1 = \begin{bmatrix} B_1 \\ 0 \\ 1 \end{bmatrix}, \quad \mathcal{D}_1 = \begin{bmatrix} 0 \\ 0 \\ -1 \end{bmatrix}.$$

We consider the case where $r_1(t)$ is a step function. Since the pair (F_1, B_1) is controllable, it is easily seen that the pair $(\mathcal{A}_1, \mathcal{B}_1)$ is also controllable. Since our problem is to regulate (30), a feedback control law

$$f(t) = Kx(t) = K_1 x_1(t) + K_2 \tilde{\theta}(t) + K_3 \tilde{\omega}_1(t) \tag{31}$$

should be sought by minimizing the performance index

$$\begin{aligned} J(f) = \int_0^\infty & [q_1 y^2(t, L) + q_2 \dot{y}^2(t, L) + q_3 \tilde{\theta}^2(t) \\ & + q_4 \tilde{\omega}_1^2(t) + r f^2(t)] e^{2\sigma t} dt, \end{aligned} \tag{32}$$

where $q_i \geq 0$ $(i = 1, \cdots, 4)$, $r > 0$, and $\sigma \geq 0$. Because $x_1(t)$ cannot be measured, we use the observer output $w_1(t)$ in place of $x_1(t)$ in (28).

If the rotation motor M_1 has an amplifier of speed-feedback type, the torque developed by the motor M_1 is given by

$$T_1(t) = k_1 (V_{ref1}(t) - k_2 \omega_1(t)), \tag{33}$$

where $\omega_1(t) = \dot{\theta}(t)$, $V_{ref1}(t)$ is the input-speed reference voltage to the amplifier, and k_1 and k_2 are gain constants. Substituting $T_1(t)$ in (8) into (33) gives

$$V_{ref1}(t) = (k_2 + \frac{\gamma_1}{k_1})[\int_0^t f(t) dt + \omega_1(0)] + \frac{J_1}{k_1} f(t) - \frac{EI}{k_1} z_1(t), \tag{34}$$

where $\omega_1(0)$ and $z_1(t)$ are known, and $f(t)$ is given by (31).

The feedback control law for the torsional vibration can be obtained in the entirely same manner as discussed above. Namely,

$$g(t) = K_4\xi_1(t) + K_5\tilde{\varphi}_m(t) + K_6\tilde{\omega}_2(t), \tag{35}$$

where ξ_1 is the finite-dimensional state vector corresponding to the first several modes of (20), $\tilde{\varphi}_m(t) = \varphi_m(t) - r_2(t)$, $\tilde{\omega}_2(t) = \dot{\varphi}_m(t) - \dot{r}_2(t)$, and $r_2(t)$ is a given reference input function for the twist angle.

Acknowledgment. The authors wish to thank Mr. T. Murachi for his cooperation.

REFERENCES

[1] Y. Sakawa and Z. H. Luo, Modeling and control of coupled bending and torsional vibrations of flexible beams, IEEE Trans. Automat. Contr., vol. 34, no. 9, pp. 970-977, 1989.

[2] Y. Sakawa, Feedback control of second order evolution equations with damping, SIAM J. Contr. Optimiz., vol. 22, no. 3, pp. 343-361, 1984.

[3] Y. Sakawa, F. Matsuno, and S. Fukushima, Modeling and feedback control of a flexible arm, J. Robotic Systems, vol. 2, no. 2, pp. 453-472, 1985.

[4] L. Meirovitch, Analytical Methods in Vibrations. New York: McGraw-Hill, 1967.

STRONG SOLUTIONS AND OPTIMAL CONTROL FOR STOCHASTIC DIFFERENTIAL EQUATIONS IN DUALS OF NUCLEAR SPACES *

Situ Rong
Department of Mathematics, Zhongshan University, Guangzhou, China

Introduction

Stochastic differential equations in a dual of a nuclear space is an appropriate stochastic model for investigating a diffusion process, which is a limit of a sequence of stochastic processes arising in very diverse fields such as neurophysiology [7], interacting particle diffusions [17], and chemical kinetics [8]. Moreover, it is also an effective model for exploring some physical problems, such as the random motion of strings, etc. ([6], [2] ,[5]). In this paper we obtain results on Girsanov theorem, existence of weak and strong solutions, martingale representation theorem, maximum principle to optimal stochastic control, and pathwise uniqueness, stability of solutions for stochastic differential equations (SDE) with discontinuous drift, which even can be greater than linear growth, and with non-Lipschitzian diffusion coefficient in the duals of nuclear spaces.

Girsanov theorem and weak solutions

Let Φ be a real nuclear Frechet space with topology generated by a countable family of increasing Hilbertian norms $\{|\cdot|_n\}$. Denote the completion of Φ under norm $|\cdot|_n$ by H^n, the strong dual of Φ by Φ', the dual of H^n by H^{-n}. Then

$$\Phi = \bigcap_{n=1}^{\infty} H^n, \quad \Phi' = \bigcup_{n=1}^{\infty} H^{-n}.$$

Now let us adopt the notation from [6], [8]. A Φ'-valued process is called a Φ' Brownian Motion process (BM), iff w_t is a Φ'-valued L^2-martingale (i.e. $w_t[\varphi]$ is a real square integrable martingale for each $\varphi \in \Phi$) such that

1° $w_0 = 0$, P - a.s.;

2° $\langle w\rangle_t[\varphi,\psi] = t\,Q[\varphi,\psi]$, for all $\varphi, \psi \in \Phi$,

where $Q \in \Phi' \times \Phi'$, which is non-random, and $\langle w\rangle_t$ is a $\Phi' \times \Phi'$ - valued predictable process, which exists, is increasing in t, and for all $\varphi, \psi \in \Phi$

$$Y_t[\varphi,\psi] = w_t[\varphi]w_t[\psi] - \langle w\rangle_t[\varphi,\psi]$$

is a real local martingale satisfying $Y_0[\varphi,\psi] = 0$, P - a.s. By [12], [6] there exists a $0 < q < \infty$ such that for each $r \geq q$ $w_{\cdot}(\omega) \in C(R_+; H^{-r})$ (the space of strong continuous maps in H^{-r}), and Q has a unique continuous extension to a nuclear form on H^{-r} such that $Q: H^r \times H^r \to R^1$, and for all $\varphi, \psi \in H^r$

* This work is supported in part by the Foundation of Zhongshan University Advanced Research Center.

$$Q[\varphi,\psi] = (\varphi, Q_r\psi)_r = (Q_r^{\frac{1}{2}}\varphi, Q_r^{\frac{1}{2}}\psi)_r, \quad |Q|_{-r,-r} = \sum_{j=1}^{\infty} Q[h_j^r, h_j^r] < \infty,$$

for a unique non-negative trace-class operator Q_r on H^r, where $\{h_j^r\}$ is a complete orthonormal system (CONS) in H^r. In the whole paper we shall fix a special index r ($\geq q$), and call Q the covariance operator of w_t. Now let us introduce the definition of a stochastic integral with respect to (w.r.t.) a Φ' BM w_t with Q. Set

L_w^2 = the totality of $f_s = f(s,\omega)$: $R_+ \times \Omega \to L(\Phi', \Phi')$, which is $\mathcal{F}_t$-adapted, and

$$E\int_0^T Q[f_s^*\varphi, f_s^*\varphi]\, ds < \infty, \text{ for all } \varphi \in \Phi, \quad T > 0,$$

where $f_s^* = f_s^*(s,\omega)$: $R_+ \times \Omega \to L(\Phi, \Phi)$ is the adjoint operator of f_s such that

$$\psi[f_s^*\varphi] = (f_s\psi)[\varphi], \quad \text{for all } \varphi \in \Phi,\ \psi \in \Phi'.$$

Definition 1. For $f \in L_w^2$ the stochastic integral $I_t^f = \int_0^t f_s dw_s$, $t \in [0,T]$, is a Φ'-valued L^2-martingale, which is strongly continuous w.r.t. t, and satisfies

1° I_t^f is linear in $f \in L_w^2$, i.e. for all a, b - real numbers, and $f, g \in L_w^2$

$$I_t^{af+bg} = aI_t^f + bI_t^g,$$

2° $\langle I_\cdot^f\rangle_t[\varphi,\psi] = \int_0^t Q[f_s^*\varphi, f_s^*\psi]\, ds$, for all $\varphi, \psi \in \Phi$.

Remark 1. By a category argument [6] there exists a sufficiently large $m \geq r$ (depending on f and T) such that $I_\cdot^f \in C([0,T]; H^{-m})$, P - a.s.

3° if $\{h_j^m\}_{j=1} \subset \Phi$ is any CINS in H^m, then as $0 \leq t \leq T$

$$I_t^f[\varphi] = \int_0^t f_s dw_s[\varphi] = \sum_{i=1}^{\infty} \int_0^t (f_s^*\varphi, h_i^m)_m\, dw_s[h_i^m], \quad \text{for all } \varphi \in \Phi,$$

where the right hand side is an L^2-convergent series of the usual Ito stochastic integrals.

We have the following Girsanov type theorem.

Theorem 1. Assume that

$$b: R_+ \times \Phi' \to \Phi', \quad \sigma: R_+ \times \Phi' \to L(\Phi', \Phi'),$$

for each $T > 0$ and sufficient large $m \geq r$ there exists an index $p \geq m$, as $t \leq T$

1° $b(t,u)$: $R_+ \times H^{-m} \to H^{-p}$, which is jointly measurable, and

$$(u, b(t,u)) \leq k_0(1 + |u|_{-p}^2 \prod_{i=1}^{\infty} g_i(u)), \quad \text{for all } u \in H^{-m},$$

where for $i = 1, 2, \ldots, k$

$$g_i(u) = 1 + \ln g_{i-1}(u), \quad g_0(u) = 1 + |u|_{-p}^{2n_0}, \quad 1 \leq n_0 \text{ - constant},$$

and $|b(t,u)|_{-p} \leq k_N$, as $|u|_{-p} \leq N$, $u \in H^{-m}$;

2° $|\sigma(t,u)v|_{-p} \leq k_0|v|_{-p}$, for all $u \in H^{-m}$, $v \in H^{-p}$.

If x_t with Φ' BM w_t satisfies Φ' - valued SDE

$$x_t = x_0 + \int_0^t \sigma(s,x_s)dw_s, \quad x_0 \in \Phi', \quad 0 \leq t \leq T, \tag{1}$$

then

$$\bar{w}_t = w_t - \int_0^t Q_p^* b(s,x_s)ds, \quad 0 \leq t \leq T,$$

is a Φ' BM with the same covariance operator Q under the probability measure

$$d\bar{P} = Z_T dP,$$

where

$$Z_t = \exp\left(\int_0^t (b(s,x_s), dw_s)_{-p} - \tfrac{1}{2}\int_0^t |Q_p^{*\frac{1}{2}} b(s,x_s)|_{-p}^2 ds\right),$$

and $p \geq m$, m comes from remark 1 such that $x_. \in C([0,T]; H^{-m})$, and p from the above assumption; and $Q_p^*: H^{-p} \to H^{-p}$ is the dual of Q_p such that

$$\psi[Q_p\varphi] = (Q_p^*\psi)[\varphi], \quad \text{for all } \varphi \in H^p, \quad \psi \in H^{-p}.$$

The proof of theorem 1 depends on the following

Lemma 1. If x_t with Φ' BM w_t satisfies (1) and $E\, Z_T = 1$, then the conclusion of theorem 1 holds.

Proof. Denote $b_t = b(t,x_t)$, $N_t = \int_0^t (b_s, dw_s)_{-p}$,

$$\mathcal{B}_t^{w,\Phi} = \sigma\{\{\omega: w_s[\varphi] \in \Gamma\},\ \Gamma \in \mathcal{B}(R^1),\ s \leq t,\ \varphi \in \Phi\}.$$

Then it is not difficult to derive that

$$\langle N_., w_.[\varphi]\rangle_t = \int_0^t Q_p^* b_s[\varphi]\, ds.$$

By Ito formula for arbitrary η - $\mathcal{B}_s^{w,\Phi}$ - measurable, bounded function it yields tha

$$E(\exp\{i\lambda(\bar{w}_t - \bar{w}_s)[\varphi]\}Z_t\eta) = \exp(-\lambda^2/2\ Q[\varphi,\varphi](t-s))\ E(Z_s\eta).$$

Hence if denote the expectation w.r.t. probability $\bar{P}$ by $E_{\bar{P}}$, then

$$E_{\bar{P}}(\exp\{i\lambda(\bar{w}_t - \bar{w}_s)[\varphi]\}|\mathcal{B}_s^{w,\Phi}) = \exp(-\lambda^2/2\ Q[\varphi,\varphi](t-s)).$$

From this $\bar{w}_t$ is a Φ'-valued L^2- martingale, and for all $\varphi, \psi \in \Phi$, $0 \leq s \leq t$

$$\langle \bar{w}\rangle_t[\varphi,\psi] = t\ Q[\varphi,\psi]. \qquad \text{Q.E.D.}$$

The proof of theorem 1 can be accomplished by lemma 1 and the technique as [14]. We omit it here. In the following we always make the assumption

(A) there exists $\{h_j\}_{j=1}^{\infty} \subset \Phi$, which is a common orthogonal system in H^m, for all $m \geq 1$. (Hence $\{h_j^m\}_{j=1}^{\infty} \subset \Phi$ is a CONS in H^m, if set $h_j^m = h_j/|h_j|_m$).

Applying lemma 1 and a result from [6] we have immediately a theorem on the existence of a weak solution for a Φ'-valued SDE with discontinuous drift, which even can be greater than linear growth as following:

Theorem 2. Assume that conditions 1° and 2° in theorem 1 hold, and

3° $\sigma: R_+ \times \Phi' \to L(\Phi', \Phi')$ is jointly continuous,

4° $\sigma(s,u)v \in H^{-m}$, if $u, v \in H^{-m}$,

5° $Q(\sigma^*(s,u)\varphi, \sigma^*(s,u)\varphi)$ is continuous in $u \in \Phi'$ for each $\varphi \in \Phi$.

Then there exists a weak solution for Φ'-valued SDE

$$x_t = x_0 + \int_0^t \sigma(s,x_s)Q^*b(s,x_s)ds + \int_0^t \sigma(s,x_s)dw_s, \quad x_0 = x \in \Phi', \quad 0 \leq t \leq T.$$

In case, if

6° σ^{-1} exists and $\|\sigma^{-1}\|_{\mathcal{L}(H^{-p},H^{-p})} \leq k_0$,

then the following Φ'-valued SDE also has a weak solution:

$$x_t = x_0 + \int_0^t Q^*b(s,x_s)ds + \int_0^t \sigma(s,x_s)dw_s, \quad x_0 = x \in \Phi', \quad 0 \leq t \leq T.$$

Remark 2. Here we define

$$Q^*\psi = Q_p^*\psi, \quad \text{as } \psi \in H^{-p}. \qquad (*)$$

Since the extension Q_p is unique, hence

$$Q_{p_2}^*|_{H^{-p_1}} = Q_{p_1}^*, \quad \text{as } p_1 \leq p_2.$$

(*) is well defined.

Proof. By [6] (1) has a weak solution. Applying theorem 1 one obtains the desired result of theorem 2. (Note that $\sigma Q_p^* \sigma^{-1} b = Q_p^* b$). Q.E.D.

Tanaka Formula and Strong Solutions

Theorem 3. (Tanaka formula). Assume that

$$b^i : R_+ \times \Phi' \to \Phi', \quad i = 1, 2; \quad \sigma : R_+ \times \Phi' \to L(\Phi', \Phi'),$$

for each $T > 0$ and sufficient large $m \geq r$ there exists an index $p \geq m$, as $t \leq T$

1° $|b^i(t,x)|_{-p}^2 + \mathrm{trac}(\sigma'\sigma Q^*) \leq g(|x|_{-p}), \quad i = 1, 2, \quad$ as $x \in H^{-m}$,

where $b(t,.) : H^{-m} \to H^{-p}, \quad \sigma(t,.) : H^{-m} \to L(H^{-p}, H^{-p}),$

$$\mathrm{trac}(\sigma'\sigma Q^*) = \sum_{j=1}^{\infty} (\sigma Q_p^{*\frac{1}{2}} h_j^{-p}, \sigma Q_p^{*\frac{1}{2}} h_j^{-p})_{-p},$$

and $g: R^1 \to R^1$ is continuous;

2° for all $x, y \in H^{-m}$, as $|x|_{-p}, |y|_{-p} \leq N, \quad N = 1, 2, \ldots$

$$\|\sigma(t,x) - \sigma(t,y)\|_{L(H^{-p},H^{-p})} \leq k_N(t)\, \rho_N(|x - y|_{-p}),$$

where $0 \leq \rho_N(u)$ ia strictly increasing, continuous, $\rho_N(0) = 0$; $k_N(t) \geq 0$, and

$$\int_0^T k_N^2(t)dt < \infty, \quad \int_{0+} du/\rho_N(u)^2 = \infty, \quad \text{for any } 0 \leq T < \infty.$$

If Φ'-valued processes x_t^i, $i = 1, 2$, satisfy

$$x_t^i = x_0^i + \int_0^t b(s,x_s^i)ds + \int_0^t \sigma(s,x_s)dw_s, \quad t \geq 0, \tag{2}$$

then for each $T > 0$, there exist $m \geq r$, and $p \geq m$ such that as $0 \leq t \leq T$

$$|x_t^1 - x_t^2|_{-p} = |x_0^1 - x_0^2|_{-p} + \int_0^t I_{(x_s^1 \neq x_s^2)} |x_s^1 - x_s^2|_{-p}^{-1} \{(x_s^1 - x_s^2, b^1(s,x_s^1) -$$
$$- b^2(s,x_s^2))_{-p} ds + (x_s^1 - x_s^2, (\sigma(s,x_s^1) - \sigma(s,x_s^2))dw_s)_{-p}\} +$$
$$+ \tfrac{1}{2} \sum_{j=1}^{\infty} \int_0^t \{ |(\sigma(s,x_s^1) - \sigma(s,x_s^2))Q^{*\frac{1}{2}} h_j^{-p}|_{-p}^2 \, |x_s^1 - x_s^2|_{-p}^2 -$$
$$- ((\sigma(s,x_s^1) - \sigma(s,x_s^2))Q^{*\frac{1}{2}} h_j^{-p}, x_s^1 - x_s^2)_{-p}^2 \} |x_s^1 - x_s^2|_{-p}^{-3} I_{(x_s^1 \neq x_s^2)} ds. \tag{3}$$

Theorem 3 can be proved similarly by the technique as [13]. We omit if here.

Applying theorem 3 one can derive the pathwise uniqueness theorem for (2):

Theorem 4. Assume that all conditions 1° and 2° in theorem 3 hold, and $b^1 = b^2 = b$, $x_0^1 = x_0^2$, moreover, as $x, y \in H^{-m}$, $|x|_{-p}, |y|_{-p} \leq N$, $t \in [0,T]$

3° $\|\sigma(t,x) - \sigma(t,y)\|^2_{L(H^{-p},H^{-p})} \leq k_N(t)\, G_N(|x - y|_{-p})\, |x - y|_{-p}$,

where $G_N(u) > 0$, as $u > 0$, it is increasing, concave and such that

$$\int_{0+} du/G_N(u) = \infty, \quad \int_0^T k_N(t)dt < \infty, \quad k_N(t) \geq 0, \; t \geq 0,$$

4° $I_{(x \neq y)}\, (x - y, b(t,x) - b(t,y))_{-p}/|x - y|_{-p} \leq k_N(t)\, G_N(|x - y|_{-p})$,

where $k_N(t)$ and $G_N(u)$ have the same property as that in 3°.

If two Φ'-valued processes with the same Φ' BM w_t satisfy (2) on the same probability space, then

$$P(\omega: x_t^1 = x_t^2, \text{ for all } t \geq 0) = 1.$$

Theorem 4 can be proved by the category argument [6] and the similar technique as [13].

Remark 3. Each of the following $G_N(u)$ satisfies condition in 3°:

$$G_N(u) = u, \; G_N(u) = u \ln(1/u), \; G_N(u) = u \ln(\ln(1/u)), \; \ldots, \text{ etc.}$$

Remark 4. $b(t,x)$, which satisfies 4°, can be discontinuous, e.g.

$$b(t,x) = -x/|x|_{-p}, \text{ as } x \neq 0; \quad b(t,x) = 0, \text{ as } x = 0;$$

where we assume that (1) has a weak solution and $m \geq r$ is such that $x_\cdot \in C([0,T]; H^{-m})$, and $p \geq m$. Then 4^0 holds for $b(t,x)$ with such p and T.
Since the Yamada - Watanabe theorem still holds for Φ'-valued SDE (see [6]), so applying it and by theorem 2 and 4 one obtains the following

Theorem 5. Assume that now $1^0 - 5^0$ in theorem 2 hold, and conditions $1^0 - 3^0$ in theorem 4 are fulfilled; moreover, 4^0 in theorem 4 holds for $\bar{b}$, where $\bar{b} = \sigma Q^* b$. Then SDE (2) with coefficients $\bar{b}$ and σ has a pathwise unique strong solutions. In case 6^0 in theorem 2 holds, and 4^0 in theorem 4 is fulfilled for $\tilde{b}$, where $\tilde{b} = Q^* b$, then SDE (2) has a pathwise unique strong solution with coefficients $\tilde{b}$ and σ.

Example 1. Assume that $\sigma(t,x)$ satisfies $2^0 - 6^0$ in theorem 2, then by theorem 2 (1) has a weak solution. For simplicity suppose that x_t with w_t satisfies (1). Furthermore, assume that conditions $1^0 - 3^0$ in theorem 4 for $\sigma(t,x)$ are fulfilled. Then by theorem 4 and the Yamada-Watanabe theorem (1) has a pathwise unique strong solution, denote it by x_t (with w_t) again. Now fix a $T > 0$, by the category argument [6] there exists a sufficiently large $m > 0$ such that the trajectories of solutions $x_\cdot \in C([0,T]; H^{-m})$, P - a.s. Take an index $p \geq m$, and set

$$b(t,x) = -|x|_{-p}^{N_0} Q_p^* x - Q_p^{*2} x / |Q_p^* x|_{-p} = b_1(x) + b_2(x),$$

where $N_0 \geq 2$ is any natural number. Then b satisfies 1^0 in theorem 1 and 4^0 in theorem 4 for this T and p. Indeed,

$$(x, b(t,x))_{-p} = -|x|_{-p}^{N_0} |Q_p^{*\frac{1}{2}} x|_{-p}^2 - |Q_p^* x|_{-p} \leq 0,$$

and as $|x|_{-p}, |y|_{-p} \leq N$, $x, y \in H^{-m}$

$$|b_1(x) - b_1(y)|_{-p} \leq \bar{k}_N |x - y|_{-p}, \quad \bar{k}_N \text{ - constant;}$$

$$(x - y, b_2(x) - b_2(y))_{-p} = -|Q_p^* x|_{-p} - |Q_p^* y|_{-p} + (Q_p^* x, Q_p^* y)_{-p} \cdot$$
$$\cdot (|Q_p^* x|_{-p}^{-1} + |Q_p^* y|_{-p}^{-1}) \leq 0.$$

Therefore theorem 5 is applied. (2) has a pathwise unique strong solution on $t \in [0,T]$ with such σ and b, where σ is non-Lipschitzian, non-monotonic, and b is discontinuous, greater than linear growth in x very much; e.g. if $e_0 \in H^{-p}$,

$$Q_p^* e_0 = \lambda_0 e_0, \quad |e_0|_{-p} = 1, \quad \lambda_0 > 0,$$

then set $x_n = n e_0$, one has

$$b(t,x_n) = -(n^{N_0+1} + 1) \lambda_0 e_0, \qquad |b(t,x_n)|_{-p} \geq \lambda_0 |x_n|_{-p}^{N_0+1}.$$

Stability Theorems

Applying theorem 3 one can derive stability theorems for solutions.

Theorem 6. Assume that for n = 0, 1, 2, ...

$$b^n = b^n(t,x,\omega): R_+ \times \Phi' \times \Omega \to \Phi', \quad \sigma = \sigma(t,x): R_+ \times \Phi' \times \Omega \to L(\Phi', \Phi')$$

for each $T > 0$ and sufficiently large $m > 0$ there exists an index $p \geq m$ such that as $x, y \in H^{-m}$, $0 \leq t \leq T$, n = 0, 1, 2, ...

1^0 $|b^n(t,x,\omega)|_{-p}^2 + \|\sigma(t,x,\omega)\|_{L(H^{-p},H^{-p})} \leq k_0(1 + |x|_{-p}^2)$,
where $b^n(t,\cdot,\omega): H^{-m} \to H^{-p}$, $\sigma(t,\cdot,\omega): H^{-m} \to L(H^{-p}, H^{-p})$;

2° 2° in theorem 3 holds for all $x, y \in H^{-m}$ with $k(t) = k_N(t)$ and $\rho(u) = \rho_N(u)$, which do not depend on N;

3° 3° in theorem 4 holds for all $x, y \in H^{-m}$ with $k(t) = k_N(t)$ and $G(u) = G_N(u)$, which do not depend on N, and in addition, G is strictly increasing and continuous;

4° 4° in theorem 4 for $b^0(t,x,\omega)$ holds as $x, y \in H^{-m}$, where $k(t)$ and $G(u)$ have the same property as that in 3°;

5° $|b^n(t,x,\omega) - b^0(t,x,\omega)|_{-p} \leq F^n(t,\omega)$, $\lim_{n\to\infty} E\int_0^T F^n(t,\omega)dt = 0$,

6° $\lim_{n\to\infty} E|x_0^n - x_0^0|_{-p} = 0$.

If Φ'-valued processes x_t^n satisfy (2) with coefficients b^n and σ, $n = 0, 1, 2, \ldots$ and for each $T > 0$ there exists a sufficiently large $m \geq r$ such that

$$x_\cdot^n \in C([0,T]; H^{-m}), \quad \text{for all } n = 0, 1, 2, \ldots, \qquad (*)$$

then for each $T > 0$ there exists an index p ($\geq m$) such that

$$\lim_{n\to\infty} E|x_t^n - x_0^0|_{-p} = 0, \quad \text{as } t \leq T.$$

Remark 5. If b^n and σ satisfy all conditions in theorem 5, then (*) holds.

Remark 6. Beware of that b^0 satisfying 4° can be discontinuous. Moreover, if set

$$G(u) = \begin{cases} u \ln(1/u), & \text{as } u < a, \\ (\ln(1/a) - 1)u + a, & \text{as } u \geq a, \end{cases}$$

where a is a constant satisfying $\ln(1/a) - 1 > 0$. Then $G(u)$ satisfies the conditions 3° and 4°. Besides, in this case $\rho(u) = u^{\frac{1}{2}}G(u)^{\frac{1}{2}}$ satisfies 2°.

Proof. Given $T > 0$, then (*) holds. By 1° applying Ito formula in H^{-p} and Gronwall inequality, we get that there exists a constant $k_T' \geq 0$ such that

$$E|x_t^n|_{-p}^2 \leq k_T', \quad \text{for all } n = 0, 1, 2, \ldots, \text{ as } t \leq T.$$

Hence as $t \leq T$

$$G(E|x_t^n - x_t^0|_{-p}) \leq G(2k_T') < \infty.$$

Therefore, by theorem 3 and Fatou lemma as $t \leq T$

$$\overline{\lim}_{n\to\infty} E|x_t^n - x_t^0|_{-p} \leq k_1 \int_0^t k(s)\, \overline{\lim}_{n\to\infty} G(E|x_s^n - x_s^0|_{-p})ds$$
$$\leq k_1 \int_0^t k(s)\, G(\overline{\lim}_{n\to\infty} E|x_s^n - x_s^0|_{-p})ds.$$

It yields that $\overline{\lim}_{n\to\infty} E|x_t^n - x_t^0|_{-p} = 0$, as $t \leq T$. Q.E.D.

We also have other stability theorems.

Theorem 7. Assume that

$$b: R_+ \times \Phi' \to \Phi', \quad \sigma: R_+ \times \Phi' \to \Phi'$$

for each $T > 0$ and sufficiently large $m \geq r$ there exists an index $p \geq m$ such that as $u, v \in H^{-m}$, $t \leq T$

1° $\sigma(t,u)v \in H^{-m}$, $|\sigma(t,u)v|_{-p} \leq k_0\, \bar{g}_{k+1}(u)^{\frac{1}{2}}\, |v|_{-p}$;

where $\bar{g}_{k+1}(u) = g_{k+1}(u) - 1$, and $g_{k+1}(u)$ is defined in 1° of theorem 1;

2° $b(t,\cdot): H^{-m} \to H^{-p}$, $|b(t,u)|_{-p} \leq k_0(1 + |u|_{-p} \prod_{i=1}^k g_i(u))\, \bar{g}_{k+1}(u)^{\frac{1}{2}}$;

3° conditions 3° - 6° in theorem 2 hold.

Then SDE (2) with coefficients σ and $\tilde{b}$, where $\tilde{b} = Q*b$, has a pathwise unique strong solution on $t \geq 0$ provided that $\tilde{b} = \tilde{b}^1 = \tilde{b}^2$, $x_0 = x_0^1 = x_0^2 \in H^{-m_0}$ is non-random. Moreover, if we denote this solution by $x(t,x_0)$, then for each $T > 0$ there exists an index p ($\geq m_0 \vee r$) such that for arbitrary $\varepsilon > 0$

$$\lim_{|x_0|_{-p}\to 0} P(\sup_{0\leq t\leq T} |x(t,x_0)|_{-p} > \varepsilon) = 0. \tag{4}$$

Remark 7. Since by assumption $b(t,0) = 0$, $\sigma(t,0) = 0$, 0 is evidently a pathwise unique strong solution of (2) with initial value $x_0 = 0$. Therefore (4) means that the solution of (2) is stable in probability on each finite interval in terms of [3].

The proof of theorem 7 needs some technique as that of theorem 1 etc. We omit it.

Theorem 8. Assume that all conditions in theorem 5 hold, where, in addition, assume that $G_N(u)$ in 3° and 4° of theorem 4 is increasing strictly, continuous; and $k_N = k_N(t)$ in 3° and 4° of theorem 4 does not depend on t, which is a constant depending on N only. Moreover, assume that

7° $b(t,0) = \theta$, $\sigma(t,0) = 0$, for all $t \geq 0$;

then the conclusion of theorem 7 holds.

Theorem 8 can be proved by theorem 3 and a little bit more discussion. We omit it here.

Martingale Representation Theorems and Maximum Principle

Let us give now a real-valued martingale representation theorem with respect to a Φ' - diffusion process for our purpose.

Theorem 9. Assume that all conditions in theorem 1 and 3 hold, and 3° in theorem 4 and 6° in theorem 2 also hold. If ξ_t is a real $\mathcal{F}_t^x$ - adapted square integrable martingale, where x_t satisfies Φ' - valued SDE (2) with $\tilde{b}$ and σ, where $\tilde{b} = Q^*b$, then for each $T > 0$ there exists a sufficiently large $m \geq r$ such that $x_. \in C([0,T]; H^{-m})$, and there exists an index $p \geq m$ and there exists a $\mathcal{F}_t^x$ - adapted H^{-p}-valued process $f(t,\omega)$ satisfying

$$\xi_t = \xi_0 + \int_0^t (f(s,\omega), dw_s)_{-p}, \quad \text{as } t \in [0,T],$$

where

$$E\int_0^T |Q_p^{*\frac{1}{2}} f(t,\omega)|_{-p}^2 \, dt < \infty.$$

The proof of theorem 9 depends on the following

Lemma 2. If N_t is a real $\mathcal{F}_t^w$ - adapted square integrable martingale with $N_0 = 0$, where w_t is a Φ' BM with the covariance operator Q, then for any $T > 0$ and $p \geq r$ there exists a $\gamma_t(\omega)$, $t \leq T$, which is $\mathcal{F}_t^w$-adapted and H^{-p}-valued satisfying

$$N_t = \int_0^t (\gamma_s, dw_s)_{-p}, \quad \text{for all } t \in [0,T],$$

where

$$E\int_0^T |Q_p^{*\frac{1}{2}} \gamma_s|_{-p}^2 \, ds < \infty.$$

The proof of lemma 2 and theorem 9, even it is similar to [10], needs more discussion. For saving pages we omit it here.

Now we are in a position to discuss the stochastic optimal control problem. From now on we shall always make the following assumpation:

(I) The conditions of theorem 5 (including 6° of theorem 2) for $\sigma(t,x)$ itself hold.

Applying theorem 5 we obtain that (1) has a pathwise unique strong solution x_t, which is $\mathcal{F}_t^w$ - adapted. By the category argument for given $T > 0$ there exists a sufficiently large $m \geq r$ such that $x(.) \in C([0,T]; H^{-m})$. In the following we shall fix this m. Consider now SDE (2) with drift coefficient as $\tilde{b} = Q^*b(t,x,u(t,x))$, where $u: [0,T] \times \Phi' \to V$ is jointly measurable, and V is some Hilbert space. We also make the following assumption:

(II) For given $T > 0$ and $m \geq r$ there exists an index $p \geq m$ such that as $t \leq T$

$$b(t,.,.): H^{-m} \times V \to H^{-p},$$

$$|b(t,x,u)|_{-p} \leq k_0(1 + |x|_{-p} \prod_{i=1}^{k} g_i(x)), \quad \text{as } x \in H^{-m},\ u \in V.$$

Then by theorem 2 (2) has a weak solution (x_t, w_t^u) for coefficients $\tilde{b}$ and σ on probability space $(\Omega, \mathcal{F}, (\mathcal{F}_t^w), P^u)$ as $t \in [0,T]$, where P^u is a probability measure defined by

$$dP^u = Z_T^u \, dP,$$

Z_T^u is defined in theorem 1 but with $b(t,x)$ substituted by $\sigma^{-1}b(s,x_s,u(s,x_s))$, and w_t^u is a Φ' BM under P^u such that (denote $\bar{b} = \sigma^{-1}b$)

$$w_t^u = w_t - \int_0^t Q_p^*\bar{b}(s,x_s,u(s,x_s))ds.$$

Denote now by $\mathcal{U}$ the totality of u mentioned above, i.e.

$$\mathcal{U} = \{u = u(t,\omega) = u(t,x_t(\omega)): \text{it is } \mathcal{F}_t^w \text{ - adapted, and } u(t,\omega) \in V, \text{ for all } t \in [0,T]\}$$

Introduce a metric d on $\mathcal{U}$ by setting

$$d(u,v) = (m \times P)\{(t,\omega) \in [0,T] \times \Omega : u(t,\omega) \neq v(t,\omega)\}, \quad \text{for } u, v \in \mathcal{U},$$

where m is the Lebesgue measure on $[0,T]$. Then d is a complete distance on $\mathcal{U}$. (See [1]). Consider the minimization of functional

$$F(u) = E_u\, g_T(x(.)) = \int_\Omega g_T(x(.))\, dP^u(\omega),$$

where

$$g_T(x(.)) = \int_0^T c(t,x_t)dt + h(x_T),$$

among all $u \in \mathcal{U}$. Let us make the following assumption:

(III) $c(t,x)$ and $h(x)$ are real, jointly measurable, and as $t \in [0,T]$, $x \in H^{-m}$

$$|c| + |h| + |b(t,x,u)|_{-p} + \|\sigma(t,x)\|_{L(H^{-p},H^{-p})} \leq k_0.$$

We have the following Maximum Principle.

Theorem 10. Under assumption (I) - (III) then

1) for any $\varepsilon > 0$ there exists a control $u \in \mathcal{U}$ such that

$$F(u^\varepsilon) \leq \inf_{u \in \mathcal{U}} F(u) + \varepsilon,$$

and for all $v \in \mathcal{U}$

$$F(v) \geq F(u^\varepsilon) - d(u^\varepsilon, v);$$

2) there exists a $\mathcal{F}_t^w$ - adapted process $\gamma_t^\varepsilon(\omega)$, which is H^{-p}-valued, and satisfies

$$P(\int_0^T |\gamma_t^\varepsilon|_{-p}^2 dt < \infty) = 1, \qquad (5)_1$$

and P^u - a.s. for all $t \in [0,T]$

$$E_{u^\varepsilon}(g_T(x(.))|\mathcal{F}_t^w) = F(u^\varepsilon) + \int_0^t (\gamma_s^\varepsilon, dw_s^{u^\varepsilon})_{-p}, \qquad (5)_2$$

such that for all $v \in \mathcal{U}$ (denote $\rho_0^t(b) = Z_t$, where Z_t is defined in theorem 1)

$$\rho_0^t(\bar{b}^{u^\varepsilon})(\gamma_t^\varepsilon, Q_p^*\bar{b}^v)_{-p} \geq \rho_0^t(\bar{b}^{u^\varepsilon})(\gamma_t^\varepsilon, Q_p^*\bar{b}^{u^\varepsilon})_{-p} - \varepsilon, \quad m \times P - \text{a.e.} \quad (5)_3$$

3) if u^ε with γ_t^ε verifies $(5)_1 - (5)_3$, then for all $v \in \mathcal{U}$

$$F(v) \geq F(u^\varepsilon) - \exp(2k_1T)\,\varepsilon T,$$

where $k_1 = k_0^2 \| Q_p \|_{L(H^{-p},H^{-p})}$, and k_0 comes from (III) and 6° of theorem 2;

4) if u^0 is an optimal control, then 2) holds for $\varepsilon = 0$;

5) if $u^0 \in \mathcal{U}$ and $\gamma_t^0(\omega)$ is a $\mathcal{F}_t^w$ - adapted process such that

$$E_{u^0}(g(x(.))|\mathcal{F}_t^w) = J(u^0) + \int_0^t(\gamma_s^0, dw_s^0)_{-p}, \quad (\text{set } w_t^0 = w_t^{u^0}),$$

i.e. γ_t^0 is the integrand in the martingale representation for the $E_{u^0}(g_T(x(\cdot))|\mathcal{F}_t^w)$

and if for all $v \in \mathcal{U}$ $m \times P$ - a.s.

$$(\gamma_t^0(\omega), Q_p^* \bar{b}(t,x_t(\omega),v_t(\omega)))_{-p} \geq (\gamma_t^0(\omega), Q_p^* \bar{b}'(t,x_t(\omega),u_t^0(\omega)))_{-p},$$

then u^0 is an optimal control.

To prove theorem 10 one has to apply the Ekeland lemma [1] and theorem 9. But for applying the Ekeland lemma one needs the following

lemma 3. $F(u)$: $(\mathcal{U},d) \to V$ is a continuous map under assumption (I) - (III).

Remark 8. The proof of lemma 3 of Elliot and Kohlmann [1] is incomplete. Since the following incorrect fact was used: d - convergence implies convergence a.e. with respect to m x P. A counter example is given by [16]. Hence it is doubtful that their lemma 3 is true under condition

$$\sigma^{-1}(t,x)b(t,x,u) \leq k_0(1 + |x|), \quad \text{as } t \in [0,T],\ x \in R^n,$$

where b is a n - dimensional vector, and σ is a n x n matrix.

Lemma 3 can be proved as [15]. To show our method let us just prove 3) of theorem 10 only.

Proof of 3). Note that for any $v \in \mathcal{U}$

$$\int_0^T (\gamma_t^\varepsilon, dw_t^v)_{-p} = \int_0^T (\gamma_t^\varepsilon, dw_t - Q_p^* \bar{b}_t^v)_{-p},$$

$$E\rho_0^t(\bar{b}^v - \bar{b}^{u^\varepsilon}) \exp(\int_0^t (|Q_p^* \bar{b}^{u^\varepsilon}|_{-p}^2 - (Q_p^* \bar{b}^v, Q_{pp}^{*} \bar{b}^{u^\varepsilon})_{-p})ds) \leq e^{2k_1T},$$

where $k_1 = k_0^2 \| Q_p \|_{L(H^{-p},H^{-p})}$. Therefore by $(5)_3$

$$E_v(F(u^\varepsilon) + \int_0^T (\gamma_t^\varepsilon, dw_t^{u^\varepsilon})_{-p} = F(u^\varepsilon) + E_v\int_0^T (\gamma_t^\varepsilon, Q_p^*(\bar{b}^v - \bar{b}^{u^\varepsilon}))_{-p}dt$$

$$\geq F(u^\varepsilon) - \varepsilon\int_0^T E\rho_0^t(\bar{b}^v)\rho_0^t(\bar{b}^{u^\varepsilon})^{-1}dt \geq F(u^\varepsilon) - e^{2k_1T}\varepsilon T.$$

Hence

$$F(v) = E_v(g_T(x(\cdot))) = E_v(E_{u^\varepsilon}(g_T(x(\cdot))|\mathcal{F}_T^w)) \geq F(u^\varepsilon) - e^{2k_1T}\,T\varepsilon. \quad \text{Q.E.D.}$$

REFERENCES

[1] Elliot, R.J. and M. Kohlmann. (1980). The variational principle and stochastic optimal control. Stochastics, 3, 229-241.

[2] Funaki, T. (1980). Random motion of strings and related stochastic evolution equations. Nagoya Math. J., 89, 129-193.

[3] Hazminskii, R.Z. (1980). Stochastic Stability of Differential Equations. Netherlands and Rockville, MD.

[4] Ikeda, N. and S. Watanabe. (1981). Stochastic Differential Equations and Diffusion Processes. North-Holland.

[5] Ito, K. (1984). Foundations of Stochastic Differential Equations in Infinite Dimensional Spaces. SIAM, Philadelphia.

[6] Kallianpur, G., Mitoma, I. and R.J. Wolpert. (Preprint). Diffusion equations in duals of nuclear spaces.

[7] Kallianpur, G. and R. Wolpert. (1987). Weak convergence of stochastic neuronal models. Stochastic Methods in Biology, Lecture Notes in Biomathematics 70, Eds. M. Kimura, G. Kallianpur and T. Hida, 116-145.

[8] Kallianpur, G. and V. Perez-Abreu. (1988). Stochastic evolution equations driven by nuclear-space-valued martingales. Appl. Math. Optim., 17, 237-272.

[9] Korezlioglu, H. and C. Martias. (1988). Stochastic integration for operator valued processes on Hilbert spaces and nuclear spaces. Stochastics 14, 171--219.

[10] Liptser, R.S. and A.N. Shiryayev. (1977). Statistics of Random Processes. Vol. 1. Springer-Verlag.

[11] Mitoma, I. (1981). Martingales of random distributions. Mem.Fac. Sci. Kyushu Univ. Ser. A., 35, 185-197.

[12] Perez-Abreu, V. (1985). Product stochastic measures, multiple stochastic integrals and their extensions to nuclear space valued Wiener processes. Tech. Rept. 107, Center for Stochastic Processes, Univ. Of North Carolina at Chapel Hill.

[13] Situ Rong. (1984). An application of local time to stochastic differential equations in m-dimensional space. Acta Sci. Naturali Univ. Sunyatseni, 3, 1-12. (In Chinese).

[14] Situ Rong. (1987). On weak, strong solutions and pathwise Bang-Bang control for non-linear degenerate stochastic system. IFAC Stochastic Control, USSR, 1986, Proceedings Series (1987), No. 2, Eds. N.K. Sinha and L. Telksnys, Pergamon Press, 145-150.

[15] Situ Rong and W.L. Chan. Existence of solutions and optimal control for reflecting stochastic differential equations with applications to population control theory. (1990). (To appear in Stochastic Analysis and Applications).

[16] Situ Rong. (1983). Non-convex stochastic optimal control and maximum principle. IFAC 3rd Symposium, Control of Distributed Parameter System 1982, Eds. J.P. Babary and L.L. Letty, Pagamon Press, 401-407.

[17] Tanaka, H. and M. Hitsuda. (1981). Central limit theorem for a simple diffusion model of interacting particles. Hiroshima Math. J. 11, 415-423.

SOME NEW RESULTS ON APPROXIMATE CONTROLLABILITY FOR SEMILINEAR SYSTEMS

H. W · SUN

GUNANGDONG INDUSTRY COLLEGE, GUANGZHOU, CHINA

Y. ZHAO

ZHONGSHAN UNIVERSITY, GUNANZHOU, CHINA

ABSTRACT

In this paper, the existence of the local or global solutions for an abstract semilnear time—variant system is discussed under weaker conditions. The approximate controllabity of the system is obtaied in two case respectively and the results have improved that in some papers published recently. Some examples are also given to illustrate the applications of the results.

I. INTRODUCTION

In this paper, we consider an abstrat semiliear system in a reflexive Banach X as follows

$$x(t) = U(t,0)x_0 + \int_0^t U(t,s)(N(x(s),u(s)) + Bu(s))ds \quad t \geqslant 0 \qquad (1.1)$$

where $x_0 \in X$, $U(t,s)$ is a mild evolution opoerator, control space U is another reflexive Banach space, $u(\cdot) \in L^p([0,\infty),U)(1<p<+\infty)$, $B \in L(U,X)$, $N(x,u)$ is a X—valued function from $X \times U$ to X.

Denote linear system correspoding to (1.1) as N=0 by

$$x(t) = U(t,0)x_0 + \int_0^t u(t,s)Bu(s)ds \qquad (1.2)$$

In the case of time—invariant systems, $U(t,s)$ is reduceed to be a C_0 semigroup $S(t-s)$ and the controllability of the system was discussed by some papers (4)—(10), but the conditions needed are restricted somewhat. That is, in addition to uniformly Lip condition for $N(x,u)$, $BU=X$(see(4)) or that $N(x,u)$ is uniformly bounded(see(5)—(8)) or that $S(t)$ is a compact(see(10)) is proposed. The results of these papers are mainly applied to discuss the approximate controllability with out regard to terminal time. We attempt to obtain the existence of the solution under weaker conditions and the results of approximate controllabilty under two cases: N is uniformly Lip or uniformly bounded for time—variant system(1.1)

Ⅱ. THE EXISTENCE OF SOLUTIONS

First, we list some hypotheses which will be used in the sequel:

(H1) : $U(t,s)$ is mild evolution operator on X and for any given $T<+\infty,(T>0)$, exist $M(T)>0$ such that

$$\max_{0\leqslant s\leqslant t\leqslant T} \| U(t,s) \| \leqslant M(T) \tag{2.1}$$

(H2) : (H_1) holds and (i) $U(t,s)(0\leqslant s<t<+\infty)$ is a compact opreator on X;

(ii) there exists $M>0$ such that

$$\sup_{0\leqslant s<t<\infty} \|U(T,s)\| \leqslant M \tag{2.2}$$

(N1): $N(x,u)=N(x)$ is unfiormly Lip with respect to x, that is, there exists $K>0$ such that

$$\|N(x_1) - N(x_2)\| \leqslant K\|x_1 - x_2\| \qquad for\ x_1,x_2 \in X \tag{2.3}$$

(N2): for any given $u\in L^p([0,\infty);U)$, $N(x,u(t))\triangleq F(f,x)$ satisfies

(i) $F(\cdot,x) \in L^p([0,\infty),X)$ for $x\in X$

(ii) for any given $c>0$ and $T>0$, there exists $L(c,T)>0$ such that

$$\|F(t,x_1) - F(t,x_2)\| \leqslant L(c,T)\|x_1 - x_2\| \qquad for\ t \in [0,T] \tag{2.4}$$

for any $x_1,x_2\in Bc(0)=\{x;x\in X, \|x\|\leqslant c\}\subset X$

(N3): For any given $u\in L^p([0,\infty);U)$, $N(u(t),x)=F(t,x)$ satisfies (i) in (N2) and

(i) it is continuous at any $x\in X$ unfiormly for $t\in[0,T]$;

(ii) it is L^p integrable on $[0,T]$ for any $x\in X$,

(iii) it maps bounded subset in $[0,\infty]\times X$ to bounded subset in X.

(N4): $N(x,u)$ is uniformly bounded with respect to $x\in X$ and $u\in U$ jointly.

In what follows, we present a series of results on the existence of local or global solutions for (1.1) under different hypotheses.

Theorem 2.1 If (H1) and (N1) hold, then for any given $u\in L^p([0,\infty);U)$ (1.1) has unique solution $x\in C([0,\infty);X)$.

Theorem 2.2 If (H2) and (N2) hold, then for any given $u\in L^p([0,\infty);U)$. There exists $\bar{t}\in[0,\infty)$ such that (1.1) has unique solution $x\in C([0,\bar{t});x)$. Moreover, if $\bar{t}\neq\infty$ and thes solutions of (1.1) can not be extended to the right side from, $\bar{t}$, then

$$\lim_{t\to\bar{t}-0} \|x(t)\| = \infty.$$

Corollary 2.1 If (H1), (N2) and (N4) hold, then for any given $u\in L^p([0.\infty);U)$ (1.1) has unique solution $x\in C([0,\infty);X)$.

Theorem 2.3 If (H2), (N3) hlod. The conclusions of theorem2.2 hold.

Corollary2.2 If (H2), (N3) and (N4) hold, The conclusions of corollary 2.1 hold.

The thorem 2.1 can be proved using generalized Banach contraction mapping principle by means of the methods similar to that in (1).

Proof of theorem 2.2: For any $t \geqslant t_0 \geqslant 0$, (1.1) can be rewritten as follows

$$x(t) = U(t,t_0)x(t_0) + f(t) - U(t,t_0)f(t_0) + \int_{t_0}^{t} U(t,s)F(s,x(s))ds$$

$$\triangleq g(t,t_0,x(t_0)) + \int_{t_0}^{t} U(t,s)F(s,x(s))ds \tag{2.5}$$

where $f(t) = U(t,0)x_0$.

Since $F(t) \triangleq (\int_0^t \|\tilde{F}(s,0)\|^p ds)^{\frac{1}{p}}$ is continuous with respect to t, so for any given $T < \infty$, It exists $M_F(T) > 0$) such that

$$\max_{t \in [0,T]} \|\tilde{F}(t)\| \leqslant M_F(T) \tag{2.6}$$

Also, it exists $M_f(T) > 0$ such that

$$\|f(\cdot)\|_{C([0,T],X)} \leqslant M_f(T) \tag{2.7}$$

beacuse $f \in C([0,T],X)$

Consequently, we have

$$\|g(t,t_0,x(t_0))\| \leqslant M_f(T)(1 + M(t_0+1) + M(t_0+1)\|x(t_0)\|) \triangleq N(\|x(t)\|) \tag{2.8}$$

for $t \in [t_0, t_0+1]$ by (2.1) and (2.7)

Let

$$K(t_0) = 2N(\|x(t_0)\|) \tag{2.9}$$

and take

$$\delta = \min(1, (N(\|x_0\|)/M(t_0+1)(L(K(t_0),t_0+1)K(t_0) + M_F(t+1))^q) \tag{2.10}$$

where $1/p + 1/q = 1$.

Define mapping $Q: Y = C([t_0, t_0+\delta]; B_{K(t_0)}(0)) \to C([t_0,t_0+\delta]; X)$ as follows

$$Q(x)(t) = g(t,t_0,x(t_0)) + \int_{t_0}^{t} U(t,s)F(s,x(s))ds \tag{2.11}$$

According to (2.4), (2.6), (2.8), (2,9) and (2.10), we see that $\|Q(x)(t)\| \leqslant K(t_0)$ for any $X \in Y$ which implies that Q maps Y into itself.

On the other hand, for $x_1, x_2 \in Y$, one can deduce that

$$\|(Q^n(x_1)(\cdot) - Q^n(x_2)(\cdot)\|_Y \leqslant ((M(t+1))L(K(t),T+1)^n/n!)$$
$$(\|x_1(\cdot) - x_2(\cdot)\|_Y) \triangleq m\|x_1(\cdot) - x_2(\cdot)\|_Y$$

As n is taken large enough we have $m < 1$, So Q has fixed point in Y by Banach generized contraction principle and (1.1) has unique solution $\in Y$.

The reainder can be proved by the methods similar to that in (1).

The coroll ary 2.1 follows theorem 2.2 and (N4) immediately.

Proof of theorem 2.3: We start to work with (2.5) and assume that it exists $x(t_0)$ For any given $\rho > 0$, let $B\rho(t_0) = S(\rho; x(t_0)) \subset X$ and $t \in [t_0, t_0+1]$. By (N3) it exists $N(t_0) > 0$ such that

$$\|F(t,x)\| \leqslant N(t_0) \quad \text{for} \quad (t,x) \in [0, t_0+1] \times B\rho(t_0)$$

Noting that $g(t,t_0,x(t_0))$ is continuous at $t=0$ and $g(t_0,t_0,x(t_0))=x(t_0)$ one can see that there exists $\delta'>0$ such that

$$\|g(t,t_0,x(t_0)) - x(t_0)\|_X \leqslant \frac{\rho}{2} \tag{2. 12}$$

Let

$$\delta = min(1,\delta',\rho/2 \cdot MN(t_0)) \tag{2. 13}$$

where M is as in (H2).

Let

$$Y = C([t_0,t_0+\delta];X), Y_0 = (z; z \in Y, z(t_0) = x(t_0), z(t) \in B\rho(t_0) \ for \ t \in [t_0,t_0+\delta])$$

It is easy to check that Y_0 is a bounded convex subset in Y. Define mapping $P: Y \to Y$ as follows

$$(Px)(t)) = g(t,t_0,x(t_0)) + \int_{t_0}^{t} U(t,s)F(s,x(s))ds \tag{2. 14}$$

One can work out that (2. 14)

$$\|(px)(t) - x(t_0)\| \leqslant \rho$$

for any $x \in Y_0 \subset Y$. One can see that P is a continunous mapping from Y_0 to Y_0 by (N3) and $PY_0 = \tilde{Y}(Px; x \in Y_0)$ is compact in X and consists of equicontinous family in Y by regular conputation and estimation.

Hence $\tilde{Y}$ is a compact subest in Y by Arzeia—Ascoli theorem. Furthermore, P has a fixed point in Y_0 by Scharder's fixed point theorem. It implies that (2. 5) has unique solution $x \in Y_0$.

The remainder can proved by methods similar to that in (1).

The corollary 2. 2 follows theorem 2. 3 and (N4) immediateiy.

Ⅲ. THE APPROXIMATE COMTROLLABILITY OF (1. 1)
——THE CASE OF N (x,u) BEING UNIFORMLY BOUNDED

Introduce the "blocking" system correopoding to (2. 1) as follows

$$y(t) = y(t,t_0,y_0,u) = U(t,t_0)y_0 + \int_{t_0}^{t} U(t,s)Bu(s)ds \qquad t \geqslant t_0 \tag{3. 1}$$

and express the solution of (1. 1) by $x(t)=X(t,x_0,u)$.

Let

$$G(t_1,t_2)p(\cdot) = \int_{t_1}^{t_2} U(t_2,s)p(s)ds \quad : \quad L^p([0,\infty),X) \to X \tag{3. 2}$$

Where $t_2>t_1$, $t_1,t_2 \in [0,\infty)$

$$K(t_0,T,y_0) = (\xi_T : \xi_T = y(T,t_0,y_0,u), u \in L^p([t_0,T);U))$$

$$K_N(T,x_0) = (\xi_T = x(T,x,u), u \in L^p(0,T;U))$$

Definition 3. 1 (3. 1) is called approximately controllable on $[t_0,T]$ with initial value y if $K(t,T,y_0)=X$; (1. 1) is called approximately controllable on $[t\ ,T]$ with initial value x_0 if $K_N(T,x_0)=X$

As well known we have

Lemma 3.1 $K(t_0,T,y_0)=X$ iff $R(G(t_0,T)B)=X$

Now, introduce some hypotheses as follows

(B_1) : $R(G(t_1,t_2)B)=X$ for $t_2>t_1\geqslant 0$

(B2) : $\| G(t_1,t_2)N(x,u) \|_x=0(1)$ with respect to $(t_2-t_1)(t_2>t_1\geqslant 0)$ uniformly for $x \in X$ and $u\in U$.

(F) : (1.1) has unique solution $x\in C([0,\infty);X)$ for nay given $u\in L^p([0,\infty);U)$ where $N(x(\cdot),u(\cdot))\in L^p([0,\infty);X)$.

Theorem 3.1 Assume that (B1), (B2) and (F) hold, then $K_N(T,x_0)=X$ for any gaven $T>0$ and $x_0\in X$.

Proof : It is only necesary to prove that for any given $\eta\in X$ and $\varepsilon>0$, there exists $u\in L^p(,T;U)$ such that

$$\| x(T,x_0,u) - \eta \| < \varepsilon \tag{3.3}$$

Let $t_n=(1-2^{-n})T$ so that $t_n\to T$ as $n\to\infty$. We have that

(i) There exists $v_0(\cdot)\in L^p(0,T;U)$ such that

$$y(T,t_0,x_0,v_0(\cdot)) - \eta\|_X < \frac{\varepsilon}{2}$$

by (B1).

(ii) After taking $x_1=x(t_1,x_0,v_0(\cdot))$, there exists $v_1(\cdot)\in L^p(t_1,T;U)$ such that

$$\|y(T,t_1,x_1,v(\cdot)) - \eta\| < \frac{\varepsilon}{2}$$

by (B1). Let

$$u(t) = \begin{cases} v_0(t) & 0\leqslant t< t_1 \\ v_1(t) & t_1\leqslant t\leqslant T \end{cases}$$

and we can see that $u_1(\cdot)\in L^p(0,T;U)$

(iii) After taking $x_2=x(t_2,x_0,u(\cdot))$, there exists $v_2(\cdot)\in L^p(t,T;U)$ such that

$$\| y(T,t_2,x_2,v_2(\cdot)) - \eta \| < \frac{\varepsilon}{2}$$

by (B1). Let

$$u_2(t) = \begin{cases} u_1(t) & 0\leqslant t< t_2 \\ v_2(t) & t_2\leqslant t\leqslant T \end{cases} \qquad u_2(\cdot) \in L^p(0,T;U)$$

(iv) Repeating above procedures we have that $v_n(\cdot)\in L^p(t,T;U)$ such that

$$y(T,t_n,x_n,v_n) - \eta \| < \frac{\varepsilon}{2} \tag{3.4}$$

and

$$u_n(t) = \begin{cases} u_{n-1}(t) & 0\leqslant t< t_n \\ v_n(t) & t_n\leqslant t\leqslant T, \end{cases} \qquad u_n(\cdot) \in L^p(0,T;U) \tag{3.5}$$

(v)According to (B2), we have

$$\lim_{n\to\infty} G(t_n, T)N(x(\cdot), u(\cdot)) = 0$$

so that there exsts N such that

$$\|G(t_n, T)N(x(\cdot), u_n)\| < \frac{\varepsilon}{2} \quad (3.6) \qquad as\ n > N.$$

(vi) Taking $u_{N+1}(\cdot) \in \{u_n(\cdot)\} \subset L^p(0, T; U)$ we have

$$\|x(T, x_0, u_{N+1}(\cdot)) - \eta\| \leqslant \|x(T, x_0, u_{N+1}(\cdot) - y(T, t_{N+1}, x_{N+1}, v_{N+1}\|$$
$$+ \|y(T, t_{N+1}, x_{N+1}, v_{N+1}) - \eta\|$$

By(1. 1), (3. 1), and (3, 6) we can deduce that (3. 3) hold as $u(\cdot) = u_{N+1}(\cdot)$ which completes the proof.

Theorem 3. 2 If (H1), (N4) and (B1)hold then $K_N(T, x_0) = X$ for any given $x_0 \in X$ and $T > 0$.

Proof : Noting that (B2) is valid as (N4) hold, the conçlusion follows from corollary 2. 1 and theorem 3. 1.

Similary, one has

Theorem 3. 3 If (H2), (H3), (H4)and (B_1) hold then $K_N(T, x_0) = X$ for any given $x_0 \in X$ and $T > 0$.

Example 3. 1[6]: Let $X = L^2(0, \pi)$, $e_n(x) = \sqrt{2/\pi}\sin nx$, $n = 1, 2, \cdots$, where $\{e_n\}$ is a family of orthogonal basis in X and

$$U\{u; u = \sum_{n=2}^{\infty} u_n e_n, \sum_{n=2}^{\infty} u_n^2 < \infty\}, \|u\|_u = (\sum_{n=1}^{\infty} u_n^2)^{1/2}$$

Consider a heat—conduction equation as follows

$$\begin{cases} \dfrac{\partial y(t,x)}{\partial t} = \dfrac{\partial^2 y(t,x)}{\partial x^2} + F(t, y(t,z)) + Bu(t,x) & t \in (0,T) \\ & x \in (0,\pi), \\ y(t,0) = y(t,\pi) = 0 \qquad t \in [0,T] \\ y(0,x) = y_0(x) \qquad x \in [0,\pi] \end{cases} \quad (3.7)$$

Let $Ay = -(y'')$ so that $(-A)$ generates a compact semigroup of contraction S(t).

The mild solution of (3. 6) is expressed by

$$y(t) = S(t)y_0 + \int_0^t S(t-s)(F(y(s)) + Bu(s))ds \quad (3.8)$$

Let

$$Bu = 2u_1 e_1 + \sum_{n=2}^{\infty} u_n e_n \quad (3.9)$$

It follows from above hypotheses that(H2) and (B1) hold. If F(y) is continious and bounded uniformly for x, then (N3) and (N4) hold so that (3. 7) is approxemately controllable for any given $T > 0$ by theorem3. 3 without uniformly Lip condition even local Lip condition for F. Say,

$$F(y) = F(\sum_{n=1}^{\infty} y_n e_n) = \begin{cases} \sin y_1 \sin \frac{1}{y_1} e_1 & y_1 \neq 0 \\ 0 & y_1 = 0 \end{cases} \tag{3,10}$$

This result can not be obtained from (5),(8) or(10).

Example 3. 2 If S(t) is only a semigroup in example 3. 1 and

$$F(y,u) = F(\sum_{n=1}^{\infty} y_n e_n, \sum_{n=2}^{\infty} u_n e_n) = \cos y_1^2 e_1 + \cos u_2^2 e_2$$

instead of F(y), it is easy to chech that (H1),(H2),(N4) and (B1) hold. So,the system is approximalely controllable by Theorem 3. 2 without uniformly Lip condition for F(x,u), This can not be obtained from(4),(8),or(10).

VI. THE APPROXIMATE CONTROLLABILITY OF(1. 1) ——THE CASE OF N(x,u) SATISFIYING UNIFORMLY Lip CONDITION

In this case,N(x,u)=N(x) satisfies(N1) and (1. 1)becomes

$$x(t) = U(t,0)x_0 + \int_0^t U(t,s)(N(x) + Bu(s))ds \tag{4. 1}$$

Theorem 4. 1 If (N1) and (B1) hold,then (4. 1) is approimately controllable for any given $x_0 \in X$ and $T>0$.

Proof: Similar to the proof in theorem 3. 1 from the begining to the end of (iv). We have(3. 4) and (3. 5). Consequently, we have

$$\begin{aligned} \| x(t)-\eta \| &\leqslant \| y(t,t_n,x_n,v_n)-\eta \| + \| \int_{t_n}^t U(t,s)Nx(s))ds \| \\ &\leqslant \| y(t,t_n,x_n,v_n)-\eta \| \\ &\quad + M(T)(\| N(0) \| + K \| \eta \|)(t-t_n) + M(t)K\int_{t_n}^t \| x(s)-\eta \| \end{aligned}$$

ds

for $t \in [t_n,T]$ where K is Lip constant and

$$M(T) = \max_{0 \leqslant s \leqslant t \leqslant T} \| U(t,s) \|$$

Furthemore,by Growall's inequality one has

$$\| x(t) - \eta \| \leqslant e^{M(T)Kt}[\| (y(t,t_n,x_n,v_n) - \eta \| + M(T)(\| N(0) \| + K \| \eta \|)(t - t_n)] \tag{4. 2}$$

Taking t=T in (4. 2),it follows that

$$\| x(T) - \eta \| \leqslant e^{M(T)KT}(\frac{\varepsilon}{2} + M(T)(\| N(0) \| \eta \|)(T - t_n))$$

So,there exists $N>0$ and $u(t)= u_{N+1}(t)$ such that

$$\|x(T) - \eta\| \leqslant e^{M(T)KT}(\frac{\varepsilon}{2} + \frac{\varepsilon}{2}) = e^{M(T)KT}\varepsilon$$

that complets the proof.

In example 3. 1 if F(y) satisties (N1), the system is approximately controllable without any other conditions. This rasult can not be obtained from(5)—(10).

REFERENCE

1. Pazy. A., Semigroups of linear Operators to Partial Dfferential Equations, Lecture Notes 10, Dept. of Math, University of Maryland, 1974.

2. Curtain. R. F., and Prirchard . A. T., Infinte Dimensional Linear Syetems Theory, Springer—verlag, New York, 1978.

3. Russell. D. L. Controllability and Stabilizability Theory for Linea Partical Differential E-quations; Recent Progress and Open Question. SIAM Rievew 20(1978) PP639—739

4. Henry, J; Etudc ole La Controlabilite Eole Certains Equtains Pareboliques Non — Lin-eaires. These, detat, Paris, June, 1978.

5. Zhou. H. X, A note on Approxmate Controllability for Semilinear one dimensional Heat Equation; Appl. math. option 8 (1982), pp275—285.

6. ——, Applroximate Controllability for a class of Semilinear Abstract Equation, SIAM, J. Control and Option, (21)(1983)pp551—565.

7. ——, Controllability Properties of Linear and Semilinear Abstract Control Systems, I bid 22(1984), pp405——422.

8. ——, Approximate Controllability on Aastract Control systems with a Nonliear Ditur-bance, IFAC 84.

9. H. Zhou, andY. Zhao. A Survey of Conrtollability Theory for Nonlinear systems, Con-trol Theory and Applications, Vol. 5. No2. 1988.

10. Naito. K. Cntrollability of Semilinear Control Systems dominted by the linear part, SIAM, J, Control and Optim, 25 (1987) pp715—722.

OPTIMAL CONTROL FOR A CLASS OF SYSTEMS AND ITS APPLICATIONS IN THE POWER FACTOR OPTIMIZATION OF THE NUCLEAR REACTOR†

Wang Miansen Kuang Zhifeng
Department of Mathematics , Xi'an Jiaotong University
Xi'an, Shaanxi, 710049, China
Zhu Guangtian
Institute of Systems Sciences, Acadimia Sinica
Beijing 100080, China

1. Introduction

On the background of the power factor optimization of the nuclear reactor, the optimal control for a class of systems governed by the eigenequation

$$L\varphi + \sigma\varphi = \frac{1}{\lambda}K\varphi$$

has been presented in [1]. Under the variance index, [1] shows the existence of optimal control and gives the corresponding optimality conditions. But these results can't be applied to those systems in which the neutron density is relative with the direction of the speed. In this paper, we put these results into so general case that they can be applied to all stable transport systems.

We start from a example.

Example 1. Consider the following energy-dependant steady-state transport equation in a slab with generalized reflexive boundary conditions:

$$\begin{cases} v\mu\dfrac{\partial N(x,v,\mu)}{\partial x} + v\Sigma(x,v)N(x,v,\mu) = \dfrac{1}{\lambda}\displaystyle\int_0^{v_m}\int_{-1}^{1} k(x,v,v',\mu,\mu')N(x,v',\mu')dv'd\mu' & (1.1) \\ |x|< a, \quad |\mu|< 1, \quad v\in(0,v_m), \quad v_m<+\infty & \\ N(-a,v,\mu) = \alpha(v,\mu)N(-a,v,-\mu), \quad 0\le\mu\le 1 & (1.2) \\ N(a,v,-\mu) = \beta(v,\mu)N(a,v,\mu), \quad 0\le\mu\le 1 & (1.3) \end{cases}$$

where $N(x,v,\mu)$ denotes the neutron density with the velocity v and direction μ at the point x; $\Sigma(x,v)$ is the macro-absorption cross section; $k(x,v,v',\mu,\mu')$ is the transfer kernel; $\alpha(x,v)$ and $\beta(x,v)$ are the reflexive coefficients.

If we suppose that $\Sigma(x,v)$ is composed by two parts: $\Sigma(x,v) = \Sigma_a(x,v) + \Sigma_c(x,v)$, where $\Sigma_a(x,v)$ is a fixed part, and $\Sigma_c(x,v)$ is a controllable part realized by regulating the control bars, and denote the following:

$$\Omega_1 = (-a,a)\times(0,v_m), \Omega_2 = (-1,1), \Omega = \Omega_1\times\Omega_2$$

$$L\cdot = v\mu\frac{\partial}{\partial x}\cdot + v\Sigma_a(x,v).$$

$$D(L) = \{\varphi\in L^2(\Omega) \mid \varphi \textit{ is absolutely continuous in } x, L\varphi\in L^2(\Omega), \varphi \textit{ satisfies boundary conditions } (1.2) \textit{ and } (1.3)\}$$

$$K\cdot = \int_0^{v_m}\int_{-1}^{1} k(x,v,v',\mu,\mu')\cdot dv'd\mu'$$

$$\sigma\cdot = v\Sigma_c(x,v)\cdot, \qquad D(K) = D(\sigma) = L^2(\Omega)$$

†This work is supported by the National Natural Science Fundation of China

then above system becomes

$$L\varphi + \sigma\varphi = \frac{1}{\lambda}K\varphi \tag{1.4}$$

According to the reactor theory[2], in order that reactor operator safety and efficiently, the equation (1.4) should be in a critical state (i.e. $\lambda(\sigma) = r = 1$) meanwhile the power factor

$$J_0(\sigma) = \max_{\Omega} \frac{(K\varphi)(x, v, \mu)}{(h, \varphi)} \tag{1.5}$$

takes the minimum and total power mains constant (i.e. $(h, \varphi) = p$). This is the so-called power factor opimization problem. For simplicity, instead of the index (1.5), we adopt the variance index

$$J(\sigma) = \frac{1}{2}\int_{\Omega} | \varphi(\sigma) - M\varphi(\sigma) |^2 d\Omega$$

where

$$M\varphi = \frac{1}{m(\Omega)}\int_{\Omega} \varphi(\sigma)(x, v, \mu)d\Omega$$

$m(\Omega)$ denotes the Lebesgue measure of Ω.

From this example, we present the optimal control problem for a class of systems. Let's state our problem more precisely as follows.

Let $\Omega_1 \subset R^n$ and $\Omega_2 \subset R^m$ be bounded Lebesgue measurable domains, $\Omega = \Omega_1 \times \Omega_2$, $L^2(\Omega)$ be a real Hilbert space with inner product

$$(f, g) = \int_{\Omega} f(x)g(x)dx$$

K and L be linear operators in $L^2(\Omega)$. Consider the systems with the state $(\lambda(\sigma), \varphi(\sigma))$

$$\begin{cases} L\varphi + \sigma\varphi = \frac{1}{\lambda}K\varphi & (1.6) \\ (k, \varphi) = p & (1.7) \end{cases}$$

where $\lambda(\sigma)$ is the critical eigenvalue of the operator $T(\sigma) = (L + \sigma)^{-1}K$ and σ is the controllable variation. The control set is defined by

$$U = \{\sigma \in L^{\infty}(\Omega_1) \mid 0 \leq a(x) \leq \sigma(x) \leq b(x) a.e. in \Omega_1\}$$

with the designated function $a(x)$ and $b(x)$ in $L^{\infty}(\Omega_1)$ and $b(x) \neq ma(x)$.

Given the cost functional

$$J(\sigma) = \frac{1}{2}\int_{\Omega} | \varphi(\sigma) - M\varphi(\sigma) |^2 \, d\Omega$$

We consider the following problem (P): Given a positive number r, we want to find a control $\sigma_0 \in U$ such that

$$\lambda(\sigma_0) = r, \qquad J(\sigma_0) = \inf_{\sigma \in U} J(\sigma)$$

In the case that $\Omega = \Omega_1$, the problem has been solved in [1]. But $\Omega \neq \Omega_1$ in example 1, and their methods is of useless.

2. Existence of the Optimal Control

In this section, we prove the existence of the optimal control by use of compact imbedding theorem[3]. First we give following hypertheses:

(H1) Ω_1, Ω are the domains of the class $J_{\alpha_1}, J_\alpha (1-\frac{1}{n} \le \alpha_1 \le 1, 1-\frac{1}{n+m} \le \alpha \le 1)$[3] , respectively;

(H2) p is a positive number, h is a positive function in $L^2(\Omega)$, L is a closed linear operator, and K bounded linear operator under the norm $\|\cdot\|_1 = \int_\Omega |\cdot(x)|\, dx$;

(H3) for any $\sigma \in U, (L+\sigma)^{-1}$ exists and is a positive operator, $T(\sigma) = (L+\sigma)^{-1}K$ is a positive compact operator in $L^2(\Omega)$; and the critical eigenvalue $\lambda(\sigma)$ of $T(\sigma)$ exists.

Lemma 2.1 For any $\sigma \in U$, there exixts a unique positive function which satisfies (1.6) and (1.7).

Proof This is a direct consequence of (H3) and Theorem 2.1 in [4].

Denoting $U_0 = \{\sigma \in U \mid \lambda(\sigma) = r\}$ as the admissible control set, we have

Lemma 2.2 U_0 is an infinite set, if $\lambda(a(x)) > r > \lambda(b(x))$.

Proof For any σ_1 and σ_2 in U, since

$$T(\sigma_1) - T(\sigma_2) = (L+\sigma_1)^{-1}K - (L+\sigma_2)^{-1}K = (L+\sigma_1)^{-1}(\sigma_2-\sigma_1)(L+\sigma_2)^{-1}K \le 0$$
$$as \;\; \sigma_1(x) \ge \sigma_2(x) \;\; a.e. \;\; in \;\; \Omega_1$$

By Theorem 4.2 in [4], $\lambda(\sigma_1) \le \lambda(\sigma_2)$. If we define $\lambda_1(t) = \lambda(ta(x))$, $(t \ge 1)$, then $\lambda_1(t)$ is a nonincreasing continuous function by Lemma 3.2.

Therefore, there exists a sequence $\{\delta_k\}_1{}^\infty, \delta_k \to 1^+$, such that $\lambda_1(\delta_k) = \lambda(\delta_k a(x)) > r$ and for every δ_k, we consider the nonincreasing continuous function:

$$\lambda_2(t) = \lambda(\delta_k a(x) + t(b(x) - \delta_k a(x))), \qquad t \in [0,1]$$

It's easy to know that there exists a $t_0 \in (0,1)$ or $\sigma_{k_0} \in U$ such that $\lambda(\sigma_{k_0}) = r$ and $\sigma_{k_0} \ne \sigma_{k_0'}$, if $\sigma_k \ne \sigma_{k'}$. Thus the Lemma is right.

Define a subset in $L^2(\Omega)$:

$$W = \{\varphi \in L^2(\Omega) \mid \varphi > 0, there \;\; exists \;\; \sigma \in U \;\; such \;\; that \;\; \varphi \;\; is \;\; the$$
$$solution \;\; to \;\; L\varphi + \sigma\varphi = \frac{1}{r}K\varphi \qquad (h,\varphi) = p\}$$

Then by Lemma 2.2, we have

Lemma 2.3 If $\lambda(a(x)) > r > \lambda(b(x))$, then W is a infinity set.

Note W isn't a convex set here, but W is convex in [1].

It's easy to know the problem (P) is equivalent to the minimization problem:

$$\begin{cases} inf \hat{J}(\varphi) \\ s.t. \varphi \in W \end{cases}$$

where $\hat{J}(\varphi)$ is defined as

$$\hat{J}(\varphi) = \frac{1}{2}\int_\Omega |\varphi(x) - M\varphi|^2\, dx$$

Lemma 2.4[1] $J(\varphi)$ is a continuous convex functional in $L^2(\Omega)$.

Lemma 2.5 $J(\varphi)$ is a weakly lower semi-continuous functional.

This is the conclusion of Lemma 2.4 and Cor. 1.8.6 in [5].

Lemma 2.6 U is a bounded weakly closed subset in $L^2(\Omega_1)$.

Proof Obviously, U is bounded.

It's enough to show that U is strongly closed because of the convexity.

Let $\{\sigma_n\}$ be an arbitrary strongly convergent sequence in $U \subset L^2(\Omega_1)$ and $\sigma_n \to \sigma_0$ strongly in $L^2(\Omega_1)$. Then, by Cauchy inequelity, for any $\varepsilon > 0$

$$\sqrt{m(\Omega_1)}\left(\int_{\Omega_1} |\sigma_n(x) - \sigma_0(x)|^2 dx\right)^{\frac{1}{2}} \geq \int_{\Omega_1} |\sigma_n(x) - \sigma_0(x)| dx$$
$$\geq \int_{\Omega_1(|\sigma_n - \sigma_0| \geq \varepsilon)} |\sigma_n(x) - \sigma_0(x)| dx \geq \varepsilon m(\Omega_1(|\sigma_n - \sigma_0| \geq \varepsilon)).$$

Let $n \to \infty$, $m(\Omega_1(|\sigma_n - \sigma_0| \geq \varepsilon)) \to 0$. By F. Riesz Theorem, there exists a convergent subsequence $\{\sigma_{n_k}\} \subset \{\sigma_n\}$, such that

$$|\sigma_{n_k}(x) - \sigma_0(x)| \to 0 \qquad a.e. \quad in\ \Omega_1, \quad as\ k \to \infty \tag{2.1}$$

Since $a(x) \leq \sigma_{n_k}(x) \leq b(x)$, and hence $a(x) \leq \sigma_0(x) \leq b(x)$ or $\sigma_0 \in U$.

Lemma 2.7 W is a weakly closed subset in $L^2(\Omega)$.

Proof Let $\{\varphi_n\}$ be a weakly convergent sequence in W, and $\varphi_n \to \varphi_0$ weakly in $L^2(\Omega)$. By the definition of W, there exists a sequence $\{\sigma_n\}$ in U, such that

$$\begin{cases} L\varphi_n + \sigma_n\varphi_n = \frac{1}{r}K\varphi_n & (2.2)\\ (h, \varphi_n) = p & (2.3)\end{cases}$$

It's obvious that $(h, \varphi_0) = p$ from (2.3).

By Lemma 2.6, there exists a subsequence $\{\sigma_{n_k}\} \subset \{\sigma_n\}$ such that $\sigma_{n_k} \to \sigma_0$ weakly in $L^2(\Omega_1)$.

Because the embedding operators $I_1 : L^2(\Omega) \to L^1(\Omega)$, and $I_2 : L^2(\Omega_1) \to L^1(\Omega_1)$ are compact[3],

$$\varphi_n \to \varphi_0 \qquad strongly\ in \quad L^1(\Omega) \tag{2.4}$$

$$\sigma_{n_k} \to \sigma_0 \qquad strongly\ in \quad L^1(\Omega_1) \tag{2.5}$$

By the process of Lemma 2.6, there exists a subsequence of $\{\sigma_{n_k}\}$, written still $\{\sigma_{n_k}\}$, such that (2.1) is right. Then $|(\sigma_{n_k}(x) - \sigma_0(x))\varphi_0(x,y)| \to 0$ a.e. (in Ω) and $|(\sigma_{n_k}(x) - \sigma_0(x))\varphi_0(x,y)| \leq 2b(x)|\varphi_0(x,y)|$, by Lebesgue Theorem,

$$\iint_{\Omega_1 \times \Omega_2} |(\sigma_{n_k}(x) - \sigma_0(x))\varphi_0(x,y)| dxdy \to 0, \qquad as\ k \to \infty$$

or $\|(\sigma_{n_k} - \sigma_0)\varphi_0\|_1 \to 0$, as $k \to \infty$. From (2.2),

$$L\varphi_{n_k} = \frac{1}{r}K\varphi_{n_k} - \sigma_{n_k}\varphi_{n_k} = \frac{1}{r}K\varphi_{n_k} - \sigma_{n_k}(\varphi_{n_k} - \varphi_0) - (\sigma_{n_k} - \sigma_0)\varphi_0 - \sigma_0\varphi_0$$
$$\to \frac{1}{r}K\varphi,\varphi_0 - \sigma_0\varphi_0 \qquad strongly\ in\ L^1(\Omega) \tag{2.6}$$

Combining (2.4), (2.6) and (H2), we have

$$\varphi_0 \in D(L) \qquad L\varphi_0 = \frac{1}{r}K\varphi_0 - \sigma_0\varphi_0$$

so $\varphi_0 \in W$. That is W is weakly closed.

Lemma 2.8[1] Each minimizing sequence of $\hat{J}$ in W is bounded.

Theorem 2.1 If $\lambda(a(x)) > r > \lambda(b(x))$, there exists $\varphi_0 \in W$ such that $\hat{J}(\varphi_0) = \inf_{\varphi \in W} \hat{J}(\varphi)$.

Proof Let $\{\varphi_n\}$ be a minimizing sequence of $\hat{J}$ in W, that is $\lim_{n\to\infty} \hat{J}(\varphi_n) = inf_{\varphi \in W} \hat{J}(\varphi)$.

According to Lemma 2.7 and Lemm 2.8, there exists a weakly convergent subsequence $\{\varphi_{n_k}\} \subset \{\varphi_n\}$, such that $\varphi_{n_k} \rightarrow \varphi_0 \in W$ *weakly in* $L^2(\Omega)$.

By Lemma 2.5, $\hat{J}(\varphi_0) = \inf_{\varphi \in W} \hat{J}(\varphi)$.

Corollary 2.1 If $\lambda(a(x)) > r > \lambda(b(x))$, then the optimal solution to (P) exists.

3. Necessary Conditions for Optimal Control

In [1], the authors gives the necessary conditions for the optimal control by use of Duboviskii-Milyutin Theorem[6]. In their process, the main difficulty is to prove the conclusion that " $\lambda(\sigma)$ and $\varphi(\sigma)$ are Frechet differentialbe mappings from U into R and $L^2(\Omega)$, respectively". But the conclusion is not an immediate consequence of [7] as they say. In this section, first we give a strict verification of the conclusion, then give the same necessary condition by use of generalized Kuhu-Tucker Theorem[6].

In this section, we further assume that

(H4) L is densely defined, so the adjoint operator L^* exists and is unique.

(H5) for any $\sigma_\alpha \in$ U, there exists a neighbourhood $\Delta(\sigma_\alpha)$ of σ_α (it may not be contained in U) such that (H3) is satisfied in $\Delta(\sigma_\alpha)$, and $\| (L+\sigma)^{-1} \| \leq M$ for any σ in $\Delta(\sigma_\alpha)$.

Lemma 3.1 For any $\sigma_\alpha \in U$, there exists a neighbourhood $\Delta(\sigma_\alpha)$ of σ_α and a commom Jordan curve Γ such that the eigen-projection operator P_σ connecting with $\lambda(\sigma)$ can be represented by

$$P_\sigma = \frac{-1}{2\pi i} \int_\Gamma R_\sigma(z) dz \quad for \ any \ \sigma \ in \ \Delta(\sigma_\alpha)$$

and $P_\sigma L^2(\Omega)$ is the eigenspace of $T(\sigma)$ corresponding to the critical eigen-value $\lambda(\sigma)$. And P_σ is Fréchet differentiable operator, where $R_\sigma(z)$ is the resolvent of $T(\sigma)$.

Proof According to the Theorem IV.1.5 and it's notes in [8], we have

$$\begin{aligned}
&\| T(\sigma) - T(\sigma_\alpha) \| = \| (L+\sigma)^{-1}K - (L+\sigma_\alpha)^{-1}K \| \\
&\leq \| K \| \| (L+\sigma)^{-1} - (L+\sigma_\alpha)^{-1} \| \leq \frac{\| K \| \| (L+\sigma_\alpha)^{-1} \|^2 \| \sigma - \sigma_\alpha \|}{1 - \| (L-\sigma_\alpha)^{-1} \| \| \sigma - \sigma_\alpha \|} \\
&\rightarrow 0, \qquad as \quad \| \sigma - \sigma_\alpha \| \rightarrow 0
\end{aligned}$$

and

$$\begin{aligned}
&\| R_\sigma(z) - R_{\sigma_\alpha}(z) \| = \| (z - T(\sigma))^{-1} - (z - T(\sigma_\alpha))^{-1} \| \\
&\leq \frac{\| (z - T(\sigma_\alpha))^{-1} \|^2 \| T(\sigma) - T(\sigma_\alpha) \|}{1 - \| (z - T(\sigma_\alpha))^{-1} \| \| T(\sigma) - T(\sigma_\alpha) \|} \rightarrow 0, \quad as \quad \| \sigma - \sigma_\alpha \| \rightarrow 0
\end{aligned}$$

Therefore there exists a neighbourhood $\Delta(\sigma_\alpha)$ of σ_α and a bounded function $M(z)$ such that $\| R_\sigma(z) \| \leq M(z)$.

Similar to the definition in [9], $T(\sigma)$ is a strongly stable approximation to $T(\sigma_\alpha)$.

By the similar process with proposition 5.6 in [9], we can conclude that there exists a common Jordan curve Γ such that

$$P_\sigma = \frac{-1}{2\pi i} \int_\Gamma R_\sigma(z) dz \qquad for \ every \ \sigma \ in \ \Delta(\sigma_\alpha)$$

and $P_\sigma L^2(\Omega)$ is the eigenspace of $T(\sigma)$.

Let \$ denote $R_{\sigma_\alpha}(z)(L+\sigma_\alpha)^{-1}\ \sigma\ (L+\sigma_\alpha)^{-1}KR_{\sigma_\alpha}(z)$, since

$$
\begin{aligned}
&P_{\sigma_\alpha+\sigma}-P_{\sigma_\alpha}-\frac{-1}{2\pi i}\oint_\Gamma \$dz \\
&=\frac{-1}{2\pi i}\oint_\Gamma (R_{\sigma_\alpha+\sigma}(z)-R_{\sigma_\alpha}(z))dz-\frac{-1}{2\pi i}\oint_\Gamma \$dz \\
&=\frac{-1}{2\pi i}\oint_\Gamma R_{\sigma_\alpha+\sigma}(z)(L+\sigma_\alpha)^{-1}\sigma(L+\sigma_\alpha+\sigma)^{-1}KR_{\sigma_\alpha}(z)dz-\frac{-1}{2\pi i}\oint_\Gamma \$dz \\
&=\frac{-1}{2\pi i}\oint_\Gamma (R_{\sigma_\alpha+\sigma}(z)-R_{\sigma_\alpha}(z))(L+\sigma_\alpha)^{-1}\sigma(L+\sigma_\alpha+\sigma)^{-1}KR_{\sigma_\alpha}(z)dz \\
&+\frac{-1}{2\pi i}\oint_\Gamma R_{\sigma_\alpha}(z)(L+\sigma_\alpha)^{-1}\sigma[(L+\sigma_\alpha+\sigma)^{-1}-(L+\sigma_\alpha)^{-1}]KR_{\sigma_\alpha}(z)dz
\end{aligned}
$$

thus

$$\| P_{\sigma_\alpha+\sigma}-P_{\sigma_\alpha}-\frac{-1}{2\pi i}\oint_\Gamma \$dz \| / \| \sigma \| \to 0 \qquad as \qquad \| \sigma \| \to 0.$$

So P_σ is Fréchet differentiable operator, for σ_α is an arbitrary point.

Theorem 3.1 For any $\sigma_\alpha \in U$ there exists a neighbourhood $\Delta(\sigma_\alpha)$ of σ_α such that the solution to (1.6) and (1.7) can be represented by

$$\varphi(\sigma)=\frac{p}{(1,P_\sigma\varphi(\sigma_\alpha))}P_\sigma\varphi(\sigma_\alpha)$$

and P_σ is Fréchent defferentiable in $\Delta(\sigma_\alpha)$.

Proof By lemma 3.1,

$$\frac{p}{(1,P_\sigma\varphi(\sigma_\alpha))}P_\sigma\varphi(\sigma_\alpha)$$

is the positive solution to (1.6) and (1.7). By Lemma 2.1,

$$\varphi(\sigma)=\frac{p}{(1,P_\sigma\varphi(\sigma_\alpha))}P_\sigma\varphi(\sigma_\alpha)$$

Since P_σ is Fréchet differentiable in $\Delta(\sigma_\alpha)$, it is easy to show $\varphi(\sigma)$ is, too.

Lemma 3.2 λ_α is a continuous functional from U into R.

Proof By [7], for any bounded linear operator T, the spectrum set $\sigma(T)$ is upper semi-continuous function, that is for any $\varepsilon > 0$, there exists a $\delta > 0$, such that

$$sup_{\lambda\in\sigma(S)} dist(\lambda,\sigma(T)) < \varepsilon, \qquad as \qquad \| S-T \| < \delta$$

First given a $0<\delta_0<1$, we can take

$$\delta' \le min(\delta_0/\| (L+\sigma_\alpha)^{-1} \|, \quad (1+\delta_0)\delta/\| (L+\sigma_\alpha)^{-1} \|^2 \| K \|)$$

From (3.1), $\| T(\sigma)-T(\sigma_\alpha) \| < \delta$, as $\| \sigma-\sigma_\alpha \| < \delta'$. Since $\lambda(\sigma)\in\sigma(T(\sigma))$, $dist(\lambda(\sigma),\sigma(T(\sigma_\alpha))) < \varepsilon$.

Considering the compactness of $\sigma(T(\sigma_\alpha))$, we know there exists $\lambda^* \in \sigma(T(\sigma_\alpha))$ such that $|\ \sigma(\sigma)-\lambda^*\ | < \varepsilon$. Thus (since $\lambda(\sigma_\alpha) > |\ \lambda^*\ |$)

$$\lambda(\sigma)-\lambda(\sigma_\alpha)\le\lambda(\sigma)-|\ \lambda^*\ |=|\ \lambda(\sigma)\ |-|\ \lambda^*\ |\le|\ \lambda(\sigma)-\lambda^*\ |<\varepsilon$$

By the same process, $\lambda(\sigma_\alpha)-\lambda(\sigma)<\varepsilon$, and so $|\ \lambda(\sigma_\alpha)-\lambda(\sigma)\ |<\varepsilon$, as $\| \sigma_\alpha-\sigma \| \to 0$, or $\lambda(\sigma)$ is continuous at σ_α.

Lemma 3.3 For any $\sigma_\alpha \in U$, the Gateaux derivative $\dot{\lambda}_\alpha$ of $\lambda(\sigma)$ at σ_α exists and is given by

$$\dot{\lambda}_\alpha = (\dot{\lambda}(\sigma_\alpha), \sigma) = -\lambda_\alpha{}^2 \frac{(\sigma\varphi_\alpha, \psi_\alpha)}{(K\varphi_\alpha, \psi_\alpha)}, \qquad \sigma \in L^\infty(\Omega_1) \tag{3.2}$$

where $\lambda_\alpha = \lambda(\sigma_\alpha), \varphi_\alpha = \varphi(\sigma_\alpha)$ and ψ_α is an eigenfunction of $(L^* + \sigma_\alpha)^{-1}K^*$ associated with λ_α, i.e.

$$(L^* + \sigma_\alpha)\psi_\alpha = \frac{1}{\lambda_\alpha} K^* \psi_\alpha$$

Proof Denote $\hat{T}(t) = T(\sigma_\alpha + t\sigma)$, then it's easy to show that $\hat{T}(t)$ is a holomorphic function at a neighbourhood of 0.

By [7], $\lambda(t) = \lambda(\sigma_\alpha + t\sigma)$ is analytic at 0, that is $\lambda(\sigma)$ is Gateaux differentiable at σ_α in any direction σ in $L^\infty(\Omega_1)$.

According to Lemma 3.2 in [1], we obtain (3.2).

Theorem 3.2 For any $\sigma_\alpha \in U$ there exists a neighbourhood $\Delta(\sigma_\alpha)$ such that $\lambda(\sigma)$ is continuously Fréchet differentiable in $\Delta(\sigma_\alpha)$.

Proof Combining (3.2) with Lemma 3.2 and Theorem 3.1, and considering ψ_α sharing the same properties with φ_α, we can conclude the conclusion after a complexive estimation.

Similar to Lemma 3.3 in [1], we obtain

Lemma 3.4 The Fréchet derivative $\dot{\varphi}_\alpha$ of $\varphi(\sigma)$ at σ_α satisfies

$$(L + \sigma_\alpha)\dot{\varphi}_\alpha = \frac{1}{\lambda_0} K\dot{\varphi}_\alpha - \frac{(\sigma\varphi_\alpha, \psi_\alpha)}{(K\varphi_\alpha, \psi_\alpha)} K\varphi_\alpha - \sigma\varphi_\alpha \tag{3.3}$$

$$(\dot{\varphi}_\alpha, h) = 0 \tag{3.4}$$

where the notation is as Lemma 3.3.

Lemma 3.5 For $\sigma_\alpha \in U(\sigma_0)$, $J(\sigma_0)$ is Fréchet differentiable at σ_α and the derivative is given by

$$J'(\sigma_\alpha)(\sigma) = (\varphi_\alpha - M\varphi_\alpha, \dot{\varphi}_\alpha(\sigma))$$

where $\dot{\varphi}_\alpha, \varphi_\alpha$ are defined as Lemma 3.4, $U(\sigma_0)$ is a neighbourhood of σ_0.

Proof It's easy to show $\hat{J}'(\varphi_\alpha)(\varphi) = (\varphi_\alpha - M\varphi_\alpha, \varphi)$. Because $J(\sigma) = (\hat{J} \circ \varphi)(\sigma)$,

$$J'(\sigma_\alpha)(\sigma) = (\hat{J}'(\sigma_\alpha) \circ \dot{\varphi}_\alpha)(\sigma) = (\varphi_\alpha - M\varphi_\alpha, \dot{\varphi}_\alpha(\sigma))$$

Theorem 3.3 If $\lambda(a(x)) > r > \lambda(b(x))$, and σ_0 is the optimal solution to (P), then there exist $r_1 \geq 0, r_2 \in R$, not all zero, and adjoint state $\psi \in D(L^*)$ such that

$$\begin{cases} L\varphi_0 + \sigma_0\varphi_0 = \dfrac{1}{\lambda(\sigma_0)} K\varphi_0, \qquad \lambda(\sigma_0) = r \\ (h, \varphi_0) = p \\ L^*\varphi + \sigma_0\varphi = \dfrac{1}{r} K^*\varphi - r_1(\varphi_0 - M\varphi_0) + \dfrac{r_1}{p}(\varphi_0 - M\varphi_0, \varphi_0)h \\ \displaystyle\int_\Omega (\sigma - \sigma_0)\varphi_0(\psi + r_2\psi_0)dx \geq 0 \quad for \quad every \quad \sigma \in U \end{cases}$$

Proof After the preparation above, the Kuhn-Tucker theorem can be applied to (P), and if σ_0 is the solution,then there exist $r_1 \geq 0, \mu_1 \in R$, which are not equal to zero simultaneously, such that

$$r_1(\varphi_0 - M\varphi_0, \varphi_0'(\sigma - \sigma_0)) + \mu_1((\sigma - \sigma_0)\varphi_0, \psi_0) \geq 0 \qquad \forall \sigma \in U \tag{3.5}$$

In order to simplify (3.5), we introduce the adjoint state ψ which is described by

$$L^*\psi + \sigma_0\psi = \frac{1}{r}K^*\psi - r_1(\varphi_0 - M\varphi_0) + \alpha h \tag{3.6}$$

where α is a constant to be defined. Similar to [1], $\alpha = \frac{r_1}{p}(\varphi_0 - M\varphi_0, \varphi_0)$, so

$$r_1((\varphi_0 - M\varphi_0), \varphi_0'(\sigma - \sigma_0)) = -(L^*\psi + \sigma_0\psi - \frac{1}{r}K^*\psi - \alpha h, \varphi_0'(\sigma - \sigma_0))$$

$$= -(\psi, (L\varphi_0' + \sigma_0\varphi_0' - \frac{1}{r}K\varphi_0')(\sigma - \sigma_0)) = -(\psi, -\frac{((\sigma - \sigma_0)\varphi_0, \psi_0)}{(K\varphi_0, \psi_0)}K\varphi_0 - (\sigma - \sigma_0)\varphi_0)$$

$$= \frac{(\psi, K\varphi_0)}{(K\varphi_0, \psi_0)}((\sigma - \sigma_0)\varphi_0, \psi_0) + (\psi, (\sigma - \sigma_0)\varphi_0)$$

Let

$$r_2 = \mu_1 + \frac{(\psi, K\varphi_0)}{(K\varphi_0, \psi_0)},$$

then (3.5) becomes $r_2((\sigma - \sigma_0)\varphi_0, \psi_0) + ((\sigma - \sigma_0)\varphi_0, \psi) \geq 0$, that is

$$((\sigma - \sigma_0)\varphi_0, \psi + r_2\psi_0) \geq 0, \quad \forall \sigma \in U$$

Similar to [1], we have

Theorem 3.4 We can take $r_1 = 1$ in Theorem 3.3, and the last formula can be replaced by

$$\iint_{\Omega_1 \times \Omega_2} (\sigma(x) - \sigma_0(x))\varphi_0(x, y)\psi(x, y)dxdy \geq 0, \qquad \forall \sigma \in U^0$$

where $U^0 = \{\sigma \in U \mid \quad (\sigma\varphi_0, \psi_0) = (\sigma_0\varphi_0, \psi_0)\}$.

4. Applications

Let's return to the example 1, we give the following hypertheses:

1) $\alpha(v, \mu), \beta(v, \mu)$ are positive measurable functions and $0 < \alpha(v, \mu) \leq 1, 0 < \beta(v, \mu) \leq 1$, $\alpha(v, \mu) = \alpha(v, -\mu)$, $\beta(v, \mu) = \beta(v, -\mu)$;

2) $v\Sigma_\alpha(x, v), k(x, v', v, \mu', \mu)$ are nonnegative bounded measurable functions and

$$\lambda_0{}^* = essinf\, v\Sigma_\alpha(x, v) > 0$$

3) $k(x, v', v, \mu', \mu) > 0$ a.e. in $[-a, a] \times (0, v_m] \times (0, v_m] \times [-1, 1] \times [-1, 1]$ and there exest $x_1, x_2 \in [-a, a]$, $x_1 < x_2$ and $v_1, v_2 \in (0, v_m], v_1 < v_2$, such that $k(x, v', v, \mu', \mu) \geq k_0 > 0$ a.e. in $[x_1, x_2] \times [v_1, v_2] \times [v_1, v_2] \times [-1, 1] \times [-1, 1]$.

With the above hypertheses, basing on [10], we can imitate the process in [11] to show that all the conditions stated in section 2 and section 3 are satisfied. So we obtain

Theorem 4.1 If $\lambda(c_1) > 1 > \lambda(c_2)$, the optimal control exists, where $c_1 = a(x, v)v$, and $c_2 = b(x, v)v$.

Let $S = \{\Sigma_c \in L^\infty(\Omega_1) \quad | \quad 0 \leq a(x, v) \leq \Sigma_c(x, v) \leq b(x, v) \quad a.e. \quad in \;\; \Omega_1\}$, we have

Theorem 4.2 The optimal macro-absorption cross section $\Sigma_c{}^0 \in U$ is given by

$$\begin{cases} v\mu\dfrac{\partial\varphi_0(x, v, \mu)}{\partial x} + v(\Sigma_\alpha(x, v) + \Sigma_c{}^0(x, v))\varphi_0(x, v, \mu) = \displaystyle\int_0^{v_m}\int_{-1}^1 k(x, v', v, \mu', \mu)\varphi_0(x, v', \mu')dv'd\mu' \\ \varphi_0(-a, v, \mu) = \alpha(v, \mu)\varphi_0(-a, v, -\mu) \\ \varphi_0(a, v, -\mu) = \beta(v, \mu)\varphi_0(a, v, \mu) \\ (h, \varphi_0) = p \\ \displaystyle\int_\Omega (v\Sigma_c(x, v) - v\Sigma_c{}^0(x, v))(\psi(x, v, \mu) + r_2\psi_0(x, v, \mu))dxdvd\mu \geq 0 \quad \forall \Sigma_c(x, v) \in S \end{cases} \tag{4.1}$$

where $r_2 \in R$, and ψ is directed by (4.2):

$$\begin{cases} -v\mu\dfrac{\partial\psi(x,v,\mu)}{\partial x}+v(\Sigma_a(x,v)+\Sigma_c{}^0(x,v))\psi(x,v,\mu) \\ =\displaystyle\int_0^{v_m}\int_{-1}^{1}k(x,v',v,\mu',\mu)\psi(x,v',\mu')dv'd\mu'-r_1(\varphi_0-M\varphi_0)+\frac{r_1}{p}(\varphi_0-M\varphi_0,\varphi_0)h \\ \psi(-a,v,-\mu)=\alpha(v,\mu)\psi(-a,v,\mu) \\ \psi(a,v,\mu)=\beta(v,\mu)\psi(a,v,-\mu) \end{cases} \tag{4.2}$$

and ψ_0 is a solution to the homogeneous equation of (4.2).

Theorem 4.3 In Theorem 4.2, we can take $r_1 = 1$, and (4.1) can be replaced by

$$\int_\Omega (v\Sigma_c(x,v)-v\Sigma_c{}^0(x,v))\varphi_0(x,v,\mu)\psi(x,v,\mu))dxdvd\mu \geq 0 \quad \forall\Sigma_c \in S^0$$

where

$$S^0=\{\Sigma_c\in S \mid (v\Sigma_c\varphi_0,\psi_0)=(v\Sigma_c{}^0\varphi_0,\psi_0)\}$$

Similar to example 1, the obtained results in section 2 and section 3 can be applied to the 3-dimensional stable transport equation with isotropic scattering:

$$\begin{cases} v\vec{\Omega}\cdot grad_{\vec{r}}\psi(\vec{r},v,\vec{\Omega})+v\Sigma(\vec{r},v)\psi(\vec{r},v,\vec{\Omega})=\dfrac{1}{\lambda}\displaystyle\int_{v_m}^{v_M}dv'\int_{V_{\vec{\Omega}}}k(\vec{r},v,v')\psi(\vec{r},v',\vec{\Omega}')d\vec{\Omega}' \\ \psi(\vec{r},v,\vec{\Omega})=0, \quad \vec{r}\in\Gamma_V, \quad \vec{n}\cdot\vec{\Omega}<0 \end{cases}$$

where V is a bounded convex domain in R^3.

References

[1] Dexing Feng and Guangting Zhu: J. Math. Pures et. Appl. 1984(2), page 169-186
[2] A. M. Wernberg, and E. Wigner: The physical Theory of Neutron Chain Reactor, Univ. of Chicago Press 1958
[3] Maz. Ja. V. G.: Sobolev Space, Berlin-Springer, 1984 (Trans. from Russion)
[4] Marek. I.: SIAM J. Appl. Math. 1970(19) 607-628
[5] Balakrishnann A. V.: Applied Functional Analysis, Springer-Verlag 1976
[6] Zeidler, E.: Nonlinear Functional Analysis and its Applications
[7] Kato. T.: Perturbation Theory of Linear Operators, Springer-Verlag 1984
[8] Taylor A. E.: Introduction to Functional Analysis, N. Y. 1980
[9] Chatelin F.: Spectral Approximation of Linear Operators, N. Y. 1983
[10] Song Degong, Wang Miansen and Zhu Guangtian: Systems Science and Mathematical Sciences, China, 1990(2), 102-125
[11] Yang Minzhu and Zhu Guangtian: Acta Mathematica, Scientia, China, Vol. 1, No. 1, 1981, 1-12

SINGLE INPUT CONTROLLABILITY FOR SPECTRAL SYSTEMS IN BANACH SPACES

JINGBO WU

Department of Computer & System Sciences

Nankai University, Tianjin, China

Consider the discrete-time system defined by

$$x_{n+1} = Ax_n + Bu_n$$

where A is a bounded linear operator on a Banach space X and B is a linear operator from the input space C^m to X. This system is denoted by $\{A, B\}$. We say that the system is controllable if $X = \bigvee_{k=0}^{\infty}$ ran $A^k B$. If A is a scalar type spectral operator, then $\{A, B\}$ is called a scalar type spectral system. In the case where A is a normal operator on a Hilbert space, the system $\{A, B\}$ is called a normal system. We can define self-adjoint system in an analogous way. We say that the normal system $\{A, B\}$ is bilaterally controllable if $X = \bigvee_{k,l=0}^{\infty}$ ran $A^{*k} A^l B$.

In [7], Wonham showed that if X is finite dimensional and A is cyclic, then the controllability of $\{A, B\}$ implies that $\{A, b\}$ is controllable for some $b \in$ran B. There have been many papers concerned with the single input controllability for infinite dimensional state space. In [5], Fuhrmann proved that this property holds if A is self-adjoint. Feintuch [4] pointed out that Fuhrmann's proof in the slef-adjoint case remains valid for reductive normal operators. Lubin [6] gave an example to show that this result may fail in general for normal operators.

Recall that a controllable system $\{A, B\}$ is single input controllable if $\{A, b\}$ is controllable for some $b \in$ran B. The main purpose of this note is to discuss the following problem: when is a controllable scalar type spectral system on a Banach space single input controllable? The main results will be given in Theorem 5 and Corollary 6.

Throughout, by an operator we mean a bounded linear transformation unless it is otherwise stated. The Banach algebra of operators on X is denoted by $L(X)$. The σ-algebra of Borel subsets of the complex plane is denoted by Σ. Let A be a scalar type spectral operator with resolution of the identity $E(\cdot)$. For fixed $x_0 \in X$, denote by $M(x_0)$ the closed subspace spanned by $E(\sigma)x_0$ for all $\sigma \in \Sigma$. We say that x_0 is a cyclic vector of $E(\cdot)$ if $X = M(x_0)$ and that x_0 is a cyclic vector of A if $X = V_{k=0}^{\infty} A^k x_0$. For normal operator A, we say that x_0 is a star-cyclic vector of A if $X = V_{k,l=0}^{\infty} A^{*k} A^l x_0$. Clearly, x_0 is star-cyclic for A if and only if it is cyclic for $E(\cdot)$. Given $x_0 \in X$, there exists $x_0^* \in X^*$ such that $x_0^* E(\sigma)x_0 \geq 0$ for all $\sigma \in \Sigma$ and such that $x_0^* E(\sigma)x_0 = 0$ implies $E(\sigma)x_0 = 0$ (cf. Theorem 3.1 of [1]). Write $\mu_0(\cdot) = x_0^* E(\cdot)x_0$. For any bounded Borel function f, the integral

$$S(f) = \int f(\lambda)E(d\lambda)$$

exists in the uniform operator topology. When f is unbounded, define $\sigma_n = \{\lambda | |f(\lambda)| \leq n\}$, $n = 1, 2, \cdots$. The unbounded operator $S(f)$ has domain

$$D(S(f)) = \{x | \lim_{n\to\infty} \int_{\sigma_n} f(\lambda)E(d\lambda)x \text{ exists}\}$$

and is given by the formula

$$S(f)x = \lim_{n\to\infty} \int_{\sigma_n} f(\lambda)E(d\lambda)x, \quad x \in D(S(f)).$$

The operator $S(f)$ (not necessarily bounded) corresponding to a Borel function f is a densely defined closed operator. For $x_0 \in X$,

$$M(x_0) = \{S(f)x_0 | x_0 \in D(S(f))\}$$

(cf. [2], §4). Note that A has the single-valued extension property (cf. Theorem 5.31 of [3]). Denote by $\rho_A(x)$ the local resolvent of A at x and by $\sigma_A(x)$ the local spectrum of A at x.

Lemma 1. Let X be a Banach space and let A be a scalar type spectral operator with resolution of the identity $E(\cdot)$. Assume that x_0, in X, is a cyclic vector of $E(\cdot)$. If f is a Borel function such that $x_0 \in D(S(f))$, then $S(f)x_0$ is cyclic for $E(\cdot)$ if and only if $f(\lambda) \neq 0$ almost everywhere with respect to μ_0.

Proof. If $f(\lambda) = 0$ on a Borel subset σ_0 of the complex plane with $\mu_0(\sigma_0) > 0$, then $E(\sigma_0)x_0 \neq 0$. It follows that $E(\sigma_0)x_0 \notin M(S(f)x_0)$ and so $S(f)x_0$ is not cyclic for $E(\cdot)$. Conversely, assume that $f(\lambda) \neq 0$ a.e. with respect to μ_0. Let σ be a closed subset of the complex plane such that $|f(\lambda)|$ is bounded above and possesses a positive lower bound on σ. Define g by letting $g(\lambda) = f(\lambda)^{-1}$ for $\lambda \in \sigma$ and $g(\lambda) = 0$ elsewhere. We have

$$E(\sigma)x_0 = S(gf)x_0 = S(g)S(f)x_0.$$

Using the countable additivity of $E(\cdot)$ in the strong operator topology together with the cyclicity of x_0, we have $X \subset \mathcal{M}(S(f)x_0)$ and so $S(f)x_0$ is cyclic for $E(\cdot)$.

Lemma 2. Let the assumptions of Lemma 1 be satisfied. If f and g are Borel functions such that $x_0 \in D(S(f)x_0) \cap D(S(g)x_0)$ and such that $X = \mathcal{M}(S(f)x_0) V \mathcal{M}(S(g)x_0)$, then $\sigma(A) = \sigma_A(S(f)x_0) \cup \sigma_A(S(g)x_0)$ and there exists a complex number α such that $S(f + \alpha g)x_0$ is cyclic for $E(\cdot)$.

Proof. Since $X = \mathcal{M}(S(f)x_0) V \mathcal{M}(S(g)x_0)$, we have

$$X = [E(\sigma_A(S(f)x_0))X] V [E(\sigma_A(S(g)x_0))X]$$

(cf. Theorem 5.33 of [3]) and so $\sigma(A) = \sigma_A(S(f)x_0) \cup \sigma_A(S(g)x_0)$. For fixed ε, we write

$$\sigma_\varepsilon = \{\lambda \in \sigma(A) | f(\lambda) + \varepsilon g(\lambda) = 0\}.$$

For $\alpha \neq \beta$, it is clear that $f(\lambda) = g(\lambda) = 0$ on $\sigma_\alpha \cap \sigma_\beta$. Suppose that $\mu_0(\sigma_\alpha \cap \sigma_\beta) > 0$. We have $E(\sigma_\alpha \cap \sigma_\beta)x_0 \neq 0$. It follows that $E(\sigma_\alpha \cap \sigma_\beta)x_0 \notin \mathcal{M}(S(f)x_0) V \mathcal{M}(S(g)x_0)$. But this is a contradiction and so $\mu_0(\sigma_\alpha \cap \sigma_\beta) = 0$. We have first $\mu_0(\sigma_\varepsilon) \geq n^{-1}\mu_0(\sigma(A))$ for a finite set of subscripts and then $\mu_0(\sigma_\varepsilon) > 0$ for a countable set of subscripts only. This implies that there exists an α such that $\mu_0(\sigma_\alpha) = 0$ and hence $f(\lambda) + \alpha g(\lambda) \neq 0$ a.e. with respect to μ_0. Using Lemma 1, $S(f + \alpha g)x_0$ is cyclic for $E(\cdot)$.

Theorem 3. Let the assumptions of Lemma 1 be satisfied. If $X = V_{x \in Y} \mathcal{M}(x)$ where Y is a finite dimensional subspace of X, then $X = \mathcal{M}(y_0)$ for some $y_0 \in Y$.

Proof. There exist unit vectors $x_1, \cdots, x_m$ in Y such that $X = V_{k=1}^m \mathcal{M}(x_k)$. By Theorem 4.5 of [2], there exist Borel functions $f_1, \cdots, f_m$ such that $x_k = S(f_k)x_0$, $k = 1, \cdots, m$. Using Lemma 2 $m-1$ times, we have $\sigma(A) = \cup_{k=1}^m \sigma_A(x_k)$. Further, there exist $\alpha_1, \cdots, \alpha_{m-1}$ such that

$$\mathcal{M}(S(f_1 + \alpha_1 f_2 + \cdots + \alpha_{m-1} f_m)x_0) = V_{k=1}^m \mathcal{M}(x_k) = X.$$

This completes the proof by letting $y_0 = x_1 + \sum_{k=1}^{m-1} \alpha_k x_{k+1}$.

The proof of the following theorem is straightforward and will be omitted.

Theorem 4. Let $\{A, B\}$ be a bilaterally controllable normal system. If A has a star-cyclic vector, then there exists $b \in \text{ran } B$ such that $\{A, b\}$ is bilaterally controllable.

Theorem 5. Let $\{A, B\}$ be a controllable scalar type spectral system. Suppose that A has a cyclic vector. If the resolution of the identity $E(\cdot)$ for A leaves invariant every closed subspace of X invariant under A, then there exists $b \in \text{ran } B$ such that $\{A, b\}$ is controllable.

Proof. Note that a cyclic vector of A is necessarily cylcic for $E(\cdot)$. By Theorem 3, there exists $b \in \text{ran } B$ such that $X = \mathcal{M}(b)$. Write $Y = V_{n=0}^{\infty} A^n b$. It is clear that Y is invariant under A and so by the assumptions it is invariant under $E(\cdot)$. Hence $Y = X$ and so $\{A, b\}$ is controllable.

Corollary 6. Let $\{A, B\}$ be a controllable scalar type spectral system. Suppose that A has a cyclic vector. If $\sigma(A)$ is nowhere dense and $\rho(A)$ is connected, then there exists $b \in \text{ran } B$ such that $\{A, b\}$ is controllable.

Proof. Let Y, a closed subspace of X, be invariant under A. Then Y is invariant under $p(A)$ where p is a polynomial. Define

$$\alpha(f) = \int_{\sigma(A)} f(\lambda) E(d\lambda)$$

where $f \in C(\sigma(A))$. Note that $\alpha(f)$ is a bicontinuous algebra isomorphism from $C(\sigma(A))$ into $L(X)$ (cf. Proposition 5.9 of [3]). By Lavrentieff's theorem (cf. Theorem 5.37 of [3]), Y is invariant under $\alpha(f)$ for all $f \in C(\sigma(A))$. Using Proposition 12.13 of [3], Y is invariant under $E(\cdot)$. This completes the proof by using Theorem 5.

References

1. W.G. Bade, On Boolean algebras of projections and algebras of operators, Trans. Amer. Math. Soc. 80 (1955), 345-360.

2. W.G. Bade, A multiplicity theory for Boolean algebras of projections in Banach spaces, Trans. Amer. Math. Soc. 92 (1959), 508-530.

3. H.R. Dowson, Spectral theory of linear operators, Academic Press, London, 1978.

4. A. Feintuch, On single input controllability for infinite dimensional linear systems, J. Math. Anal. Appl. 62 (1978), 538- 546.

5. P.A. Fuhrmann, Some results on controllability, Ricerche di Automatica 5 (1974), 1-5.

6. A.R. Lubin, A note on single input controllability for normal systems, Math. Systems Theory, 15 (1982), 371-373.

7. W.M. Wonham, Linear multivariable control, Lecture Notes in Economics and Mathematical Systems, 101, Springer-Verlag, Berlin, 1974.

Distributed Parameter Systems with Measure Controls *

Jiongmin Yong
Department of Mathematics, Fudan University, Shanghai 200433, China

§1. Introduction.

Let us first give a motivation of our optimal control problem which is to be studied in this paper.

Let X be a Banach space and $A : \mathcal{D}(A) \subset X \to X$ be the infinitesimal generator of some C_0-semigroup e^{At} on X. We consider a controlled evolution system

$$
(1.1) \qquad \begin{cases} \dot{x}(t) = Ax(t) + g(t, x(t), u(t)), & t \in [0, T], \\ x(0) = x_0. \end{cases}
$$

The state of the system is usually understood as the mild solution of (1.1). Now, suppose we have another control action—an impulse control, i.e., at times $t = \tau_i$, $i \geq 1$, we make an impulse ξ_i to the state $x(\tau_i - 0)$. We refer $\{(\tau_i, \xi_i) \mid i \geq 1\}$ as an "impulse control". Thus, one has

$$
x(\tau_i) = x(\tau_i - 0) + \xi_i, \qquad i \geq 1.
$$

Then, the state $x(\cdot)$ formally satisfies the following evolution equation:

$$
(1.2) \qquad \begin{cases} \dot{x}(t) = Ax(t) + g(t, x(t), u(t)) + \sum_{i \geq 1} \xi_i \delta(t - \tau_i), & t \in [0, T], \\ x(0) = x_0, \end{cases}
$$

where $\delta(\cdot)$ is the δ-function. Similarly, we understand the state of the system $x(\cdot)$ as the mild solution of the system (1.2). It is more reasonable that, in general, the impulse ξ_i at time $t = \tau_i$ should also depend on the state $x(\tau_i - 0)$, which is the state of the system before making impulse ξ_i. Thus, it is more natural to consider the following state equation (compare with (1.2))

$$
(1.3) \qquad \begin{cases} dx(t) = \big(Ax(t) + g(t, x(t), u(t))\big)dt \\ \qquad\qquad + g_0(t, x(t-0), u(t))d\xi(t), & t \in [0, T], \\ x(0) = x_0, \end{cases}
$$

where

$$
\xi(t) = \sum_{i \geq 1} \xi_i \chi_{[\tau_i, \infty)}(t), \qquad t \in [0, \infty).
$$

Hence, in general we have the following type evolution equation:

$$
(1.4) \qquad \begin{cases} dx(t) = Ax(t)dt + F(t, x(t-0), u(t))d\mu(t), & t \in [0, T], \\ x(0) = x_0, \end{cases}
$$

* This work was partially supported by the Chinese NSF under Grants 0188416.

with some $\mathcal{L}(Z,X)$-valued function F and some Z-valued vector measure $\mu(\cdot)$, which together with $u(\cdot)$ will be considered as control actions. Associated with (1.4), we are given a cost functional

$$J(x(\cdot),u(\cdot),\mu(\cdot)) = \int_0^T \langle f(t,x(t-0),u(t)),\mu(dt)\rangle, \tag{1.5}$$

with some X^*-valued function f. In the case there is no impulse, or the measure $\mu(\cdot)$ is fixed and is absolutely continuous with respect to the Lebesgue measure, the problem is reduced to a classical semilinear distributed parameter systems with Lagrange form cost functional.

The control problem we will study in this paper is to minimize functional (1.5) over some class of admissible controls, subject to the state equation (1.4) and an end constraint for the state of the following type:

$$(x(0),x(T)) \in \Omega \subset X \times X. \tag{1.6}$$

In [14,15], a similar problem in finite dimensional spaces was studied. It is immediate that the major difference between this paper and [14,15] is that whether the coefficient of $\mu(\cdot)$ depends on the state $x(\cdot)$. Secondly, we are in infinite dimensional space and we have a general end constraint (1.6), which is different from the separated end constraint case (see [12] for comments). On the other hand, we should point out that the operator-valued function $F(t,x,u)$ has to be assumed Frechet differentiable in x, which is a little more restrictive than [15] for the finite dimensional case (in [15], only the Lipschitz continuity in x was assumed). The main results of this paper consist of the study of the infinite dimensional Volterra-Stieljes integral equations and the Pontryagin type maximum principle for the related optimal control problems.

§2. Evolution Equations.

Let us first introduce the so-called Young integral of operator valued functions with respect to vector measures. To this end, let X and Z be Banach spaces and $T>0$ be given. For any metric space V, we denote

$$BV_0([0,T];V) = \{v(\cdot):[0,T]\to V \mid v(\cdot) \text{ is of bounded variation, } v(0)=0\}.$$

For any $\mu(\cdot)\in BV_0([0,T];Z)$, there exists a unique vector measure associated with it. We denote it by $\bar\mu(\cdot)$. Next, by noticing the fact that any BV function $\mu(\cdot)$ has at most countably many discontinuity points, we may define

$$\begin{cases} \mu_b(t) = \sum\limits_{0\le\tau<t} [\mu(\tau+0)-\mu(\tau-0)] + \mu(t)-\mu(t-0), & \forall t\in[0,T], \\ \mu_c(t) = \mu(t)-\mu_b(t), & \forall t\in[0,T]. \end{cases} \tag{2.1}$$

Now, we let $\bar\mu_b(\cdot)$ and $\bar\mu_c(\cdot)$ be the vector measures induced by $\mu_b(\cdot)$ and $\mu_c(\cdot)$, respctively.

Definition 2.1. An operator-valued function $F(\cdot):[0,T]\to\mathcal{L}(Z,X)$ is said to be uniformly Borel measurable if there exists a sequence of Borel measurable simple functions $F_n(\cdot):[0,T]\to\mathcal{L}(X,Z)$, such that

$$\lim_{n\to\infty}\|F_n(t)-F(t)\|_{\mathcal{L}(Z,X)}=0,\qquad \forall t\in[0,T]. \tag{2.2}$$

Then, we may define Bochner integral for operator-valued functions with respect to vector measures. Next, we would like to introduce the extended Young integral.

Definition 2.2. Let $\mu(\cdot)\in BV_0([0,T];K)$ and $F(\cdot):[0,T]\to\mathcal{L}(Z,X)$ be uniformly Borel measurable. We say that $F(\cdot)$ is μ-Young integrable if $F(\cdot)$ is $\bar{\mu}_c$-Bochner integrable and

$$\sum_{0\le\tau\le T}\|F(\tau)\|_{\mathcal{L}(Z,X)}|\mu(\tau+0)-\mu(\tau-0)|<\infty. \tag{2.3}$$

In this case, we define the Young integral of $F(\cdot)$ with respect to $\mu(\cdot)$ by the following

$$\begin{cases}\displaystyle\int_s^t F(\tau)d\mu(\tau)=\int_{[s,t]}F(\tau)\bar{\mu}_c(d\tau)+\sum_{s<\tau<t}F(\tau)[\mu(\tau+0)-\mu(\tau-0)]\\ \qquad +F(s)[\mu(s+0)-\mu(s)]+F(t)[\mu(t)-\mu(t-0)],\qquad \forall 0\le s<t\le T,\\ \displaystyle\int_t^t F(\tau)d\mu(\tau)=0,\qquad \forall t\in[0,T].\end{cases} \tag{2.4}$$

We let $Y_\mu([0,T];\mathcal{L}(Z,X))$ be the set of all μ-Young integrable functions $F(\cdot):[0,T]\to\mathcal{L}(Z,X)$. Next, we let

$$\rho_\mu(s,t)=|s-t|+\big||\mu|(s)-|\mu|(t)\big|,\qquad \forall s,t\in[0,T]. \tag{2.5}$$

Then, for any metric space V, we define

$$\begin{aligned}C_\mu([0,T];V)=\{v(\cdot):[0,T]\to V\mid v(\cdot)\text{ is uniformly}\\ \text{continuous from }[0,T]\text{ with metric }\rho_\mu(\cdot)\text{ to }V\}.\end{aligned} \tag{2.6}$$

We should note that, in general, $[0,T]$ is not necessarily compact under metric $\rho_\mu(\cdot)$ (see [8] for relevant remarks for the scalar valued case).

Now, let $\mu(\cdot)\in BV_0([0,T];K)$ and $F(\cdot)\in Y_\mu([0,T];\mathcal{L}(Z,X))$. Then, by (2.4), for any $t\in[0,T]$, one has

$$\begin{aligned}\int_0^t F(\tau)d\mu(\tau)=\int_{[0,t]}F(\tau)\bar{\mu}_c(d\tau)+\sum_{0<\tau<t}F(\tau)[\mu(\tau+0)-\mu(\tau-0)]\\ +F(0)[\mu(0+0)-\mu(0)]+F(t)[\mu(t)-\mu(t-0)].\end{aligned} \tag{2.7}$$

We refer the above the indefinite Young integral. Let

$$\bar{\Delta}\equiv\{(t,\tau)\in[0,T]\times[0,T]\mid t\ge\tau\},$$

and let $G(\cdot,\cdot):\bar{\Delta}\to\mathcal{L}(Z,X)$ and consider the following integral

$$g(t)\equiv\int_0^t G(t,\tau)d\mu(\tau),\qquad t\in[0,T]. \tag{2.8}$$

We have the following

Proposition 2.3. *Let $\mu(\cdot)\in BV_0([0,T];K)$ and $G(\cdot,\cdot):\bar{\Delta}\to\mathcal{L}(Z,X)$. We assume the following:*

(i) *$G(\cdot,\cdot)$ is uniformly Borel measurable. There exists a $|\bar{\mu}|$-integrable function $a(\cdot):[0,T]\to\mathbf{R}$, such that*

$$\|G(t,\tau)\|_{\mathcal{L}(Z,X)}\le a(\tau),\qquad \forall(t,\tau)\in\bar{\Delta}. \tag{2.9}$$

(ii) *There exist functions $\omega_G:\mathbf{R}^+\to\mathbf{R}^+$ and $\theta:[0,T]\to\mathbf{R}^+$, with the properties that ω_G is nondecreasing, $\omega_G(0)=0$ and $\theta(\cdot)$ is $|\bar{\mu}|$-integrable, such that*

$$\|G(t,\tau)-G(s,\tau)\|_{\mathcal{L}(Z,X)}\le\omega_G(\rho_\mu(t,s))\theta(\tau),\qquad\forall(t,\tau),(s,\tau)\in\bar{\Delta}. \tag{2.10}$$

Then, the function $g(\cdot)$ defined by (2.8) is μ-continuous and

$$g(t+0)=\int_0^t G(t+0,\tau)d\mu(\tau)+G(t+0,t)[\mu(t+0)-\mu(t)],\quad t\in[0,T], \tag{2.11}$$

$$g(t-0)=\int_0^t G(t-0,\tau)d\mu(\tau)-G(t-0,t)[\mu(t)-\mu(t-0)],\quad t\in[0,T], \tag{2.12}$$

where

$$G(t-0,t)\equiv G(t,t),\qquad t\in[0,T], \tag{2.13}$$

$$G(T+0,T)=G(T,T). \tag{2.14}$$

In particular, if there exists a constant C, such that

$$\omega_G(r)=Cr,\qquad\forall r\in\mathbf{R}, \tag{2.15}$$

then, $g(\cdot)\in BV_0([0,T];X)$.

Now let us consider the following integral equation

$$x(t)=\varphi(t)+\int_0^t G(t,\tau,x(\tau-0))d\mu(\tau),\qquad t\in[0,T]. \tag{2.16}$$

We state the following assumptions.

(I1) The function $\varphi(\cdot)\in C_\mu([0,T];X)$.

(I2) The function $G:\bar{\Delta}\times X\to\mathcal{L}(Z,X)$ satisfies

(i) There exists a constant L, such that

$$\|G(t,\tau,x)\|_{\mathcal{L}(Z,X)}\le L(1+|x|),\qquad\forall(t,\tau)\in\bar{\Delta},x\in X, \tag{2.17}$$

$$\|G(t,\tau,x)-G(t,\tau,\widehat{x})\|_{\mathcal{L}(Z,X)}\le L|x-\widehat{x}|,\qquad \forall(t,\tau)\in\bar{\Delta},x,\widehat{x}\in X. \tag{2.18}$$

(ii) For any $\tau\in[0,T]$ and $x\in X$, the map $G(\cdot,\tau,x)$ is continuous on $[\tau,T]$.

(I3) There exist functions $\omega_G:[0,T]\times\mathbb{R}^+\to\mathbb{R}^+$ and $\theta:[0,T]\to\mathbb{R}+$ with the properties that $\omega_G(0,r)=0,\forall r\in\mathbb{R}^+$, $\omega_G(t,\cdot)$ and $\omega_G(\cdot,r)$ are nondecreasing and $\theta(\cdot)$ is $|\bar{\mu}|$-integrable, such that

$$\begin{aligned}&\|G(t,\tau,x)-G(s,\tau,x)\|_{\mathcal{L}(Z,X)}\le\omega_G(\rho_\mu(t,s),|x|)\theta(\tau),\\&\qquad\forall\tau\in[0,T],\ t,s\in[\tau,T],\ x\in X.\end{aligned} \tag{2.19}$$

Following result gives the existence, uniqueness and some properties of the solution to (2.16).

Theorem 2.4. *Let $\mu(\cdot)\in BV_0([0,T];K)$ and let (I1)-(I2) hold. Then, (2.16) has a unique solution $x(\cdot)\in C_\mu([0,T];X)$. If $\varphi(\cdot),\widehat{\varphi}(\cdot)\in C_\mu([0,T];X)$ and $x(\cdot)$ and $\widehat{x}(\cdot)$ are the solutions of (2.16) corresponding to $\varphi(\cdot)$ and $\widehat{\varphi}(\cdot)$, respectively. Then, for any $\beta>1$,*

$$|x(t)-\widehat{x}(t)|\le\frac{\beta}{\beta-1}[\sup_{0\le\tau\le t}|\varphi(\tau)-\widehat{\varphi}(\tau)|]e^{\beta L|\mu|(t)},\qquad t\in[0,T], \tag{2.20}$$

$$|x(t)|\le\frac{\beta}{\beta-1}[\sup_{0\le\tau\le t}|\varphi(\tau)|+L|\mu|(t)]e^{\beta L|\mu|(t)},\qquad t\in[0,T]. \tag{2.21}$$

Moreover, if (I3) holds, then, for the unique solution $x(\cdot)$ of (2.16), one has

$$\begin{aligned}|x(t)-x(s)|&\le|\varphi(t)-\varphi(s)|+L(1+\sup_{0\le\tau\le t}|x(\tau)|)(|\mu|(t)-|\mu|(s))\\&\quad+\omega_G(\rho_\mu(t,s),\sup_{0\le\tau\le t}|x(\tau)|)\int_0^s\theta(\tau)d|\mu|(\tau),\\&\qquad\forall 0\le s\le t\le T.\end{aligned} \tag{2.22}$$

In particular, if $\varphi(\cdot)\in BV([0,T];X)$ and

$$\omega_G(t,r)=t\bar{\omega}(r),\qquad\forall(t,r)\in[0,T]\times\mathbb{R}, \tag{2.23}$$

for some nondecreasing function $\bar{\omega}:\mathbb{R}^+\to\mathbb{R}^+$ with $\bar{\omega}(0)=0$, then, $x(\cdot)\in BV([0,T];X)$.

We should point out that in the case we will study, the function $\varphi(\cdot)$ is not necessarily bounded variational and (2.23) does not hold (in general) either. Thus, we are not expected to obtain bounded variational solutions (even though they are μ-continuous). However, in finite dimensional case, we do have $x(\cdot)\in BV([0,T];X)$, provided (2.19) and (2.23) hold, which is not too restrictive. Also, we should note that assuming (I1) is just for simplicity. It can be replaced by

$$\varphi(\cdot)\in D([0,T];X)\equiv\{\varphi:[0,T]\to X\mid\varphi(t+0)\text{ and }\varphi(t-0)\text{ exist for all }t\in[0,T]\},$$

and we look for ths solutions of (2.16) in $D([0,T];X)$ instead. The results will be the same.

The rest of this section is devoted to the study of following linear equation:

$$
x(t) = e^{At}x_0 + \int_0^t e^{A(t-\tau)}B(\tau, x(\tau-0), d\mu(\tau)) + \int_0^t e^{A(t-\tau)}\bar{G}(\tau)d\nu(\tau), \qquad t \in [0,T]. \tag{2.24}
$$

We make the following assumptions:

(A1) The operator $A : \mathcal{D}(A) \subset X \to X$ generates a C_0-semigroup e^{At} on X and for some constants $M \geq 1$ and $\omega \in \mathbf{R}$,

$$
\|e^{At}\|_{\mathcal{L}(X)} \leq Me^{\omega t}, \qquad t \geq 0. \tag{2.25}
$$

(A2) Function $B : [0,T] \to \widetilde{\mathcal{B}}(X \times Z; X) \equiv \{B : X \times Z \to X \mid B \text{ is bilinear }\}$ is uniformly Borel measurable and $\|B(\tau)\|_{\widetilde{\mathcal{B}}}$ is bounded by L_0.

(A3) $\nu(\cdot) \in BV_0([0,T]; Y)$ and $\bar{G}(\cdot) \in Y_\nu([0,T]; \mathcal{L}(Y, X))$, with Y being some Banach space.

Theorem 2.5. *Let (A1)–(A2). Then, exists a unique solution $x(\cdot) \in C_\mu([0,T]; X)$ of (2.16). Moreover, there exist well-defined evolution operators $\Phi, \Theta : \bar{\Delta} \to \mathcal{L}(X)$ given as follows:*

$$
\begin{cases} \Phi(t,s)x_0 = e^{A(t-s)}x_0 + \int_s^t e^{A(t-\tau)}B(\tau, \Phi(\tau-0,s)x_0, d\mu(\tau)), \\ \qquad\qquad 0 \leq s \leq t \leq T, \quad x_0 \in X, \\ \Phi(s-0,s) = I, \qquad 0 \leq s \leq T, \end{cases} \tag{2.26}
$$

$$
\begin{cases} \Theta(t,s)x_0 = \int_s^t e^{A(t-\tau)}B(\tau, \Theta(\tau-0,s)x_0, d\mu(\tau)), \\ \qquad\qquad 0 \leq s \leq t \leq T, \quad x_0 \in X, \\ \Theta(s-0,s) = -I, \qquad 0 \leq s \leq T. \end{cases} \tag{2.27}
$$

such that and $x(\cdot)$ can be represented by

$$
x(t) = \Phi(t,0)x_0 + \int_0^t (\Phi(t,s) + \Theta(t,s))\bar{G}(s)d\nu(s), \qquad t \in [0,T]. \tag{2.28}
$$

From (2.27), we see that

$$
\begin{cases} \Theta(s,s) = 0, \\ \Theta(s+0,s) = B(s, -I, \mu(s+0) - \mu(s)). \end{cases} \tag{2.29}
$$

Thus, if $\mu(\cdot)$ is right-continuous at some point $s \in [0,T]$, then,

$$
\Theta(t,s) = 0, \qquad \forall 0 \leq s \leq t \leq T. \tag{2.30}
$$

§3. Optimal Control Problem and Maximum Principle.

In this section, we state our main results concerning the control problem of this paper. We let X and Z be two Banach spaces, K be a convex and closed cone in Z and U be a metric space. We define

$$\mathcal{U} = \{u(\cdot) : [0,T] \to U \mid u(\cdot) \text{ is Borel measurable } \},$$

and

$$\mathcal{M} = \{\mu_0(\cdot)\} + \mathcal{M}_0([0,T]; K),$$

where

$$\mathcal{M}_0([0,T]; K) = \{\mu(\cdot) \in BV_0([0.T]; K) \mid \bar{\mu}(E) = \int_E \theta(\tau)|\bar{\mu}|(d\tau),$$
$$\forall E \in \mathcal{B}([0,T]), \text{ for some } \theta(\cdot), \text{ with } |\theta(\tau)|_Z = 1, \quad |\bar{\mu}| - \text{a.e.}\},$$

and $\mu_0(\cdot) \in BV_0([0,T]; Z)$. We can prove that $\mathcal{M}$ is convex. Next, let us make the following hypothesis:

(H0) The dual X^* of X is strictly convex.

(H1) The same as (A1).

(H2) Maps $F : [0,T] \times X \times U \to \mathcal{L}(Z, X)$ and $f : [0,T] \times X \times U \to Z^*$ satisfy the follwoing:

(i) For any $(x, u) \in X \times U$, the maps $F(\cdot\,, x, u) : [0,T] \to \mathcal{L}(Z, X)$ and $f(\cdot\,, x, u) : [0,T] \to Z^*$ are uniformly Borel measurable; For any $(t, x) \in [0,T] \times U$, the maps $F(t, \cdot\,, u) : X \to \mathcal{L}(Z, X)$ and $f(t, \cdot\,, u) : X \to Z^*$ are continuously Frechet differentiable; For any $(t, x) \in [0,T] \times X$, the maps $F(t, x, \cdot\,) : U \to \mathcal{L}(Z, X)$ and $f(t, x, \cdot\,) : U \to Z^*$ are continuous.

(ii) There exists a constant $L > 0$, such that

$$\|F(t,x,u)\|_{\mathcal{L}(Z,X)} \le L(1 + |x|), \qquad (t,x,u) \in [0,T] \times X \times U, \tag{3.1}$$

$$\|F(t,x,u) - F(t,\hat{x},u)\|_{\mathcal{L}(Z,X)} \le L|x - \hat{x}|, \qquad (t,u) \in [0,T] \times U,\ x, \hat{x} \in X. \tag{3.2}$$

(H3) Set Ω is convex and closed in $X \times X$.

Now, for any $x_0 \in X$ and a pair $(u(\cdot), \mu(\cdot)) \in \mathcal{U} \times \mathcal{M}$, the response of the controlled system is defined to be the unique solution of the following integral equation:

$$x(t) = e^{At}x_0 + \int_0^t e^{A(t-\tau)}F(\tau, x(\tau - 0), u(\tau))d\mu(\tau), \qquad t \in [0,T]. \tag{3.3}$$

From §2, we see that the above equation makes sense. Moreover, by Theorem 2.4, we know that under (H1)–(H2), there exists a unique solution $x(\cdot) \in C_\mu([0,T]; X)$ of (3.3) corresponding to the triplet $(x_0, u(\cdot), \mu(\cdot))$. System (3.3) is a mild form of (1.4). Thus, sometimes, we refer the soultion $x(\cdot)$ of (3.3) the mild solution (1.4). We call $x(t)$ the state of our system at time t, $x(\cdot)$ the trajectory of the system and x_0, $u(\cdot)$ and $\mu(\cdot)$ the intial state, continuous control and measure control, respectively. We see that the measure control is an extended notion of the so-called impulse control ([2,16,17]). Assumption (H0) is technical for proving the maximum principle. For the case X is reflexive or separable, we can always change the norm of X to another equivalent one so that (H0) holds (see [12] for comments).

Remark 3.1. It is very important to notice that by using the Young type integral in (3.3), the trajectory of the system inherits the discontinuity of the measure control $\mu(\cdot)$. In fact, if we let $x(\cdot)$ be the solution of (3.3) corresponding to $(x_0, u(\cdot), \mu(\cdot))$, then, we have

$$\begin{cases} x(t) = x(t-0) + F(t, x(t-0), u(t))[\mu(t) - \mu(t-0)], & t \in (0,T], \\ x(t+0) = x(t) + F(t, x(t-0), u(t))[\mu(t+0) - \mu(t)], & t \in (0,T]. \end{cases}$$

While, the Bochner type integral does not have such a property! As in [32], the Bochner (or Lebesgue) type integral was used and as a result, the trajectories had to be restricted to right-continuous functions. In our case, the trajectories are just of elements in the space $D([0,T]; X) \equiv \{v(\cdot) : [0,T] \to X \mid v(t+0), v(t-0) \text{ exist } \forall t \in [0,T]\}$.

The payoff of our control problem is given by (1.5). The meaning of the right hand side of (1.5) is now clear from §2 if we regard the map f as a $\mathcal{L}(Z, \mathbb{R})$-valued function.

Now, we are ready to state our optimal control problem.

Problem C. *Minimizing functional (1.5) over all $(u(\cdot), \mu(\cdot)) \in \mathcal{U} \times \mathcal{M}$, subject to system (3.3) and the end constraint (1.6).*

Next, let $(x^{\#}(\cdot), u^{\#}(\cdot), \mu^{\#}(\cdot))$ be an optimal solution to Problem C. We define

$$\mathcal{U}^{\#} = \{u(\cdot) \in \mathcal{U} \mid u(\tau) = u^{\#}(\tau), \text{ if } \mu^{\#}(\cdot) \text{ jumps at } \tau\}. \tag{3.4}$$

Note that $F : [0,T] \times X \times U \to \mathcal{L}(Z, X)$, thus, the Frechet derivative F_x of F is a map from $[0,T] \times X \times U$ to $\widetilde{B}(Z \times X; X)$. We let

$$\begin{aligned} B(\tau, \xi, z) &= F_x(\tau, x^{\#}(\tau-0), u^{\#}(\tau); \xi)z \\ &\equiv \lim_{\varepsilon \to 0} \frac{F(\tau, x^{\#}(\tau-0) + \varepsilon\xi, u^{\#}(\tau))z - F(\tau, x^{\#}(\tau-0), u^{\#}(\tau))z}{\varepsilon}, \\ &\forall (\tau, \xi, z) \in [0,T] \times X \times Z. \end{aligned} \tag{3.5}$$

Then, we may define $B(\cdot,\cdot)^* : [0,T] \times Z \to \mathcal{L}(X^*)$ as follows:

$$\langle B(\tau, z)^* x^*, \xi \rangle = \langle B(\tau, \xi, z), x^* \rangle, \qquad \forall (\tau, \xi, z, x^*) \in [0,T] \times X \times Z \times X^*. \tag{3.6}$$

As in §2, we let $\Phi(\cdot,\cdot)$, $\Theta(\cdot,\cdot)$ and $\Psi(\cdot,\cdot)$ be the evolution operators associated with the operator A and the bilinear form valued function $B(\tau, \xi, z)$. Then, we define

(3.7)
$$\begin{aligned} \mathcal{R} = \{ &\int_0^T \Psi(T,\tau)[F(\tau, x^{\#}(\tau-0), u(\tau)) - F(\tau, x^{\#}(\tau-0), u^{\#}(\tau))]d\mu^{\#}(\tau) \\ &+ \int_0^T \Psi(T,\tau)F(\tau, x^{\#}(\tau-0), u^{\#}(\tau))d(\mu(\tau) - \mu^{\#}(\tau)) \mid (u(\cdot), \mu(\cdot)) \in \mathcal{U}^{\#} \times \mathcal{M} \}, \end{aligned}$$

and

$$Q = \{y_1 - \Phi(T,0)y_0 \mid (y_0, y_1) \in \Omega\}. \tag{3.8}$$

We introduce the Z^*-valued Hamiltonian: $\forall(\tau, x, u, \psi^0, \psi) \in [0,T] \times X \times \mathbf{R} \times X^*$

$$H(\tau, x, u, \psi^0, \psi) = \psi^0 f(\tau, x, u) + F(\tau, x, u)^* \psi. \tag{3.9}$$

Our main result for the Problem C is the following

Theorem 3.2. (Maximum Principle) *Let (H0)–(H3) hold. Let* $(x^\#(\cdot), u^\#(\cdot), \mu^\#(\cdot))$ *be an optimal solution to Problem C. Let the set*

$$\mathcal{R} - Q = \{ r - q \mid r \in \mathcal{R}, \quad q \in Q \}$$

be of finite codimensional in X ([7,11,12]). Then, there exists a pair $(\psi(\cdot), \psi^0) \in D([0,T]; X^*) \times \mathbf{R}$, *such that*

$$\begin{aligned} \psi(t) = & e^{A^*(T-t)}\psi(T) + \int_t^T e^{A^*(\tau-t)} B(\tau, d\mu^\#(\tau))^* \psi(\tau) \\ & + \psi^0 \int_t^T e^{A^*(\tau-t)} f_x(\tau, x^\#(\tau - 0), u^\#(\tau))^* d\mu^\#(\tau), \qquad \forall t \in [0,T], \end{aligned} \tag{3.10}$$

$$\psi^0 \le 0, \qquad (\psi(\cdot), \psi^0) \ne 0. \tag{3.11}$$

$$\begin{aligned} & \int_0^T \langle H(\tau, x^\#(\tau - 0), u^\#(\tau), \psi^0, \psi(\tau)), d\mu^\#(\tau) \rangle \\ & = \max_{\mu(\cdot) \in \mathcal{M}} \int_0^T \langle H(\tau, x^\#(\tau - 0), u^\#(\tau), \psi^0, \psi(\tau)), d\mu(\tau) \rangle, \end{aligned} \tag{3.12}$$

$$\begin{aligned} & \int_0^T \langle H(\tau, x^\#(\tau - 0), u^\#(\tau), \psi^0, \psi(\tau)), d\mu^\#(\tau) \rangle \\ & = \max_{u(\cdot) \in \mathcal{U}^\#} \int_0^T \langle H(\tau, x^\#(\tau - 0), u(\tau), \psi^0, \psi(\tau)), d\mu^\#(\tau) \rangle, \end{aligned} \tag{3.13}$$

$$\begin{aligned} & \langle \psi(0), y_0 - x^\#(0) \rangle - \langle \psi(T), y_1 - x^\#(T) \rangle \\ & \quad - \langle \psi(T), \Theta(T,0)(y_0 - x^\#(0)) \rangle \le 0, \qquad (y_0, y_1) \in \Omega. \end{aligned} \tag{3.14}$$

The proof is based on the Ekeland's variational principle ([6]) and the analysis of the variation of trajectories under spike purturbation of controls ([11,12]).

Remark 3.3. Condition (3.14) is the transversality condition for the optimal solution of our problem. The appearance of term

$$\langle \psi(T), \Theta(T,0)(y_0 - x^\#(0)) \rangle$$

is unexpected. This is caused by a possible jump of the measure control $\mu(\cdot)$ at $t = 0$. It is interesting to notice that in the case

$$\mu^\#(0+0) = \mu^\#(0), \tag{3.15}$$

i.e., there is no jump of $\mu^{\#}(\cdot)$ appeared at $t = 0$, then, from (2.29), we see that condintion (3.14) is reduced to the familiar form

$$(3.16) \qquad \langle \psi(0), y_0 - x^{\#}(0) \rangle - \langle \psi(T), y_1 - x^{\#}(T) \rangle \leq 0, \qquad (y_0, y_1) \in \Omega.$$

References

[1] G. Barles, *Deterministic impulse control problems, SIAM J. Control & Optim.*, 23 (1985), 419–432.

[2] A. Bensoussan and J. L. Lions, *Impulse Control and Quasi-Variational Inequalities*, Bordes, Paris, 1984.

[3] L. D. Berkovitz, *Optimal Control Theory*, Springer-Verlag, New York, 1974.

[4] P. C. Das and R. R. Sharma, *On optimal controls for measure delay-differential equations, SIAM J. Control & Optim.*, 9 (1971), 43–61.

[5] J. Diestel and J. J. Uhl, Jr., *Vector Measures*, AMS Providence, 1977.

[6] I. Ekeland,, *Nonconvex minimization problems, Bull. Amer. Math. Soc. (New Series)*, 1(1979), 443–474.

[7] H. O. Fattorini, *A unified theory of necessary conditions for nonlinear nonconvex control systems, Appl. Math & Optim.*, 15 (1987), 141–185.

[8] J. Groh, *A nonlinear Volterra-Stieltjes integral equation and a Gronwall inequality in one dimension, Illinois J. Math.*, 34 (1980), 244–263.

[9] T. H. Hildebrandt, *On systems of linear differentio-Stieltjes-integral equations, Illinois J. Math.*, 3 (1959), 352–373.

[10] C. S. Hönig, *Volterra Stieltjes Integral Equations*, North-Holland, Amsterdam, 1975.

[11] X. Li and Y. Yao, *Maximum principle of distributed parameter systems with time lags, Distributed Parameter Systems, Lecture Notes in Control and Information Sciences*, Springer-Verlag, New York Vol.75, 410–427, 1985.

[12] X. Li and J. Yong, *Necessary conditions of optimal control for distriduted parameter systems*, submitted.

[13] L. S. Pontryagin, V. G. Boltyanskii, R. V. Gamkrelidze and E. F. Mischenko, *Mathematical Theory of Optimal Processes*, Wiley, New York, 1962.

[14] R. W. Rishel, *An extended Pontryagin principle for control systems whose control law contain measures, SIAM J. Control & Optim.*, 3 (1965), 191–205.

[15] R. B. Vinter and F. M. F. L. Pereira, *A maximum principle for optimal processes with discontinuous trajectories, SIAM J. Control & Optim.*, 26 (1988), 205–229.

[16] J. Yong, *Optimal switching and impulse controls for distributed parameter systems, System Sci. & Math. Sci.*, 2 (1989), 137–160.

[17] J. Yong, *Systems governed by ordinary differential equations with continuous, switching and impulse controls, Appl. Math. Optim.*, 20 (1989), 223–236.

[18] W. H. Young, *On integration with respect to a function of bounded variation, Proc. London Math. Soc.*(2), 13 (1914), 109–150.

The Existence and the Uniqueness of Optimal Control of Population Evolution Systems

Yu Jingyuan

(Beijing Insititute of Information and control, Beijing)

Gao Ling

(Insititute of Population Research , People's University of China, Beijing)

Zhu Guangtian

(Insititute of System Science, Academia Sinica, Beijing)

Abstract

In this paper, we give a conclusion of the existence and the uniqueness of optimal control of the population evolution equation with the specific fertility rate of female $\beta(t)$ as parameter.

Recently the population control is an important topic for population workers[1]. How to realize the population control is practically how to control population to approach the population distributed state expected by humanity. Controllability of the population system was made out in [2] In [3] J.Yu etal sketched the necessary conditions for the optimal control with a contral domain consisting of continuous functions . In this paper we shall come to the conclusion that the optimal control is exist uniquely for the control domain consisting of bounded measurable functions .

We shall discuss the equation of population evolution process

$$\begin{cases} \dfrac{\partial p(r,t)}{\partial t} + \dfrac{\partial p(r,t)}{\partial r} = -\mu(r,t)p(r,t) & 0<r<r_m,\ t>0 \\ p(r,0) = p_0(r) & 0\le r\le r_m \\ p(0,t)=\beta(t)\int_{r_1}^{r_2} k(s)h(s)p(s,t)ds & 0\le t\le T \end{cases} \qquad (1).$$

Where t denotes time , r denotes age , r_m is the highest age ever attained by individuals of the population , p(r,t) is called age density function , $\mu(r)$ is the relative mortality function , $p^0(r)$ is an initial age density of the population, $\beta(t)$ is the specific fertility rate of female at time t, k(r) is the female sex ratio at age r , h(r) is the fertility pattern and $[r_1,r_2]$ is the fecundity period of females .

Suppose that $k(r)$, $h(r)$ and $\mu(r)$ are measurable functions on $[0,r_m]$, $\beta(t)$ is a measurable function on $[0,T]$, and conditions

(a) $0 \le k(r) \le 1$, $\quad r \in [0, r_m]$

$0 < k_0 \le k(r) < 1$, $\quad r \in [r_1,r_2]$

(b) $0 < h(r) \le h_0$, $\quad r \in (r_1,r_2)$

$\int_{r_1}^{r_2} h(r)dr = 1$; $\quad h(r) = 0 \quad r \notin (r_1, r_2)$

(c) $\int_0^r \mu(\rho)d\rho < \infty$, $\quad r < r_m$, $\quad \int_0^{r_m} \mu(\rho)d\rho = \infty$

(d) $p_0(r)$ is measurable on $[0,r_m]$ and

$p_0(r) > 0$, for $r \in [0, r_2]$

Where k_0, h_0 are two positive constants.

For two given positive constants $\alpha_1 < \alpha_2$ we define the control domain

$$U_{\alpha_2}^{\alpha_1} = \{ \beta(t) \mid \alpha_1 \le \beta(t) \le \alpha_2 \text{ a.e., } \beta(t) \text{ is measurable}.\}$$

The cost functional of our population control system is given by

$$J(\beta) = \int_0^T \int_0^{r_m} [p(r,t;\beta)-p^0(r,t)]^2 drdt \qquad \text{for all } \beta \in U$$

where $p^0(r,t)$ is given a non-negative L^2 goal function, on $[0,r_m] \times [0,T]$, $p(r,t;\beta)$ is the solution to the equation (1) conesponding to β which has following representation

$$p(r,t;\beta) = \begin{cases} p_0(r-t)e^{-\int_{r-t}^{r} \mu(\rho)d\rho}, & \text{for } r \ge t \\ \beta(t-r)\int_{r_1}^{r_2} k(s)h(s)p(s,t-r;\beta)ds\, e^{-\int_0^r \mu(\rho)d\rho} & \text{for } r < t \end{cases} \tag{2}$$

It is obvious that $p(r,t;\beta) \ge 0$ and

$$p(r,t;\beta) \in L^2([0,r_m] \times [0,T])$$

$p(r,t;\beta)$ may be expended into[4]

$$p(r,t;\beta)=p_0(r-t)e^{-\int_{r-t}^{r}\mu(\rho)d\rho} + \sum_{k=0}^{\infty} \phi_k(t-r)e^{-\int_0^r \mu(\rho)d\rho} \tag{3}.$$

here

$$\phi_0(t)=\beta(t)\int_{r_1}^{r_2} k(s)h(s)e^{-\int_{s-t}^{s}\mu(\rho)d\rho} p_0(s-t)ds \ ,$$

$$\phi_k(t)=\beta(t)\int_{r_1}^{r_2} k(s)h(s)e^{-\int_{0}^{s}\mu(\rho)d\rho} \phi_{k-1}(t-s)ds \ , \quad k=1,2....$$

If for $s < 0$ we define $p_0(s)=0$, then $\phi_k(t)$, $k=1,2,...,$are well defined on R and

$$\text{Supp } \{\phi_k\} = [kr_1, (k+1)r_2]$$

Recall that the optimal control of the population evolution system (1) is to find a $\beta^* \in U = U_{\alpha_2}^{\alpha_1}$ such that

$$J(\beta^*) = \min_{\beta\in U} J(\beta) \qquad (4).$$

In this paper we shall prove that there exists such a $\beta^* \in$ U uniguely .

Lemma 1. *If $\beta \in U$, then*

$$p(r,t;\beta) > 0 \ ,$$

fot $t \in [0,T]$ and $r \in [0,r_2+t] \cap [0,r_m]$.

Proof . By (3) we have

$$p(r,t;\beta)=p_0(r-t)e^{-\int_{r-t}^{r}\mu(\rho)d\rho} >0 \quad \text{for } r \geq t$$

If $r < t$, then there exists a integer k with $t-r \in [kr_1,(k+1)r_2]$. Since the support set of ϕ_k is $[kr_1,(k+1)r_2]$, we have $\phi_k(t-r) > 0$. So $p(r,t;\beta(\cdot)) > 0$.

Lemma 2 . *If β_1, $\beta_2 \in U$, and in $L^2([0,T])$, $\beta_1 \neq \beta_2$, then in $L^2([0,r_m] \times [0,T])$*

$$p(r,t;\beta_1) \neq p(r,t;\beta_2) \ .$$

Proof . If the conclusion is false , i.e. there exist β_1 and β_2 in U , $\beta_1 \neq \beta_2$, such that

$$p(r,t;\beta_1) = p(r,t;\beta_2)$$

it follows for all $t \geq r$

$$p(r,t;\beta_1) - p(r,t;\beta_2)=$$

$$= \beta_1(t-r)\int_{r_1}^{r_2} k(s)h(s)p(s,t-r;\beta_1)ds\,e^{-\int_0^r \mu(\rho)d\rho} -$$

$$- \beta_2(t-r)\int_{r_1}^{r_2} k(s)h(s)p(s,t-r;\beta_2)ds\,e^{-\int_0^r \mu(\rho)d\rho}$$

$$= [\beta_1(t-r)-\beta_2(t-r)]\int_{r_1}^{r_2} k(s)h(s)p(s,t-r;\beta_2)ds\,e^{-\int_0^r \mu(\rho)d\rho} = 0$$

By Lemma 1 and $k(s) \geq k_0 > 0$, $s \in [r_1,r_2]$,

$$\int_{r_1}^{r_2} k(s)h(s)p(s,t-r;\beta_2)ds\,e^{-\int_0^r \mu(\rho)d\rho} > 0$$

hence

$$\beta_1(t) = \beta_2(t) \ , \quad \text{a.e.} \quad \text{for all } t \geq 0 \ .$$

It contradicts our supposition . Lemma 2 is proved .

Lemma 3 . *Let* $\beta_n \in U$, $n = 1,2,\ldots.$ *If* β_n *converge to* β^* *in* $L^2([0,T])$, *then* $\beta^* \in U$ *and*

$$L^2\text{—}\lim_{n\to\infty} p(r,t;\beta_n) = p(r,t;\beta^*)$$

Proof. By the formula (2) , $p(r,t;\beta_n)$ and $p(r,t;\beta^*)$ can be extened into

$$p(r,t;\beta_n)=p_0(r-t)e^{-\int_{r-t}^{r} \mu(\rho)d\rho} + \sum_{k=0}^{\infty} \phi_k^{(n)}(t-r)e^{-\int_0^r \mu(\rho)d\rho}$$

$$p(r,t;\beta^*)=p_0(r-t)e^{-\int_{r-t}^{r} \mu(\rho)d\rho} + \sum_{k=0}^{\infty} \phi_k^*(t-r)e^{-\int_0^r \mu(\rho)d\rho}$$

respectively . So it is only need to prove that for any k , $\phi_k^{(n)}(t)$ L^2– converges to $\phi_k^*(t)$. Since

$$\phi_0^{(n)}(t)=\beta_n(t)\int_{r_1}^{r_2} k(s)h(s)e^{-\int_{s-t}^{s} \mu(\rho)d\rho} p_0(s-t)ds \ ,$$

$$\phi_0^*(t)=\beta^*(t)\int_{r_1}^{r_2}k(s)h(s)e^{-\int_{s-t}^{s}\mu(\rho)d\rho}p_0(s-t)ds \quad ,$$

So by $\beta^{(n)}\xrightarrow{L^2}\beta^*$,

$$\int_0^T[\phi_0^{(n)}(t)-\phi_0^*(t)]^2dt=$$

$$=\int_0^T(\beta_n(t)-\beta^*(t))^2 x \quad [\int_{r_1}^{r_2}k(s)h(s)e^{-\int_{s-t}^{s}\mu(\rho)d\rho}p_0(s-t)ds]^2dt$$

$$\leq C\int_0^T[\beta_n(t)-\beta^*(t)]^2dt \longrightarrow 0 \ .$$

If for some $k>0$, $\phi_k^{(n)}\xrightarrow{L^2}\phi_k^*$, then

$$\int_0^T[\phi_{k+1}^{(n)}(t)-\phi_{k+1}^*(t)]^2dt \quad \leq$$

$$\leq 2\int_0^T[\beta_n(t)-\beta^*(t)]^2x[\int_{r_1}^{r_2}k(s)h(s)e^{-\int_0^{s}\mu(\rho)d\rho}\phi_k^{(n)}(t-s)ds]^2dt$$

$$+2\int_0^T[\beta^*(t)]^2x \quad [\int_{r_1}^{r_2}k(s)h(s)e^{-\int_0^{s}\mu(\rho)d\rho}(\phi_k^{(n)}(t-s)ds-\phi_k^*(t-s))ds]^2dt$$

$$\leq 2\max[\int_{r_1}^{r_2}k(s)h(s)e^{-\int_0^{s}\mu(\rho)d\rho}\phi_k^{(n)}(t-s)ds]^2dt]^2x\int_0^T[\beta_n(t)-\beta^*(t)]^2dt$$

$$+\ 2\ h_0^2\ \|\ \beta^*\|^2\ \|\ \phi_k^{(n)}-\phi_k^*\ \|^2_{L^2}$$

$$\leq\ 2\ h_0(r_2-r_1)\ \|\ \phi_k^{(n)}(t)\|^2_{L^2}\ \|\ \beta_n(t)-\beta^*(t)\ \|^2_{L^2}$$

$$+\ 2\ h_0^2\ \|\ \beta^*\|^2\ \|\ \phi_k^{(n)}-\phi_k^*\ \|^2_{L^2}$$

Since $\phi_k^{(n)}(t)\geq 0$. Note

$$\|\ \phi_k^{(n)}(t)\ \|^2\ \leq\ h_0^2\ \alpha_2^2\ \|\ \phi_{k-1}^{(n)}\ \|^2_{L^2}$$

$$\leq\ \leq\ h_0^{2k+1}\ \alpha_2^{2k+1}\ \|\ p_0\|^2_{L^2}\ =\ C\ <\ \infty \quad .$$

So

$$\|\phi_{k+1}^{(n)}(t)-\phi_{k+1}^*(t)\|^2_{L^2}\leq$$

$$\leq 2Ch_0(r_2-r_1)\|\beta_n-\beta^*\|^2+2h_0^2\|\beta^*\|^2\|\phi_k^{(n)}-\phi_k^*\|^2_{L^2}\xrightarrow{n\to\infty}0.$$

Hence

$$L^2\text{—}\lim \phi_{k+1}^{(n)}(t) = \phi_{k+1}^{*}(t) .$$

The lemma is proved .

Denote

$$A = \{p(r,t) \in L^2([0,r_m] \times [0,T] \mid \text{there exists a } \beta \in U \text{ such that } p(r,t)=p(r.t;\beta)\}.$$

Lemma 4 . *A is a bound closed convex set in* $L^2([0,r_m] \times [0,T])$.

Proof . We firstly prove convexity . Let $p_1(r,t)$, $p_2(r,t) \in A$, then there are β_1 , $\beta_2 \in U$ such that

$$p_1(r,t) = p(r,t;\beta_1) ,$$
$$p_2(r,t) = p(r,t;\beta_2) .$$

For $0 < \alpha < 1$, the formular (2) implies

$$\alpha p_1(r,t) + (1-\alpha)\, p_2(r,t) =$$

$$= \begin{cases} p_0(r-t)e^{-\int_{r-t}^{r}\mu(\rho)d\rho} & r \geq t \\ \alpha\beta_1(t-r)\int_{r_1}^{r_2}k(s)h(s)p_1(s,t-r)ds\,e^{-\int_0^r\mu(\rho)d\rho} + \\ + (1-\alpha)\beta_2(t-r)\int_{r_1}^{r_2}k(s)h(s)p_2(s,t-r)ds\,e^{-\int_0^r\mu(\rho)d\rho} & r < t \end{cases}$$

$$= \begin{cases} p_0(r-t)e^{-\int_{r-t}^{r}\mu(\rho)d\rho} & r \geq t \\ \beta(t-r)\int_{r_1}^{r_2}k(s)h(s)[\alpha p_1(s,t-r)+(1-\alpha)p_2(s,t-r)]ds\,e^{-\int_0^r\mu(\rho)d\rho} & r < t \end{cases}$$

here

$$\beta(t)= \frac{\alpha\beta_1(t)\int_{r_1}^{r_2}k(s)h(s)p_1(s,t)ds+(1-\alpha)\beta_2(t)\int_{r_1}^{r_2}k(s)h(s)p_2(s,t)ds}{\int_{r_1}^{r_2}k(s)h(s)[\alpha p_1(s,t)+(1-\alpha)p_2(s,t)]ds}$$

By Lemma 1 , $p_1(r,t)$, $p_2(r,t) > 0$, for all $(r,t) \in [0,r_m] \times [0.T]$. Therefore

$$\int_{r_1}^{r_2} k(s)h(s)[\alpha p_1(s,t-r)+(1-\alpha)p_2(s,t-r)]ds > 0$$

for all $t \in [0,T]$

hence $\beta(t)$ is well defined . Obviously

$$\alpha_1 \le \beta(t) \le \alpha_2 , \qquad \text{for all } t \in [0,T] ,$$

So $\beta \in U$. It shows $\alpha p_1+(1-\alpha)p_2 \in A$,hence A is a convex set.

Now we prove that A is closed .Let $p_n(r,t) \in A$, $p_n(r,t)$ L^2– converge $p^*(r,t)$. According to the definition of the set A . there exist $\beta_n \in U$ .n =1,2,..., with $p_n(r,t) = p(r,t;\beta_n)$. Therefore there exists a subsequence $\{\beta_{n_i}\}$ of $\{\beta_n\}$ and β^* in U such that β_{n_i} weakly converge to β^* .

By

$$p_{n_i}(r,t) = \begin{cases} p_0(r-t)e^{-\int_{r-t}^{r}\mu(\rho)d\rho} & r \ge t \\ \beta_{n_i}(t-r)\int_{r_1}^{r_2}k(s)h(s)p(s,t-r;\beta)ds\,e^{-\int_0^r\mu(\rho)d\rho} & r < t , \end{cases}$$

The term on the left side weakly converges to $p(r,t)$ as $i \to \infty$ and one on the right side converges to

$$\begin{cases} p_0(r-t)e^{-\int_{r-t}^{r}\mu(\rho)d\rho} & r \ge t \\ \beta^*(t-r)\int_{r_1}^{r_2}k(s)h(s)p(s,t-r;\beta)ds\,e^{-\int_0^r\mu(\rho)d\rho} & r < t , \end{cases}$$

Hence

$$p(r,t) = p(r,t;\beta^*) ,$$

It shows that A is closed .

It is obvious that A be bounded .Hence Lemma 4 is proved.

Now we define a new cost functional

$$\bar{J}(\beta) = \int_0^T \int_0^{r_m} [p(r,t;\beta)-p^0(r,t)]^2 drdt$$

and consider the new problem of the optimal control

$$\begin{cases} \bar{J}(p^*) = \min\limits_{p \in A} \bar{J}(p) \\ p^* \in A \end{cases} \tag{5}$$

By Lemma 2, the existence and the uniqueness of solutions to (4) are equivalent to those to (5).

Lemma 5. *$\bar{J}$ is a strictly convex functional on A.*

Proof. Let $p_1, p_2 \in A$, $0 < \alpha < 1$, we have

$$\begin{aligned}
\bar{J}(\alpha p_1 + (1-\alpha)p_2) &= \int_0^T \int_0^{r_m} [\alpha p_1 + (1-\alpha)p_2 - p^0]^2 drdt \\
&= \int_0^T \int_0^{r_m} [\alpha(p_1-p^0) + (1-\alpha)(p_2-p^0)]^2 drdt \\
&= \int_0^T \int_0^{r_m} [\alpha^2(p_1-p^0)^2 + (1-\alpha)^2(p_2-p^0)^2] drdt + \\
&+ \int_0^T \int_0^{r_m} [2\alpha(1-\alpha)(p_1-p^0)(p_2-p^0)] drdt \\
&= \int_0^T \int_0^{r_m} [\alpha(p_1-p^0)^2 + (1-\alpha)(p_2-p^0)^2] drdt - \\
&- \int_0^T \int_0^{r_m} [\alpha(1-\alpha)(p_1-p_2)^2] drdt \\
&\le \int_0^T \int_0^{r_m} [\alpha(p_1-p^0)^2 + (1-\alpha)(p_2-p^0)^2] drdt \\
&= \alpha\bar{J}(p_1) + (1-\alpha)\bar{J}(p_2).
\end{aligned} \tag{6}$$

The above equality holds if and only if $p_1 = p_2$ almost everywhere.

From this the strictly convexity of $\bar{J}$ is at once obtained. Lemma 5 is proved.

Lemma 6 . *Let $p_n \in A$, $n = 1,2,\ldots,\{p_n\}$ weakly converge to p^* , then*

$$\bar{J}(p^*) \le \lim_n \inf \bar{J}(p_n) .$$

Proof . Since $\bar{J}(p_n) = \| p_n - p^o \|^2_{L^2}$, P_n weakly converge to p^* . So

$$\bar{J}(p^*) = \|p^* - p^o\|^2_{L^2} \le \liminf_n \| p_n - p^o \|^2_{L^2}$$

$$= \liminf_n \bar{J}(p_n) .$$

Lemma 6 is proved.

Theorem . *There exists a unique optimal control for the population evolution system (1) .*

Proof . In order to establish the existence of the optimal control , it is sufficient to show that there exists a $p^*(r,t) \in A$ such that

$$\bar{J}(p^*) = \inf_{p \in A} \bar{J}(p) .$$

Let $p_n \in A$, n=1,2,..., be a minimizing sequence of $\bar{J}$, i.e.

$$\lim_{n \to \infty} \bar{J}(p_n) = \inf_{p \in A} \bar{J}(p) .$$

Since A is a bounded closed convex set in $L^2([0,r_m] \times [0.T])$, so A is also bounded weakly closed , hence weakly compack , there is a subseqnence $\{p_{n_i}\}$ of $\{p_n\}$ which converges to $p^* \in A$. By Lemma 6 ,

$$\bar{J}(p^*) \le \liminf_{i \to \infty} \bar{J}(p_{n_i}) = \inf_{p \in A} \bar{J}(p)$$

This shows that the existence holds .

The uniqueness is implied in the result of Lemma 5 .The theorem is proved .

Reference

[1] . Song Jing , Tuan Chihsien and Yu Jingyuan , Population Control in China , New York , 1985 .

[2] . Song Jian , Yu Jingyuan ,Liu Changkai ,Zhang Lianping, Zhu Guangtian , The specture properties of the population evolution operator and the controllablity of the population system (in Chinese), Sciences Sinica A , 2,1986 .

[3] . Yu Jingyuan , Guo Baozhu , Zhu Guangtian , Optimal control of population system, Control Theory and Applications (in Chinese), 1,1989.

[4] . Yu Jingyuan , Zhu Guangtian , Guo Baozhu , The asymptotical properties and the controllability of the population evolution process in the L^p space ,System Sciences and Mathematics , No.2, 1987.

REACHABILITY FOR A CLASS OF NONLINEAR DISTRIBUTED SYSTEMS GOVERNED BY PARABOLIC VARIATIONAL INEQUALITIES

Y. Zhao and Y. Huang①
Zhongshan University, Guangzhou China

W. L. Chan②
The Chinese University of Hong Kong, Hong Kong

Abstract

In this paper the reachability for a class of distributed control problems governed by parabolic variational inequalities is considered. The negative reasult for the complete reachability of the system is obtained. Sufficient conditions for approximate reachability of the system are discussed under two kinds of hypotheses.

1. INTRODUCTION

The reachability of linear or semilinear distributed parameter control systems has been surveyed in[8]. But the reachability of the following system governed by the parabolic variational inequality, to our knowledge, has not been treated,

$$(1,1)\qquad \begin{cases} \langle y'(t)+Ay(t), y(t)-z\rangle+\varphi(y(t))-\varphi(z)\leqslant\langle Bu(t), y(t)-z\rangle \\ y(0)=y_0 \qquad\qquad \text{a. e. } t\in[0,T], z\in V \end{cases}$$

where V and H are real separable Hilbert spaces, $V \subset H \subset V'$ and denote the norms of H, V and V' by $|\cdot|$, $\|\cdot\|$ and $\|\cdot\|_*$ respectively and $\langle\cdot,\cdot\rangle$ the pairing between V and V'.

Assume that A is a linear continuous symmetric operator from V to V' and satisfies the coercive condition:

$$(1,2)\qquad \langle Ay, y\rangle \geqslant w\|y\|^2 \qquad \forall y\in V$$

for some $w>0$, φ: $V\to\bar{R}$ is a lower-semicontiuous convex functional. B is a linear bounded operator from U to H where U is a Hilbert space. $y_0\in V$.

Let: $\partial\varphi$: $V\to V'$ be the subdifferential of φ so that (1,1) is equivalent to

① Research supported partly by the Foundation of Zhongshan University Advance Research Centre.
② Reseaarch Supported by UPGC Direct Grant, CUHK.

(1,3) $$\begin{cases} y'(t) + A(y) + \partial\varphi(y(t)) \ni Bu(t) \\ y(0) = y_0 \end{cases}$$

We have following result

Theorem 1. 1 Under the above hypotheses if $y_0 \in V$ such that $\varphi(y_0) < +\infty$, then (1,3) has a unique solution

$$y \in W^{1,2}(0,T;H) \cap L^2(0,T;V)$$

for every given u $\in \mathcal{U} = L^2(0,T;U)$. Moreover, $(y_0,u) \to y$ is a Lipschitz mapping from $H \times L^2(0,T;U)$ to $C(0,T;H) \cap L^2(0,T;H)$ and

(1,4) $$y'(t) = (Bu(t) - Ay(t) - \partial\varphi(y(t))^0 \text{ a. e. } t \in (0,T)$$

Where M^0 denote the minimum element of the set M with respect to the norm in H.

In this sequel we will prove the negative result for complete reachability and the approximate reachability under two kinds of hyprtheses for (1,1) or equivalently (1,3) motivated by the works[4],[5]and[10].

2. THE REACHABILITY OF (1. 1) OR (1. 3)

Assume that hypotheses of theorem 1. 1 hold throughout this section. Let

$$K(y_0)(t) = \{y = y(t,y_0,u(\cdot)) : y \text{ is the solution of } (1.3) \text{ corresponding to } u, u \text{ runs over the space } L^2(0,T;U)\}$$

be the reachable set of (1,3) with initial value y_0 at moment $t \in [0,T]$.

Definition 2. 1 (1. 3) is called approximately reachable if there exists $t \in [0,T]$ such that $\overline{K(y_0)(t)} = H$.

Generally, as well know $K(y_0)(t)$ $(t \in [0,T])$ is not a closed subset in H even for linear parabolic distributed parameter systems. Similar result is also true for (1,3).

Theorem 2. 1 If the level set

(2,1) $$\{y \in H : |y|^2 + \langle Ay, y\rangle + \varphi(y) \leq c\}$$

is compact in H, then $y_n \to y$ in $C(0,T;H)$ as $u_n \xrightarrow{w} u$ in $L^2(0,T;U)$ where y_n are solutions of (1. 3) corresponding to u_n. Moreover, y is the solution of (1. 3) corresponding to u and

$$K(y_0) = \bigcup_{\substack{0 \leq t \leq T \\ u \in L^2(0,T,U)}} y(t, y_0, u)$$

is contained in the sum aggregate of countable compact subsets in H. Consequently, $K(y_0)$ has dense complementary set in H when $\dim H = \infty$.

Proof: Let $u_n \xrightarrow{w} u$ and y_n be corresponding solutions of (1. 3), i. e.

(2,2) $$\begin{cases} y'_n(t) + Ay_n(t) + \partial\varphi(y_n(t)) \ni Bu_n(t), \text{ a. e. } t \in [0,T] \\ y_n(0) = y_0 \end{cases}$$

Acting on both sides of (2,1) by y'_n and y_n repectively and integrate over $[0,T]$ we obtain that

(2,3) $$2^{-1}\int_0^t |y'_n|^2 ds + 2^{-1} \langle Ay_n, y_n\rangle + \varphi(y_n) \leq 2^{-1}\langle Ay_0, y_0\rangle + \varphi(y_0) + 2^{-1}\int_0^t |Bu_n(t)|^2 ds \leq c_1$$

(2. 4) $$|y_n(t) - y_0|^2 + \int_0^t \| y_n(s) \|^2 ds \leq c_2 \int_0^T |Bu_n(t)|^2 ds$$

It follows from (2,1), (2,3) and (2,4) that $\{y_n\}$ is a precompact subset in $C(0,T;H)$

by the Arzela—Ascoli theorem. So, there exists a subsequence of $\{y_n\}$, still denoted by $\{y_n\}$, such that (2.5) $y_n(t) \to y(t)$ in H uniformly for $t \in [0,T]$.

Furthermore, by a well known method in [2] we can see that y(t) is a solution of (1.3).

Let

$$K_n(y_0) = \bigcup_{\substack{0 \leqslant t \leqslant T \\ \| u \| \leqslant n}} y(t, y_0, u)$$

one can see that $K_n(y_0)$ is precompact in H by the above result so that $K(y_0)$ is contained in the sum aggregate of countable compact subsets in H because

$$K(y_0) \subset \bigcup_{n=1}^{\infty} K_n(y_0).$$

Moreover, if dim $H = \infty$ then $K(y_0)$ has dense complementary set in H by Baire's Category lemma.

In order to discuss the appro ximate reachability we assume that

(F_1) $$B\mathcal{U} = L^2(0,T;H)$$

where $B\mathcal{U} = \{y : y(t) = Bu(t), u \in L^2(0,T;U)\}$

(F_2) $D(A) \subset D(\partial\varphi)$ and $\partial\varphi$ satisfies the growth condition as follows

(2.7) $\| z \|_* \leqslant c(|\varphi(x)| + |x| + 1)$ $\quad$ *a.e.* $x \in D(\varphi), z \in \partial\varphi(x)$.

THeorem 2.2 $\quad$ If(F1) and (F2) hold then

(2,8) $$\overline{K(y_0)(t)} = H \text{ for every given } t \in [0,T].$$

To prove the theorem it is sufficient to prove that $\overline{K(y_0)(T)} = H$ and we have to prove three lemmas first.

Let us introduce a linear system as follows.

(2,9) $$\begin{cases} y'(t) + Ay(t) = Bu(t) \\ y(0) = y_0 \end{cases}$$

Lemma 2.1 $\quad$ For any given $z_T \in H$ and $\eta > 0$, there exist $u_1 \in W^{1,2}(0,T;U)$ and $y_1(t)$ satisfying (2,9) and $|y_1(T) - z_T| < c_1\eta$ (c_1 is a constant), $y_1 \in C(0,T;V) \cap W^{1,2}(0,T;H)$, $y_1(t) \in D(A)$, $t \in [0,T]$.

Proof: According to the results of theorem 1.1 and (F_1) there exist $u_2 \in L^2(0,T;U)$ and $y_2(t)$ satisfying (2,9) such that

(2.10) $$|y_2(T) - z_T| < \eta$$

Since $W^{1,2}(0,T;U)$ is dense in $L^2(0,T;U)$, so there exists $u_1 \in W^{1,2}(0,T;U)$ such that

(2,11) $$\int_0^t |u_1(t) - u_2(t)|^2 dt < \eta^2$$

Consequently, we have

(2,12) $$|y_1(t) - y_2(t)|^2 \leqslant L\int_0^T |u_1(t) - u_2(t)|^2 dt \leqslant L\eta^2$$

by the results of therom 1.1 so that

$$|y_1(T) - z_T| \leqslant |y_1(T) - y_2(T)| + |y_2(T) - z_T| < (1 + L^{1/2})\eta$$

where $L > 0$ is the Lipschitz constant.

The other conclusions of the lemma can be deduced by the orem 4.3 in [2]

Lemma 2.2 $\quad$ There exists $z \in L^2(0,T;H)$ such that $z(t) \in \partial\varphi(y_1(t))$ where y_1 is given by lemma 2.1.

Proof: According to the results of lemma 2. 1 and (F_2) one has $y_1(t)\in D(A)\subset D(\partial\varphi), t\in[0,T]$. Define a set valued mapping $P(\cdot):R\rightarrow 2^{V'}$ as follows

$$(2,13)\qquad P(t)=\begin{cases}\partial\varphi(y_1(0)) & for\quad t\leqslant 0\\ \partial\varphi(y_1(t)) & for\quad o<t\leqslant T\\ \partial\varphi(y_1(T)) & for\quad t>T\end{cases}$$

One knows that for every $t\in R$, $P(t)$ is a closed convex subset in V' by the defintion of $\partial\varphi$ and it follows from the completeness of V' that $P(t)$ is complete.

On the other hand we can also prove that P is quasi—upper—semicontinuous, i. e. ,

$$(2,14)\qquad \bigcap_{\varepsilon>0} cl\{P(t);t\in N_\varepsilon(t_0)\}=\bigcap_{\varepsilon>0} P(N_\varepsilon(t_0))\subseteq P(t_0)$$

where $N_\varepsilon(t_0)$ is a ε—neighbourhood of $t_0\in R$. In fact, for $w\in\bigcap_{\varepsilon_n>0}P(N_{\varepsilon_n}(t_0))$ and $\varepsilon_n\downarrow 0$, we have $w\in P(N_{\varepsilon_n}(t_0))$. So, there exists $t_n\in N_{\varepsilon_n}(t_0)$ such that $w\in P(t_n)$ and $t_n\rightarrow t_0$, If $t_0\notin[0,T]$ then (2,14) holds obviously by (2,13). When $t_0\in[0,T]$, We have

$$(2,15)\qquad y_1(t_n)\rightarrow y_1(t_0)\ in\ V$$

because $y_1\in C(0,T;V)$. Hence $w\in\partial\varphi(y_1(t_0)=P(t_0)$ [2]. So, P is measurable [7] and there exists a measurable selection $z_1(\cdot):R\rightarrow V'$ such that $z_1(t)\in P(t)$ by the selection theorem [7] since V' is separable and R is locally compact. Let $z(t)$ be the restriction of $z_1(t)$ on $[0,T]$ and one can see that $\varphi(y_1(\cdot))$ is absolutely continuous with $y_1\in W^{1,2}(0,T;H)$ taken into account as given in lemma 2. 1. Consequently, $z\in L^2(0,T;H)$ by (F2).

Now introduce a system as follows

$$(2.16)\qquad \begin{cases}y'(t)+Ay(t)+\partial\varphi(y(t)+y_1(t))\ni Bu(t), a.\ e.\ t\in[0,T]\\ y(0)=y_0\end{cases}$$

where y_1 is given in lemma 2. 1

Lemma 2. 3 There exist $\bar{u}\in L^2(0,T;H)$ and $\bar{y}(t)$ satisfying (2. 16) and $|\bar{y}(T)|<c_2\eta$ (c_2 is a constant).

Proof: For every given $u\in L^2(0,T;U)$ the existence of the corresponding solution $y(t)$ of (2,16) is given by thorem 1. 1 and lemma 2. 1 and y satisfies the results of theorem 1. 1.

Since $\overline{B\mathcal{U}}=L^2(0,T;U)$ by (F_1) and $V\subset H\subset V'$, B is dense in $L^2(0,T;V')$ and there exists $\bar{u}\in L^2(0,T;U)$ such that

$$(2.17)\qquad \int_0^T\|B\bar{u}-z(t)\|_*^2dt\leqslant\eta$$

by lemma 2. 2 where $z(t)$ is given in lemma 2. 2. After subtracting $z(t)$ on both sides of (2. 16) and taking the scalar product on both sides by $\bar{y}(t)$ which is the solution of (2. 16) corresponding to $\bar{u}$, we can deduce that

$$|\bar{y}(t)|^2+w\int_0^T\|\bar{y}(t)\|^2dt\leqslant\int_0^T\|B\bar{u}-z(t)\|_*\|\bar{y}(t)\|dt$$

by (1. 2) and monotonicity of $\partial\varphi$. Consequently, we have

$$|\bar{y}\ (T)|<c\eta.$$

Now we come to the proof of theorem 2. 2: According to lemma 2. 1—lemma 2. 3. for any given $z_T\in H$ and $\eta>0$ there exist $u_1\in W^{1,2}(0,T;U)$ and $\bar{u}\in L^2(0,T;U)$ such that

$$(2.18)\qquad |y_1(T)-z_T|<c_1\eta,\ |\bar{y}(T)|<c_2\eta$$

where $y_1(t)$ and $\bar{y}(t)$ are the solutions of (2. 9) and (2. 16) corresponding to u_1 and $\bar{u}$ respectively.

Taking $u(t)=u_1(t)+\bar{u}(t)$ we can see that the solution of (1. 3) corrfespionding to u is given by $y(t)=y_1(t)+\bar{y}(t)$ and

$$|y(T)-z_T| \leqslant |y_1(T)-z_T| + |\bar{y}(T)| < (c_1+c_2)\eta$$

so the proof is complete.

The approximate reachability of the systems with boundary of distribution—boundary control can be obtained by the similar method above via selection theorem.

3. DEGRADATION OF (F_1)

In practical problems there exist many systems which are approximately reachable do not satisfy (F_1) in section 2[9]. So, in this section we attempt to discuss the approximate reachabillity of (1. 3) by introducing comdition (F'_1) which is weaker than (F1), but (F2) has to be supplied.

Throughtout this section we assume that φ is a lower—semicontinous convex functional from H to R.

Introduce smoothing system corresponding to (1. 3) as follows

$$(3.1) \qquad \begin{cases} y'(t) + A_H y(t) + \nabla\varphi_\varepsilon(y(t)) = Bu(t) \ a.e\ t \in (0,T) \\ y(0) = y_0 \end{cases}$$

Where A_H is the restriction of A in H, this is,

$$(3.2) \qquad A_H y = Ay \qquad \forall\, y \in D(A_H) = \{y \in V; Ay \in H\}$$

Let

$$\varphi_\varepsilon(x) = inf\{|x-y|^2/2\varepsilon + \varphi(y), y \in H\} \qquad \varepsilon > 0$$

Lemma 3. 1[1] Assume that there exists constant $c>0$ independent of ε such that

$$(Ay, \nabla\varphi_\varepsilon(y)) \geqslant -c(1+|\nabla\varphi_\varepsilon(y)|)(1+|y|) \qquad \forall\, y \in D(A_H)$$

and $y_0 \in V$. Then, for any given $u \in L^2(0,T;U)$ and $\varepsilon>0$ (3. 1) has unique solution $y_\varepsilon \in W^{1,2}([0,T];H) \cap C([0,T];V)$, $y'_\varepsilon \in L^2(0,T;D(A_H))$. Moreover, we have

$$y_\varepsilon \to y \quad in\ C(0,T;H) \cap L^2(0,T;V)$$

$$y'_\varepsilon \overset{w}{\to} y' \quad in\ L^2(0,T;H)$$

$$A_H y_\varepsilon \overset{w}{\to} Ay \quad in\ L^2(0,T;H)$$

$$\nabla\varphi_\varepsilon(y_\varepsilon) \overset{w}{\to} Bu - y' - Ay \in \partial\varphi(y) \ in\ L^2(0,T;H)$$

as $\varepsilon \to 0$ where $y \in W^{1,2}([0,T];H) \cap C([0,T];V)$ is the solution of (1. 3) with the same control u.

One can see that A_H generates a semigroup S(t) on H since A_H is linear symmetric maximal monotone operator in H. So, the mild solution of (3. 1) can be expresed by

$$(3.4) \qquad y_\varepsilon(t) = S(t)y_0 + \int_0^t S(t-s)(\nabla\varphi_\varepsilon(y_\varepsilon(s)) + Bu(s))ds$$

Let

$$(3.5) \quad G(t_1,t_2)p(\cdot) = \int_{t1}^{t2} S(t_2-s)p(s)ds; L^2(0,T;H) \to H. \ \forall\, t_2 > t_1, t_1, t_2 \in (0,T)$$

In order to discuss the approximate reachability of (3. 1) and (1. 3) the following hypotheses are proposed

$$(F'_1) \qquad \overline{R(G(t_1,t_2)B)} = H \qquad for\ any\ t_2 > t_1 \geqslant 0$$

$$(F'_2) \qquad (\partial\varphi)^\circ\ is\ uniformly\ bounded, that\ is,$$

$$|(\partial\varphi)^\circ y| \leqslant M \qquad \forall\, y \in H$$

where $(\partial\varphi)^\circ: H \to H$ is the minimum element of $\partial\varphi$;

$$(3.6) \qquad \begin{cases} (\partial\varphi)^0(x) \in \partial\varphi(x) & x \in H \\ |(\partial\varphi)^0(x)| = inf\{|y| : y \in \partial\varphi(x)\} \end{cases}$$

Theorem 3. 1 Under the hypotheses of lemma 3. 1 and (F'_1), (3. 1) is approximately reachable for every $y_\varepsilon > 0$.

Proof: Noting that $\nabla\varphi_\varepsilon$ satisfies uniform Lipschitz condition[1], the conclusion follows from (F'_1) by means of the results in[10].

Theorem 3. 2 Under the hypotheses of lemma 3. 1, (F'_1) and (F'_2), (1. 3) is approximately reachable.

To prove theorem 3. 2, the following lemma is derived first:

Lemma 3. 2 Under the hypotheses of theorem 3. 2 there exists a constant c independent of u such that

$$(3.7)\qquad |y_\varepsilon(t) - y(t)|^2 + \int_0^T \| y_\varepsilon(t) - y(t) \|^2 dt \leqslant c\varepsilon$$

Where y_ε and y is the solution of (3. 1) and (1. 3) with the same control u respectively.

Proof: For any given $\varepsilon, \lambda > 0$, it follows from (3. 1) that

$$(3.8)\qquad \begin{aligned} & y'_\varepsilon(t) - y'_\lambda(t) + A_H y_\varepsilon(t) - A_H y_\lambda(t) + \nabla\varphi_\varepsilon(y_\varepsilon(t)) - \nabla\varphi_\lambda(y_\lambda(t)) = 0 \\ & y_\varepsilon(0) - y_\lambda(0) = 0 \qquad\qquad a.e.\ t \in [0,T] \end{aligned}$$

where y_ε and y_λ is the solution of (3. 1) corresponding to ε and λ respectively.

Multiplying (scalarly in H) (3. 8) by $(y_\varepsilon(t) - y_\lambda(t))$ one has

$$(3.9)\qquad \frac{1}{2}\frac{d}{dt}|y_\varepsilon(t) - y_\lambda(t)|^2 + w \| y_\varepsilon(t) - y_\lambda(t) \|^2 + < \nabla\varphi_\varepsilon(y_\varepsilon(t)) - \nabla\varphi_\lambda(y_\lambda(t)), y_\varepsilon(t) - y_\lambda(t) > \leqslant 0$$

as (1. 2) is taken into account.

Noting that[1]

$$\nabla\varphi_\varepsilon(y_\varepsilon) = \varepsilon^{-1}(y_\varepsilon - (I + \varepsilon\partial\varphi)^{-1}y_\varepsilon) \in \partial\varphi((I + \varepsilon\partial\varphi)^{-1}y)$$

integrating (3. 9) on [0,T] we have

$$(3.10)\qquad \frac{1}{2}|y_\varepsilon(t) - y_\lambda(t)|^2 + w\int_0^t \|y_\varepsilon(s) - y_\lambda(s)\|^2 ds \leqslant \int_0^t (\nabla\varphi_\varepsilon y_\varepsilon(s) - \nabla\varphi_\lambda(y_\lambda(s)), \varepsilon\nabla\varphi_\varepsilon(y_\varepsilon(s)) - \lambda\nabla\varphi_\lambda(y_\lambda(s)))ds$$

by the monotonicity of $\partial\varphi$.

On the other hand we have[1]

$$(3.11)\qquad |\nabla\varphi_\varepsilon(x)| \leqslant |(\partial\varphi)^\circ(x)| \qquad \forall\, x \in H$$

It follows from (3. 10), (3. 11) and (F'_2) that

$$(3.12)\qquad |y_\varepsilon(t) - y_\lambda(t)|^2 + \int_0^T \| y_\varepsilon(t) - y_\lambda(t) \|^2 dt \leqslant c(\varepsilon + \lambda) \qquad \forall\, \varepsilon, \lambda > 0$$

where c is a constant independent of u.

Leting $\lambda \to 0_+$ in (3. 12) one has (3. 7) by lemma 3. 1.

Proof of theorem 3. 2:

For any given $z_T \in H$ and $\eta > 0$ it follows from (3. 7) that

$$(3.13)\qquad |y_\varepsilon(T) - y(T)| \leqslant \frac{\eta}{2} \qquad \forall\, u \in L^2(0,T;U)$$

as $0 < \varepsilon \leqslant \varepsilon_0 = \eta^2/4c$.

On the other hand, for such ε_0 there exists $\bar{u} \in L^2(0,T;U)$ such that

$$(3.14)\qquad |\bar{y}_{\varepsilon 0}(T) - z_T| \leqslant \eta/2$$

by theorem 3. 1 where y_{ε_0} is solution of (3. 1) corresponding to u. Consequently, it follows from (3. 13) and (3. 14) that

$$|\tilde{y}(T) - z_T| \leqslant |\tilde{y}_{\varepsilon_0}(T) - \tilde{y}\,(T)| + |\tilde{y}_{n0}(T) - z_T| \leqslant \eta/2 + \eta/2 = \eta$$

which completes the proof.

4. AN OPEN QUESTION

If (1.1) holds for $z \in K$ where K is a closed convex subset of H instead of $z \in V$, what are the conditions for approximate reachable? In fact, in this case we are faced with a obstacle problem and

$$\varphi(z) = I_K(z) = \begin{cases} 0 & z \in K \\ \infty & z \in K \end{cases}$$

One of the difficulties is that we do not have $D(A) \subset D(I_K)$.

REFERENCES

[1]. Barbu, V. Nonlinear Semigroups and Differential Equations in Banach Space, Noordhoff Leiden, Netherland, 1976.

[2]. Barbu. V. Optimal Control of Variational inequalities, Pitman, Boston, 1984.

[3]. Friedman, A. , Partial differential Equations of Parabolic Type, Prentice—Hall, New York, 1964.

[4]. Henry. J. Etule de La Controlabilite de Certains Equations Paraboliques Nonlineares. These. Paris, VI, June, 1978.

[5]. Hou. S. H. Controllability and feed back systems, Nonlinear Analysis, Theory, Methods & Application, Vol. 9, No. 12, pp1487—1493, 1985.

[6]. Lions. J, Optimal Control of Systems Governed by Partial Differential Equations, Springer—Verlag, New York, 1971.

[7]. Teo. K. L, and N. V. Abmed, Optimal Control of Distributed Parameter Systems, North—Holland, Amsterdam, 1981.

[8]. Zhou. H. X, and Y. Zhao. A Survey of Controllability Theory of Nonlinear Systems, Control Theory and Applications, Guangzhou, China, No. 2, pp1—14, 1988.

[9]. Hong Xing Zhou, Approximate Controllability For a Class of Semilinear Abstract Equations, SIAM. J. Control and Optim. Vol. 21, No, 4, July1983.

[10]. H. W. Sun and Y. Zhao, Some New Results of Controllability for Semilinear Systems, To Appear.

ANALYSIS OF THE BOUNDARY SINGULARITY OF A SINGULAR OPTIMAL CONTROL PROBLEM*

Wei-Tao Zhang De-Xing Feng

Institute of Systems Science, Academia Sinica, Beijing, China.

Abstract: In this paper, we consider a singular optimal control problem with cost function containing a small parameter ϵ. Using the boundary layer theory developped by Lions in [1], we give some estimates of the singular optimal control u_ϵ in Sobolev space. On the basis of the interior estimate obtained in [4], we analyse the boundary singularity of u_ϵ. According to the generalized Pohozaev identity [3], we obtain the estimation of $\|\frac{\partial u_\epsilon}{\partial \nu}\|_{L^2(\Gamma)}$.

Key words: Singular optimal control, boundary singular, boundary Layer theory.

§1. Problem Statement

Let Ω be an open bounded set in $\mathbb{R}^n (n \geq 2)$ with the boundary Γ being differentiable $n-1$ dimension manifold. Consider the control system described by the following elliptic equation

$$\begin{cases} -\Delta y(v) &= f + v \quad \text{in } \Omega, \\ y(v) &= 0 \quad \text{on } \Gamma, \end{cases} \tag{1.1}$$

with $f \in H^1(\Omega)$, $v \in \mathcal{U} = L^2(\Omega)$. Take the cost function as follows

$$J_\epsilon(v) = \int_\Omega |\nabla y(v) - Z_d|^2 \, dx + \epsilon \int_\Omega v^2 \, dx, \tag{1.2}$$

where $Z_d = (Z_{1d}, \cdots, Z_{nd})$, $Z_{id} \in H^2(\Omega)$, $i = 1, \cdots, n$, $0 < \epsilon \ll 1$.

The problem is to find $u_\epsilon \in \mathcal{U}$ satisfying

$$J_\epsilon(u_\epsilon) = \inf_{v \in \mathcal{U}} J_\epsilon(v). \tag{1.3}$$

*This research is supported by the National Natural Science Foundation of China and Partially by the Institute of Mathematics (Open), Academia Sinica.

It is well known that there exists a unique optimal control u_ϵ which satisfies the variational inequality

$$J'_\epsilon(u_\epsilon)(v-u_\epsilon)\geq 0 \quad \forall v\in \mathcal{U}, \tag{1.4}$$

where $J'_\epsilon(u_\epsilon)$ is the Frechet devivative of J_ϵ at u_ϵ.

By using the boundary conditions $y(v)|_\Gamma = 0$, $y(u_\epsilon)|_\Gamma = 0$ and Green's formala, it follows from (1.4) that

$$-\int_\Omega(\Delta y(u_\epsilon)-\operatorname{div}Z_d)(y(v)-y(u_\epsilon))\,dx+\epsilon\int_\Omega u_\epsilon(v-u_\epsilon)\,dx\geq 0 \quad \forall v\in \mathcal{U}. \tag{1.5}$$

Now define the adjoint state $p_\epsilon = p(u_\epsilon)$ by

$$\begin{cases} -\Delta p(u_\epsilon) & =-(\Delta y(u_\epsilon)-\operatorname{div}Z_d) \quad \text{in } \Omega, \\ p(u_\epsilon) & =0 \quad \text{on } \Gamma. \end{cases} \tag{1.6}$$

Then substituting (1.6) into (1.5) yields

$$p(u_\epsilon)+\epsilon u_\epsilon = 0. \tag{1.7}$$

Taking $v = u_\epsilon$ in (1.1) and substituting (1.1) into (1.6), then using (1.7), we obtain

$$\begin{cases} -\epsilon\Delta p(u_\epsilon)+p(u_\epsilon) & =\epsilon F \quad \text{in } \Omega, \\ p(u_\epsilon) & =0 \quad \text{on } \Gamma, \end{cases} \tag{1.8}$$

with $F = f + \operatorname{div}Z_d$.

In the remaining parts of this paper, we shall give some estimates involving u_ϵ in §2, and in §3 we shall analyse the boundary singularity of u_ϵ. Finally in §4, we shall obtain the estimate of $\frac{\partial u_\epsilon}{\partial \nu}$.

§2. Some Estimates of u_ϵ

Denote $a(u,v) = \int_\Omega \nabla u\cdot\nabla v\,dx$, $b(u,v) = \int_\Omega uv\,dx$, $(F,v) = \int_\Omega Fv\,dx$, $p(u_\epsilon) = p_\epsilon$. Then (1.8) is equivalent to

$$\epsilon a(p_\epsilon,v)+b(p_\epsilon,v)=\epsilon(F,v) \quad \forall v\in H_0^1(\Omega), \tag{2.1}$$

which can be also written

$$\epsilon a(p_\epsilon,v-p_\epsilon)+b(p_\epsilon,v-p_\epsilon)=\epsilon(F,v-p_\epsilon) \quad \forall v\in H_0^1(\Omega). \tag{2.2}$$

$$b(\epsilon F,v-\epsilon F)=\epsilon(F,v-\epsilon F) \quad \forall v\in L^2(\Omega). \tag{2.3}$$

If $F\in H_0^1(\Omega)$, taking $v=\epsilon F$ in (2.2) and $v=p_\epsilon$ in (2.3), then combining these two equalities, we have

$$\|p_\epsilon-\epsilon F\|_{H^1(\Omega)}\leq \epsilon\|F\|_{H^1(\Omega)}, \tag{2.4}$$

$$\|p_\epsilon - \epsilon F\|_{L^2(\Omega)} \leq c\epsilon^{1+1/2}\|F\|_{H^1(\Omega)}, \tag{2.5}$$

where and hereafter c always represents a constant independent of ϵ.

We also have

$$b(p_\epsilon - \epsilon F, p_\epsilon - \epsilon F) = b(p_\epsilon, p_\epsilon) - b(p_\epsilon, \epsilon F) - b(\epsilon F, p_\epsilon - \epsilon F). \tag{2.6}$$

Taking $v = p_\epsilon$ in (2.1), it follows that

$$b(p_\epsilon, p_\epsilon) \leq b(p_\epsilon, \epsilon F). \tag{2.7}$$

Substituting (2.7) into (2.6) yields

$$\|p_\epsilon - \epsilon F\|_{L^2(\Omega)} \leq \epsilon\|F\|_{L^2(\Omega)}. \tag{2.8}$$

Theorem 2.1 *If $F \in H^1(\Omega)$ and $F \notin H_0^1(\Omega)$, then we have*

$$\|p_\epsilon - \epsilon F\|_{L^2(\Omega)} \leq c\epsilon^{1+1/4}\|F\|_{H^1(\Omega)}. \tag{2.9}$$

Proof. Set $G_\epsilon(F) = p_\epsilon - \epsilon F$, then $G_\epsilon(F)$ is a linear mapping. From (2.5), we know $G_\epsilon \in \mathcal{L}(H_0^1(\Omega); L^2(\Omega))$. Moreover, from (2.8), it can be seen that $G_\epsilon \in \mathcal{L}(L^2(\Omega); L^2(\Omega))$.

According to [1], for each $F \in H^1(\Omega)$, F can be represented in the form $F = F_0 + F_1$ with $F_0 \in L^2(\Omega)$ and $F_1 \in H_0^1(\Omega)$ which satisfy

$$\begin{cases} \|F_0\|_{L^2(\Omega)} & \leq c\lambda^\alpha\|F\|_{L^2(\Omega)}, \\ \|F_1\|_{H^1(\Omega)} & \leq c\lambda^{-\alpha}\|F\|_{H^1(\Omega)}, \end{cases} \tag{2.10}$$

where λ and α are constants with $0 < \lambda < 1$, $\alpha > 0$.

By using (2.10), it follows that

$$p_\epsilon - \epsilon F = G_\epsilon(F_0) + G_\epsilon(F), \tag{2.11}$$

Then using (2.5) (2.8) and (2.10), from (2.11), we obtain

$$\|p_\epsilon - \epsilon F\|_{L^2(\Omega)} \leq c\epsilon\lambda^\alpha\|F\|_{H^1(\Omega)} + c\epsilon^{1+1/2}\lambda^{-\alpha}\|F\|_{H^1(\Omega)}. \tag{2.12}$$

Thus taking $\lambda = \epsilon$, $\alpha = 1/4$ in (2.12), (2.9) is derived, which finishes the proof.

Theorem 2.2 *If $F \in H^1(\Omega) \setminus H_0^1(\Omega)$, then we have*

$$\|u_\epsilon + F\|_{L^2(\Omega)} \leq c\epsilon^{1/4}\|F\|_{H^1(\Omega)}, \tag{2.13}$$

$$\| \nabla (u_\epsilon + F)\|_{L^2(\Omega)} \leq c\epsilon^{-3/8}\|F\|_{H^1(\Omega)}. \tag{2.14}$$

Proof. (2.13) can be directly obtained from (1.7) and (2.9).

In order to prove (2.14), taking $v = p_\epsilon$ in (2.1), we have

$$\epsilon\int_\Omega |\nabla p_\epsilon|^2\, dx \leq \|p_\epsilon - \epsilon F\|_{L^2(\Omega)}\|p_\epsilon\|_{L^2(\Omega)}, \tag{2.15}$$

$$\|p_\epsilon\|_{L^2(\Omega)} \leq \epsilon\|F\|_{L^2(\Omega)}. \tag{2.16}$$

Substituting (2.9) and (2.16) into (2.15) yields

$$\|\nabla p_\epsilon\|_{L^2(\Omega)} \leq c\epsilon^{5/8}\|F\|_{H^1(\Omega)}. \tag{2.17}$$

Then from $\|\nabla(p_\epsilon - \epsilon F)\|_{L^2(\Omega)} \leq \|\nabla p_\epsilon\|_{L^2(\Omega)} + \epsilon\|F\|_{H^1(\Omega)}$ and (2.17), we obtain

$$\|\nabla(p_\epsilon - \epsilon F)\|_{L^2(\Omega)} \leq c\epsilon^{5/8}\|F\|_{H^1(\Omega)}, \tag{2.18}$$

from which (2.14) is immediately obtained. The proof is then complete.

§3. Analysis of the Boundary Singularity of u_ϵ

Denote by $d(x,\Gamma)$ the distance from $x \in \Omega$ to Γ. Using Weingarten map (see chap.2 in [2]), we can prove

$$\Delta d(x,\Gamma) = -H_\delta(x), \tag{3.1}$$

where $H_\delta(x)$ is the mean curvature of the parallel hypersurface Γ_δ which passes through x, in other word, $\Gamma_\delta = \{x | x \in \Omega, d(x,\Gamma) = \delta\}$. Since Γ is regular, in some interior neighborhood of Γ, the following estimation holds

$$|D^\alpha d(x,\Gamma)| \leq c, \tag{3.2}$$

where $D^\alpha = \frac{\partial^{\alpha_1+\cdots+\alpha_n}}{\partial x_1^{\alpha_1}\cdots\partial x_n^{\alpha_n}}$.

Let $\lambda \in [\lambda_0, \infty)$ with $\lambda_0 > 0$ small enough $(0 < \lambda_0 \ll 1)$, β_0 be a fixed positive constant, β be a positive number to be defined with $\beta < \beta_0$. Denote by ν_p the outward unit normal to Γ at p, then we can define a positive function $t(p)$ such that $p_1 = p - t(p)\nu_p \in \Gamma$ and $p_1 \neq p$. Set $2s = \min_{p\in\Gamma} t(p)$. Let $\epsilon_1 = (e^{\frac{\beta_0}{\lambda_0}-1} - 1)^{-\frac{1}{\beta_0}}$, then $\epsilon^{\lambda_0}\log(1+\epsilon^{-\beta_0})$ is an increasing function of $\epsilon \in (0, \epsilon_1)$. Let $I_1 = \{\epsilon | \epsilon^{\lambda_0}\log(1+\epsilon^{-\beta_0}) < s\}$, $(0,\epsilon_2) = I_1 \cap (0,\epsilon_1)$, $\epsilon_0 = \min(1,\epsilon_2)$, and denote $F_\lambda(\epsilon) = \epsilon^\lambda \log(1+\epsilon^{-\beta})$, $\rho_\lambda(\epsilon) = \frac{1}{3}\log\frac{1/2+\epsilon^\beta}{1/3+\epsilon^\beta}$, $\Omega_{g(\epsilon)} = \{x | x \in \Omega, d(x,\Gamma) > g(\epsilon)\}$ for some positive function $g(\epsilon)$. For $\epsilon \in (0,\epsilon_0)$, from (3.2) and [4], we may construct an infinitely differentiable function ϕ with

$$\begin{cases} 0 \leq \phi \leq 1, \\ \phi(x) = 0, \quad x \in \mathbb{R}^n \setminus \Omega_{\frac{1}{3}F_\lambda(\epsilon)-\rho_\lambda(\epsilon)}, \\ \phi(x) = 1, \quad x \in \bar{\Omega}_{\frac{2}{3}F_\lambda(\epsilon)+\rho_\lambda(\epsilon)}, \\ |\nabla\phi| \leq c\epsilon^{-\epsilon}(\phi + \epsilon^\beta), \end{cases} \tag{3.3}$$

where c is a constant greater than 1.

Theorem 3.1 *Denote* $I(\lambda) = \int_{d(x,\Gamma)\geq\epsilon^\lambda\log(1+\epsilon^{-\beta})} |\nabla u_\epsilon|^2 dx$, $\alpha(\epsilon) = 1/2\log_\epsilon\frac{1}{C}$ *(where c is the constant appearing in (3.3), hence* $\lim_{\epsilon\to0}\alpha(\epsilon) = 0$*). Suppose* $\beta \geq \frac{5}{8}$, *then*

i) If $\lambda_0 \leq \lambda \leq \frac{1}{2} - \alpha(\epsilon)$, we have

$$I(\lambda) \leq c\|F\|^2_{H^1(\Omega)}. \tag{3.4}$$

ii) If $\frac{1}{2} - \alpha(\epsilon) < \lambda < \frac{5}{8}$, we have

$$I(\lambda) \leq c\epsilon^{\frac{1}{2}-2\lambda}\|F\|^2_{H^1(\Omega)}. \tag{3.5}$$

iii) If $\lambda \geq \frac{5}{8}$, we have

$$I(\lambda) \leq c\epsilon^{-\frac{3}{4}}\|F\|^2_{H_1(\Omega)}. \tag{3.6}$$

Proof. Setting $w_\epsilon = p_\epsilon - \epsilon F$, from (2.1), it follows that

$$\epsilon a(p_\epsilon, v) + b(w_\epsilon, v) = 0. \tag{3.7}$$

Using (3.3) and taking $v = w_\epsilon\phi$ in (3.7), we obtain

$$\int_\Omega \epsilon|\nabla w_\epsilon|^2\phi dx + \int_\Omega |w_\epsilon|^2\phi dx = -\epsilon\int_\Omega w_\epsilon \nabla w_\epsilon \cdot \nabla\phi\, dx - \epsilon^2\int_\Omega \nabla F\cdot\nabla(w_\epsilon\phi)\, dx. \tag{3.8}$$

Define

$$I_1 = -\epsilon\int_\Omega w_\epsilon \nabla w_\epsilon \cdot \nabla\phi\, dx,$$

$$I_2 = -\epsilon^2\int_\Omega \nabla F\cdot\nabla(w_\epsilon\phi)\, dx.$$

Using (3.3), we have

$$I_1 \leq c\epsilon^{1-\lambda}\int_\Omega |\nabla w_\epsilon||w_\epsilon|\phi\, dx + c\epsilon^{1-\lambda+\beta}\int_\Omega |\nabla w_\epsilon||w_\epsilon|\, dx. \tag{3.9}$$

In order to further eatimate the right hand of (3.9), set

$$I_{11} = c\epsilon^{1-\lambda}\int_\Omega |\nabla w_\epsilon||w_\epsilon|\phi\, dx,$$

$$I_{12} = c\epsilon^{1-\lambda+\beta}\int_\Omega |\nabla w_\epsilon||w_\epsilon|\, dx.$$

We have

$$I_{11} \leq c\epsilon^{1-\lambda}\Big(\int_\Omega |\nabla w_\epsilon|^2\phi\, dx\Big)^{\frac{1}{2}}\Big(\int_\Omega |w_\epsilon|^2\phi\, dx\Big)^{\frac{1}{2}}. \tag{3.10}$$

Noticing that

$$\epsilon^{-\lambda}xy \leq \delta x^2 + \frac{1}{4\delta}\epsilon^{-2\lambda}y^2, \tag{3.11}$$

where δ is an arbitrary positive constant, then

$$I_{11} \leq c\delta\epsilon\int_\Omega |\nabla w_\epsilon|^2\phi\, dx + \frac{c\epsilon^{1-2\lambda}}{4\delta}\int_\Omega |w_\epsilon|^2\phi\, dx. \tag{3.12}$$

Moreover,

$$I_{12} \leq c\epsilon^{1-\lambda+\beta}\Big(\int_\Omega |\nabla w_\epsilon|^2\, dx\Big)^{1/2}\Big(\int_\Omega |w_\epsilon|^2\, dx\Big)^{1/2}. \tag{3.13}$$

By substituting (2.9) and (2.18) into (3.13), it follows from (3.13) that

$$I_{12} \leq c\epsilon^{2+\frac{7}{8}-\lambda+\beta}\|F\|^2_{H^1(\Omega)}. \tag{3.14}$$

Taking $c\delta = \frac{1}{4}$ in (3.12) and from (3.14), we obtain

$$I_1 \leq \frac{\epsilon}{4}\int_\Omega |\nabla w_\epsilon|^2 + c\epsilon^{2+\frac{7}{8}-\lambda+\beta}\|F\|^2_{H^1(\Omega)} + c\epsilon^{1-2\lambda}\int_\Omega |w_\epsilon|^2\phi\, dx \tag{3.15}$$

Next we estimate I_2, first I_2 can be written

$$I_2 = -\epsilon^2\int_\Omega \phi \nabla w_\epsilon \cdot \nabla F\, dx - \epsilon^2\int_\Omega w_\epsilon \nabla F \cdot \nabla\phi\, dx. \tag{3.16}$$

Denote

$$I_{21} = -\epsilon^2\int_\Omega \phi\nabla w_\epsilon \cdot F\, dx, \quad I_{22} = -\epsilon^2\int_\Omega w_\epsilon \nabla F\cdot\nabla\phi\, dx,$$

then $I_2 = I_{21} + I_{22}$, using (3.11), we obtain

$$\begin{aligned} I_{21} &\leq \epsilon^2\epsilon^{-\frac{1}{2}}(\textstyle\int_\Omega |\nabla w_\epsilon|^2\phi\, dx)^{1/2}\epsilon^{1/2}(\int_\Omega |\nabla F|^2\phi\, dx)^{\frac{1}{2}} \\ &\leq \epsilon^2(\epsilon^{-1}\delta \textstyle\int_\Omega |\nabla w_\epsilon|^2\phi\, dx + \frac{\epsilon}{4\delta}\int_\Omega |\nabla F|^2\phi\, dx) \\ &\leq \delta\epsilon\textstyle\int_\Omega |\nabla w_\epsilon|^2\phi\, dx + \frac{\epsilon^3}{4\delta}\|\nabla F\|^2_{H^1(\Omega)}. \end{aligned} \tag{3.17}$$

$$I_{22} \leq c\epsilon^{2-\lambda}\int_\Omega |\nabla F||w_\epsilon|\phi\, dx + c\epsilon^{2-\lambda+\beta}\int_\Omega |\nabla F||w_\epsilon|\, dx. \tag{3.18}$$

For further estimating I_{22}, set

$$I_{221} = c\epsilon^{2-\lambda}\int_\Omega |\nabla F||w_\epsilon|\phi\, dx,$$

$$I_{222} = c\epsilon^{2-\lambda+\beta}\int_\Omega |\nabla F||w_\epsilon|\, dx,$$

we have

$$\begin{aligned} I_{221} &\leq c\epsilon^{2-\lambda}\epsilon^{\frac{-\lambda-1}{2}}(\textstyle\int_\Omega |w_\epsilon|^2\phi\, dx)^{\frac{1}{2}}\epsilon^{\frac{1+\lambda}{2}}(\int_\Omega |\nabla F|^2\phi\, dx)^{\frac{1}{2}} \\ &\leq c\epsilon^{2-\lambda}(\tfrac{1}{2}\epsilon^{-(\lambda+1)}\textstyle\int_\Omega |w_\epsilon|^2\phi\, dx + \frac{1}{2}\epsilon^{1+\lambda}\int_\Omega |\nabla F|^2\phi\, dx) \\ &\leq c\epsilon^{1-2\lambda}\textstyle\int_\Omega |w_\epsilon|^2\phi\, dx + c\epsilon^3\|\nabla F\|^2_{H^1(\Omega)} \end{aligned} \tag{3.19}$$

From (2.9), it follows that

$$I_{222} \leq c\epsilon^{3-\lambda+1/4+\beta}\|F\|^2_{H^1(\Omega)} \tag{3.20}$$

Taking $\delta = \frac{1}{4}$ in (3.17) and using (3.19) (3.20), we deduce

$$I_2 \leq \frac{\epsilon}{4}\int_\Omega |\nabla w_\epsilon|^2\phi\, dx + c\epsilon^3\|F\|^2_{H^1(\Omega)} + c\epsilon^{1-2\lambda}\int_\Omega |w_\epsilon|^2\phi\, dx + c\epsilon^{3-\lambda+1/4+\beta}\|F\|^2_{H^1(\Omega)} \tag{3.21}$$

Substituting (3.15) (3.21) into (3.8) and using $\epsilon^{m_2} < \epsilon^{m_1}(m_2 > m_1)$, we have

$$\frac{\epsilon}{2}\int_\Omega |\nabla w_\epsilon|^2\phi\,dx + (1-c\epsilon^{1-2\lambda})\int_\Omega |w_\epsilon|^2\phi\,dx \le c\epsilon^3\|F\|^2_{H^1(\Omega)} + c\epsilon^{2+\frac{7}{8}-\lambda+\beta}\,|F\|^2_{H^1(\Omega)},\quad (3.22)$$

If $\lambda_0 \le \lambda \le 1-\alpha(\epsilon)$, then we have $1-c\epsilon^{1-2\lambda} \ge 0$. If $2+\frac{7}{8}-\lambda+\beta > 2+\frac{7}{8}-\frac{1}{2}+\beta \ge 3$, we then have $\beta \ge \frac{5}{8}$. Therefore with $\lambda_0 \le \lambda \le 1-\alpha(\epsilon)$, and $\beta \ge \frac{5}{8}$, from (3.22) it follows that

$$\int_\Omega |\nabla w_\epsilon|^2\phi\,dx \le c\epsilon^2\|F\|^2_{H^1(\Omega)} \quad (3.23)$$

Since $\phi(x) = 1$ in $\Omega_{F_\lambda(\epsilon)}$, using (1.7) and taking account of $|\nabla u_\epsilon|^2 \le 2(|\nabla(u_\epsilon + F)|^2 + |\nabla F|^2)$, from (3.23), we deduce (3.4).

Using (2.9) (2.18), we obtain

$$c\epsilon^{1-2\lambda}\int_\Omega |w_\epsilon|^2\phi\,dx \le c\epsilon^{3+1/2-2\lambda}\|F\|^2_{H^1(\Omega)}, \quad (3.24)$$

$$\epsilon\int_\Omega |\nabla w_\epsilon|^2\phi\,dx \le c\epsilon^{2+\frac{1}{4}}\|F\|^2_{H^1(\Omega)}. \quad (3.25)$$

Finally, using (3.24) (3.25), we deduce (3.5) (3.6). The proof is then complete.

§4. Estimate of $\|\frac{\partial u_\epsilon}{\partial \nu}\|_{L^2(\Gamma)}$

Theorem 4.1 *Let Ω be a strictly starshapped domain.*
i) If $F \in H^1(\Omega)\backslash H^1_0(\Omega)$, then we have

$$\|\frac{\partial u_\epsilon}{\partial \nu}\|_{L^2(\Gamma)} = O(\epsilon^{-\frac{1}{2}}) \quad (4.1)$$

ii) If $F \in H^1_0(\Omega)$, then we have

$$\|\frac{\partial u_\epsilon}{\partial \nu}\|_{L^2(\Gamma)} = o(\epsilon^{-\frac{1}{2}}) \quad (4.2)$$

Proof. From (1.7) (1.8), we have

$$\begin{cases} -\epsilon\Delta u_\epsilon + u_\epsilon = -F & \text{in } \Omega, \\ u_\epsilon = 0 & \text{on } \Gamma. \end{cases} \quad (4.3)$$

According to (2.9), we obtain

$$\lim_{\epsilon\to 0}\|u_\epsilon + F\|_{L^2(\Omega)} = 0. \quad (4.4)$$

We now consider the following equation

$$\begin{cases} -\Delta u = g(x,u) & \text{in } \Omega, \\ u = c \text{ on } \Gamma, \end{cases} \quad (4.5)$$

where c is a constant ($c = 0$ or $c \neq 0$) and $g \ : \Omega \times \mathbb{R} \longrightarrow \mathbb{R}$ is a continuous function, differentiable with respect to x.

Using the method similar to [3], we can prove the following generalized Pohozaev identity for (4.5),

$$\begin{aligned}\int_\Gamma (x\cdot\nu)|\frac{\partial u}{\partial\nu}|^2\,ds &= (2-n)\int_\Omega ug(x,u)\,dx + 2n\int_\Omega G(x,u)\,dx \\ &+2\int_\Omega\Big(\int_0^u\sum_{i=1}^n\frac{\partial g}{\partial x_i}x_i\,dt\Big)\,dx + (2-n)c\int_\Gamma\frac{\partial u}{\partial\nu}\,ds \\ &-2\int_\Gamma G(x,c)(x\cdot\nu)\,ds,\end{aligned} \tag{4.6}$$

where $G(x,u) = \int_0^u g(x,t)\,dt$, $G(x,c) = \int_0^c g(x,t)\,dt$, ν is the outward unit normal to Γ.

Applying (4.4) and (4.6) to (4.3), we obtain

$$\lim_{\epsilon\to 0}\int_\Gamma (x\cdot\nu)|\frac{\partial u_\epsilon}{\partial\nu}|^2\,ds = \int_\Gamma (x\cdot\nu)F^2\,ds. \tag{4.7}$$

Since Ω is strictly starshapped, i. e. $(x\cdot\nu) > 0$ on Γ, from (4.7) we obtain (4.1) (4.2). The proof is then complete.

References

[1] Lions, J. L., *Perturbations Singulières dans les Problèmes aux limites et en Contrôle Optimal,* Springer, 1973.

[2] Hicks, J. N., *Note on differentiable geometry,* D. Van Nostrand Company, Inc. Toronto, 1965.

[3] Pohozaev, S. I., *Eigenfunction of the equation* $\Delta u + \lambda f(u) = 0$, Soviet Math. Doklady 6, 1965, 1408–1411 (translated from the Russian Dokl. Akd. Nauk USSR 165, 1965, 33–36).

[4] Zhang Weitao, *Analysis of boundary layer singularity,* J. Sys. Sci. & Math. Scis., 4(2), 1984, 81–96.

ANALYSIS OF THE PARABOLIC CONTROL SYSTEM WITH A PULSE-WIDTH MODULATED SAMPLER*)

Hong Xing ZHOU
Department of Mathematics, Shandong University
Jinan, Shandong, 250100, P.R.C.

1. INTRODUCTION

In design of distributed parameter control systems one of important problems is to choose controller and actuator. As the dimension of an industrial controller in actual applications is finite it restricts us to consider the distributed parameter system with a finite-dimensional output. In industrial process control systems on-off actuators have been in engineer's good graces because of the cheep prize and the high reliability. For example, time-proportional switch actuator is applied usually in the temprature control system of a large-power electric furnace and it is a typical pulse-width modulated sampler. In this paper we will be concerned with the parabolic control system coupled with an finite-dimensional dynamical controller and a pulse-width modulated sampler.

From the point of view of engineering, a pulse-width modulated sampler can be approximately seen as an equivalent pulse-amplitude modulated sampler and the lumped parameter pulse-width modulated control system could be analysed by the classical theory of sampled-data control systems (e.g. Z-transformation method etc.[3]). Because it is impossible to make the sampling period very small in an electric-magnetic actuator some essential and important properties will be neglected in the analysis by classical methods. Therefore, from the point of view of control theory some rigorous theory to analyse pulse-width sampled-data control systems is advanced, e.g. the direct analysis method [5,6,7], discontinuous control system theory [4] etc..

Here we will be concerned with a class of control systems governed by an abstract parabolic differential equation

$$\begin{aligned} \dot{y}(t)&=Ay(t)+Bu(t)+f(t) \\ z(t)&=Cy(t) \end{aligned} \tag{1.1}$$

*) This work was supported by The State Natural Science Foundation of China.

where the state y(t) takes values in a reflexive Banach space X: $y(t) \in X$, $t \geq 0$, A is the infinitesimal generator of an analytically compact semigroup S(t), $t \geq 0$, on the state space X; u(t) is an q-dimensional control: $u(t) \in R^q$ and $B \in \mathcal{L}[R^q, X]$--- the space of all bounded linear operators from R^q into X; f(t) is a step disturbance of the system: $f(t)=f \cdot 1(t)$ with $f \in X$. In (1.1) z(t) is the p-dimensional output of the system and C is a given bounded linear operator from X into R^q.

In System (1.1) we assume that the control signal u(t) is obtained from an q-dimensional pulse-width modulated sampler with an input signal v(t) which is the output of some dynamical controller

$$\dot{v}(t)=Jv(t)+Kz(t) \tag{1.2}$$

where J and K are $q \times q$ and $q \times p$ matrices respectively. In general, the matrix J is fixed by dynamical characteristics of the controller and the matrix K called to be feed-back matrix will be chosen and tuned by the designer. The output $u(t)=(u_1(t),u_2(t),\dots,u_q(t))'$ and the input $v(t)=(v_1(t),v_2(t),\dots,v_q(t))'$ of the pulse-width sampler satisfy the following dynamic relation:

$$u_i(t)=\begin{cases} \operatorname{sign} \alpha_{ni} & nT \leq t < (n+|\alpha_{ni}|)T \\ 0 & (n+|\alpha_{ni}|)T \leq t < (n+1)T \end{cases}$$

$$\alpha_{ni}=\begin{cases} v_i(nT) & |v_i(nT)| \leq 1 \\ \operatorname{sign} v_i(nT) & |v_i(nT)| \geq 1 \end{cases} \tag{1.3}$$

$$i=1,2,\dots,q, \quad n \in I^+ \triangleq \{0,1,2,\dots\} .$$

where $T>0$ is the sampling period of the pulse-width sampler. We denote the relation described by (1.3) as $u(t)=F(v)(t)$. System (1.1)-(1.3) is called to be a parabolic pulse-width sampling control system and briefly to be written as Parabolic PM System.

The purpose of this paper is to generalize the theory for the lumped parameter control system with a pulse-width modulated sampler to the parabolic control system. Section 2 gives a strict definition of the steady-state of the system and proves the existence of the steady-state of the system. Section 3 is devoted to the analysis of the steady-state stability for a parabolic pulse-width sampled-data control system. An example to illustrate the theory is presented at last.

2. STEADY-STATE ANALYSIS

DEFINITION 2.1. In parabolic PM System the q-dimensional vector $\alpha_n=(\alpha_{n1}, \alpha_{n2}, \dots \alpha_{nq})'$ defined by (1.3) is called the duration ratio of the pulse-width sampler in the n-th sampling period, $n\in I^+$.

If we defined a closed cube Ω in R^q as

$$\Omega=\{a=(a_1,a_2,\dots,a_q)'\in R^q:\ |a_i|\leq 1,\ i=1,2,\dots,q\} \tag{2.1}$$

then we have $\alpha_n\in\Omega$ for all $n\in I^+$.

DEFINITION 2.2. In System (1.1)-(1.3) if there exists an q-dimensional vector $\alpha=(\alpha_1,\alpha_2,\dots,\alpha_q)'\in\Omega$ and a corresponding periodicly rectangular-wave control signal $u(t)=u(t;\alpha)$ defined by

$$u_i(t)=u_i(t;\alpha_i)=\begin{cases}\text{sign }\alpha_i & nT\leq t<(n+|\alpha_i|)T\\ 0 & (n+|\alpha_i|)T\leq t<(n+1)T\end{cases} \tag{2.2}$$
$$i=1,2,\dots,q,\quad n\in I^+$$

such that the closed system (1.1)-(1.3) has a corresponding periodic trajectory $y(\cdot)=y(\cdot;\alpha)$: $y(t+T;\alpha)=y(t;\alpha)$, $t\geq 0$, then the control (2.2) is called to be a steady-state control (with respect to the disturbance $f\cdot 1(t)$), the periodic projectory $y(\cdot)$ is called to be a steady-state corresponding to the steady-state control $u(\cdot)$ and the constant vector $\alpha\in\Omega$ defining the steady-state control (2.2) is called to be a steady-state duration ratio.

LEMMA 2.3. Let the control signal $u(t)$ in the dynamical system (1.1) be a rectangular-wave signal $u(t;\alpha)$ with a period T defined by (2.2) for a given $\alpha\in\Omega$. Then there exists a unique solution $y(t;\alpha)$ with the period T of (1.1) corresponding to the given control $u(t;\alpha)$ under the assumption

$$j\omega_n\in\rho(A),\quad \omega_n=2n\pi/T,\quad n=0,\pm1,\pm2,\dots, \tag{2.3}$$

where $\rho(A)$ is the resolvant set of the operator A.

COLLORARY 2.4. Consider the following open-loop system with an input $u(t;\alpha)$ defined by (2.2) for a given $\alpha\in\Omega$:

$$\begin{aligned}\dot y(t;\alpha)&=Ay(t;\alpha)+Bu(t;\alpha)+f\cdot 1(t)\\ z(t;\alpha)&=Cy(t;\alpha)\\ \dot v(t;\alpha)&=Jv(t;\alpha)+Kz(t;\alpha)\end{aligned} \tag{2.4}$$

where the operator A satisfies Assumption (2.3) and the matrix J satisfies

$$j\omega_n\in\rho(J),\quad n=0,\pm1,\pm2,\dots . \tag{2.5}$$

Then System (2.4) has a unique output $v(t;\alpha)$ with the period T corresponding to the given vector $\alpha\in\Omega$ or the given periodic control $u(t;\alpha)$ defined by (2.2).

LEMMA 2.5. Under Assumptions (2.3) and (2.5) consider a nonlinear mapping $\mathcal{J}$ from $\Omega\subset R^q$ into R^q defined by

$$\mathcal{J}(\alpha)=(I-e^{JT})^{-1}\int_0^T e^{J(T-t)}KCy(t;\alpha)dt, \quad \alpha\in\Omega , \tag{2.6}$$

where $y(t;\alpha)$ is the periodic solution of (1.1) corresponding to a given open-loop control $u(t;\alpha)$ defined by (2.2) for a given $\alpha\in\Omega$. Then there exists some constant $M_0>0$ such that

$$\|\mathcal{J}(\alpha)-\mathcal{J}(\bar\alpha)\| \leqslant M_0\|K\|\|\alpha-\bar\alpha\| \qquad \text{for any } \alpha,\bar\alpha\in\Omega . \tag{2.7}$$

PROOF. Define two q×q matrices

$$P(t,\tau)=KCS(t)R[1,S(T)]\,S(T-\tau)B, \quad \tilde P(t)=KCS(t)B. \tag{2.8}$$

Let α and $\bar\alpha$ be given in Ω arbitrarily. Then

$$KC(y(t;\alpha)-y(t;\bar\alpha))=KCS(t)(y_0-\bar y_0)$$
$$+KC\int_0^t S(t-\tau)B[u(\tau;\alpha)-u(\tau;\bar\alpha)]\,d\tau$$
$$=\int_0^t P(t-\tau)[u(\tau;\alpha)-u(\tau;\bar\alpha)]\,d\tau+\int_0^t \tilde P(t-\tau)[u(\tau;\alpha)-u(\tau;\bar\alpha)]\,d\tau .$$

Since $S(t)$ is a C_0-semigroup there exist constants $M_a>0$ and λ such that

$$\|S(t)\| \leqslant M_a e^{-\lambda t}, \quad t\geqslant 0. \tag{2.9}$$

By the boundedness of the operators B and C there exists some constant $M_1>0$ such that

$$\|P(t,\tau)\|\leqslant M_1\|K\|e^{-\lambda(t-\tau)}, \quad \|\tilde P(t)\|\leqslant M_1\|K\|e^{-\lambda t}, \quad t\geqslant\tau\geqslant 0.$$

Therefore the k-th component of the vector-valued function $KC(y(t;\alpha)-y(t;\bar\alpha))$ has the following estimation:

$$\left|\,[KC(y(t;\alpha)-y(t;\bar\alpha))]_k\right| \leqslant$$
$$\leqslant 2M_1\|K\|e^{-\lambda t}\int_0^T e^{\lambda\tau}\sum_{l=1}^{q}|u_l(\tau;\alpha_l)-u_l(\tau;\bar\alpha_l)|\,d\tau \tag{2.10}$$

When $\alpha_l\bar\alpha_l>0$, without loss of any generarity, let $0<\alpha_l<\bar\alpha_l$, we have

$$\int_0^T e^{\lambda\tau}|u_l(\tau;\alpha_l)-u_l(\tau;\bar\alpha_l)|\,d\tau=\int_{\alpha_l T}^{\bar\alpha_l T} e^{\lambda\tau}\,d\tau\leqslant e^{|\lambda|T}T|\alpha_l-\bar\alpha_l|$$

When $\alpha_l\bar\alpha_l<0$, for example, $\bar\alpha_l<0<\alpha_l$ and $|\bar\alpha_l|>\alpha_l$, we have

$$\int_0^T e^{\lambda\tau}|u_l(\tau;\alpha_l)-u_l(\tau;\bar\alpha_l)|\,d\tau\leqslant 2\int_0^{|\bar\alpha_l|T} e^{\lambda\tau}\,d\tau\leqslant 2e^{|\lambda|T}T|\alpha_l-\bar\alpha_l| .$$

Thus, from (2.10) it is known that there exists some constant $M_2>0$ dependent of T and λ such that

$$\|KC(y(t;\alpha)-y(t;\bar{\alpha}))\| \leq M_2 \|Ke^{-\lambda t}\| \, \|\alpha-\bar{\alpha}\| \tag{2.11}$$

Suppose that

$$\|e^{Jt}\| \leq M_J e^{-\omega t} , \quad t \geq 0 \tag{2.12}$$

where $M_J>0$ and ω are constants. By substituting (2.11) and (2.12) into (2.6) the conclusion (2.7) follows immediatly if we take

$$M_o=M_2 M_J \|(I-e^{JT})^{-1}\| Te^{\mu T}, \quad \mu=\max(|\lambda|,|\omega|) . \tag{2.13}$$

From Lemma 2.5 it is easy to prove the following theorem.

THEOREM 2.6. Suppse that the operator A and the matrix J satisfy Assumptions (2.3) and (2.5) respectively and the matrix K and the constant M_o defined by (2.13) satisfy

$$M_o \| K \| < 1 \tag{2.14}$$

Then Parabolic PM System (1.1)-(1.3) has a steady-state for any given $f \in X$.

3. STEADY-STATE STABILITY AND EXAMPLE

DEFINITION 3.1. Parabolic PM System (1.1)-(1.3) is called to be steady-state stable with respect to the disturbance $f(t)=f \cdot 1(t)$ if there exists some $\alpha \in \Omega$ such that

$$\lim_{n \to \infty} \alpha_n = \alpha \tag{3.1}$$

In the sequel we need some assumptions and notations. At first we assume that the constants λ and ω in (2.9) and (2.12) are positive i.e.

$$\|S(t)\| \leq M_a e^{-\lambda t} , \quad \|e^{Jt}\| \leq M_J e^{-\omega t} , \quad \lambda>0, \ \omega>0. \tag{3.2}$$

This assumption is not a severe restriction since in general industrial processes like heating, diffusion and etc and controllers are initially stable.

Denote two $q \times q$ matrices

$$G(t,\eta)= \int_\eta^t e^{J(t-\tau)} KCS(\tau-\eta) B d\tau , \quad G(t)=G(t,0)= \int_0^t e^{J(t-\tau)} KCS(\tau) B d\tau$$

ASSUMPTION (G). For the matrix G(t) and the step constant disturbance $f \in X$ we assme that (1) there exists an inverse matrix $[I- \int_0^\infty G(t)dt]^{-1}$ and

$$[I- \int_0^\infty G(t)dt]^{-1} J^{-1} KCA^{-1} f \in \mathrm{Int}\Omega \ ; \tag{3.3}$$

(2) there exists some constant $\delta>0$ such that

$$\| J^{-1}KCA^{-1}f+ \int_0^\infty G(t)dt \, \alpha -v \| \geq \delta , \quad \forall \ v \in R^q - \Omega \tag{3.4}$$

where $\alpha=\mathrm{Proj}_{\Omega}(v)$ is the projection of v on the closed convex set Ω.

REMARK 3.2. Obviously, for any given $f\in X$ there exists some constant M_K such that both of (3.3) and (3.4) are satisfied as $\|K\|\leq M_K$.

LEMMA 3.3, Let G(t) be an q×q matrix function continuously on $[0,\infty)$ and satisfy $\|G(t)\|\leq Me^{-\lambda t}$ $(t\geq 0)$ where $M>0$ and $\lambda>0$. Then for any given $\varepsilon>0$ there exists some $T^*>0$ such that

$$\Big\| \int_0^\infty G(t)dt-\sum_{l=1}^{\infty} TG(lT-\beta T)\Big\|<\varepsilon \quad , \quad 0<T<T^*,$$

holds uniformly for $\beta=(\beta_1,\beta_2,\ldots,\beta_q)'$, $0\leq\beta_j\leq 1$, $j=1,2,\ldots q$.

THEOREM 3.4. Parabolic PM System (1.1)-(1.3) is steady-state stable under Assumption (3.2) and Assumption (G).

PROOF. The output v(t) of the dynamical controller (1.2) is

$$v(t)=e^{Jt}v_0+\int_0^t e^{J(t-\tau)}KCy(\tau)d\tau= v_e(t)+v_c(t) \tag{3.5}$$

where

$$v_e(t)=e^{Jt}v_0+\int_0^t e^{J(t-\tau)}KC[S(\tau)(y_0+A^{-1}f)-A^{-1}f]d\tau$$

$$v_c(t)=\int_0^t e^{J(t-\tau)}KCd\tau\int_0^\tau S(\tau-\eta)Bu(\eta)d\eta=\int_0^t G(t,\eta)u(\eta)d\eta .$$

It is easy to see that under Assumption (3.2) one has

$$\lim_{t\to\infty} v_e(t)=J^{-1}KCA^{-1}f . \tag{3.6}$$

Since

$$v_c(nT)=\int_0^{nT} G(nT,\eta)u(\eta)d\eta=\sum_{l=0}^{n-1}\int_{lT}^{(l+1)T} G(nT,\eta)u(\eta)d\eta=\sum_{l=0}^{n-1} TG(nT,lT+\beta_l T)\alpha_l ,$$

from (3.5) we have

$$\begin{aligned} 0&=\sum_{n=0}^{\infty} [v_e(nT)+v_c(nT)-v(nT)] \\ &=\sum_{n=0}^{\infty} [v_e(nT)+\sum_{k=1}^{\infty} TG(kT-\beta_n T)\alpha_n-v(nT)] \end{aligned} \tag{3.7}$$

where $\beta_n=(\beta_{n1},\beta_{n2},\ldots,\beta_{nq})'$, $0\leq\beta_{nj}\leq|\alpha_{nj}|\leq 1$, $j=1,2,\ldots,q$, $n\in I^+$. By Lemma 3.3 there exists some q×q matrix E(T,n) such that

$$\sum_{k=1}^{\infty} TG(kT-\beta_n T)=\int_0^\infty G(t)dt+E(T,n) , \tag{3.8}$$

$$\lim_{T\to 0}\|E(T,n)\|=0 \qquad \text{uniformly for } n\in I^+. \tag{3.9}$$

For any given $T>0$ from (3.7) we have

$$0=\lim_{n\to\infty}[v_e(nT)+\sum_{k=1}^{\infty} TG(kT-\beta_n T)\alpha_n-v(nT)]$$

$$=J^{-1}KCA^{-1}f+\lim_{n\to\infty}\left\{\left[\int_0^\infty G(t)dt+E(T,n)\right]\alpha_n-v(nT)\right\} \tag{3.10}$$

Let $\delta>0$ be the positive number in Assumption (G). It is not difficult to prove that there exists some $T^*>0$ and positive integral $N^*=N^*(T^*,\delta)$ such that

$$v(nT)=\alpha_n\ ,\quad 0<T<T^*\ ,\quad n\geq N^*\ . \tag{3.11}$$

Substituting (3.11) into (3.10) we have

$$\lim_{n\to\infty}\left[I-\int_0^\infty G(t)dt-E(T,n)\right]\alpha_n=J^{-1}KCA^{-1}f\ ,\qquad 0<T<T^*\ .$$

Obviously, by (3.9) one has

$$\lim_{n\to\infty}\left[I-\int_0^\infty G(t)dt\right]\alpha_n=J^{-1}KCA^{-1}f$$

and

$$\lim_{n\to\infty}\alpha_n=\alpha=\left[I-\int_0^\infty G(t)dt\right]^{-1}J^{-1}KCA^{-1}f\ . \tag{3.12}$$

REMARK 3.5. The above theorem shows that Parabolic PM System (1.1)-(1.3) has the steady-state stability by tuning the feed-back matrix K.

EXAMPLE 3.6. We now consider a heat conduction temprature control system with an one-dimensional control and an one-dimensional output

$$\begin{aligned}
&\frac{\partial y(t,x)}{\partial t}=\frac{\partial^2 y(t,x)}{\partial x^2}+b(x)u(t)+f(x)1(t), && t>0,\ 0<x<L\\
&y(t,0)=y(t,L)=0, && t\geq 0\\
&z(t)=\int_0^L c(x)y(t,x)dx && t\geq 0
\end{aligned} \tag{3.13}$$

where $b(x)$ and $c(x)$ are in the state space $X=L^2(0,L)$. The dynamical controller is designed as a proportional controller with an inertia

$$\tau_a\dot{v}(t)=v(t)+kz(t) \tag{3.14}$$

where $k>0$ is the gain of the controller.

By the semigroup theory [2] (3.13) can be written as the abstract form (1.1) in the state space $X=L^2(0,L)$ as we define

$$D(A)=\left\{y\in L^2(0,L):\ y,y''\in L^2(0,L),\ y(0)=y(L)=0\right\}$$

$$(Ay)(x)=y''(x),\quad y\in D(A)\ .$$

The semigroup $S(t)$ generated by A has the form

$$[S(t)y](x)=\sum_{n=1}^{\infty}e^{-\lambda_n t}(y,\phi_n)\ \phi_n(x) \tag{3.15}$$

where

$$\lambda_n=(n\pi/L)^2, \quad \phi_n(x)=\sqrt{2/L}\ \sin n\pi x/L$$

$$(y,\phi_n)=\int_0^L y(x)\phi_n(x)dx \qquad n=1,2,\ldots\ .$$

Obviously

$$\|S(t)\|\leqslant e^{-\lambda_1 t}\ , \quad t\geqslant 0 \tag{3.16}$$

From (3.15) one has

$$(R[1,S(T)]z)(x)=\sum_{n=1}^{\infty}\frac{(z,\phi_n)}{1-e^{-\lambda_n T}}\phi_n(x)$$

The functions $P(t,\tau)$ and $P(t)$ defined in Lemma 2.5 are

$$P(t,\tau)=\frac{k}{\tau_a}\sum_{n=1}^{\infty}\frac{e^{-\lambda_n(t+T-\tau)}(b,\phi_n)(c,\phi_n)}{1-e^{-\lambda_n T}}$$

$$P(t)=\frac{k}{\tau_a}\sum_{n=1}^{\infty}e^{-\lambda_n t}(b,\phi_n)(c,\phi_n)\ .$$

The constant M_2 in (2.11) can be obtained as $M_2=4T|(b,c)|e^{\lambda_1 T}$. Denote

$$\mu=\min\ (\lambda_1,\ 1/\tau_a) \tag{3.17}$$

then the constant M_0 in (2.13) is chosen as $M_0=4T^2|(b,c)|e^{\mu T}$. The sufficient condition that there exists a unique steady-state with respect to any step disturbance $f(x)1(t)$, $f\in L^2(0,L)$, is

$$4k\tau_a^{-1}T^2|(b,c)|e^{\mu T}<1\ . \tag{3.18}$$

For the condition of the steady-state stability of PM System (3.13),(3.14),(1.3), at first, we give the explicit expression of $G(t)$ for the concrete case:

$$G(t)=\frac{k}{\tau_a}\int_0^t e^{-(t-\tau)/\tau_a}\sum_{n=1}^{\infty}e^{-\lambda_n\tau}(b,\phi_n)(c,\phi_n)d\tau$$

$$=\frac{k}{\tau_a}\sum_{n=1}^{\infty}\frac{e^{-\lambda_n t}-e^{-t/\tau_a}}{\frac{1}{\tau_a}-\lambda_n}(b,\phi_n)(c,\phi_n)\ .$$

Therefore

$$\int_0^{\infty}G(t)dt=k\sum_{n=1}^{\infty}\frac{1}{\lambda_n}(b,\phi_n)(c,\phi_n)\ .$$

Since

$$J^{-1}KCA^{-1}f=k\sum_{n=1}^{\infty} \frac{1}{\lambda_n} (f,\phi_n)(c,\phi_n)$$

the condition (3.3) becomes

$$\left| k\sum_{n=1}^{\infty}\frac{1}{\lambda_n}(f,\phi_n)(c,\phi_n)\right| \leqslant \left|1-k\sum_{n=1}^{\infty}\frac{1}{\lambda_n}(b,\phi_n)(c,\phi_n)\right| \tag{3.19}$$

and the condition (3.4) becomes

$$\left| k\sum_{n=1}^{\infty} \frac{(c,\phi_n)}{\lambda_n}\left[(f,\phi_n)+(b,\phi_n)\text{sign } v\right]-v\right| \geqslant \delta \tag{3.20}$$

where $|v|>1$. If we assume

$$k\sum_{n=1}^{\infty}\frac{1}{\lambda_n}\left[|(f,\phi_n)(c,\phi_n)| + |(b,\phi_n)(c,\phi_n)|\right]<1 \tag{3.21}$$

then both of (3.20) and (3.19) hold and PW System (3.13),(3.14),(1.3) is steady-state stable by Theorem 3.4.

REFERENCES

1. R.F.Curtain & A.J.Pritchard, Infinite Dimensional Linear Systems Theory, Springer-Verlag, 1978.
2. A.Pazy, Semigroups of Linear Operators and Applications to PDEs, Springer-Verlag, 1983.
3. B.C.Kuo, Degital Control Systems, Holt, Rinehart & Winston Inc., New York,1980.
4. V.I.Utkin, Discontinuous control systems: State of the art in theory and applications, Preprints of 10th World Congress on Automatic Control of IFAC, Vol.1,pp.75-94, 1987.
5. Hong Xing ZHOU, Steady-state analysis for control systems with a pulse-width sampler, Acta Automatica Sinica, 13(1987), No.1, pp.66-69.
6. ---, Steady-state stability for pulse-width modulated sampled-data control systems, Control Theory & Appl., 4(1987), No.3, pp.57-65.
7. ---, Analysis of pulse-width modulated sampled-data control systems, J. System Science and Math. Science, 8(1988), No.1, pp.11-18.

Lecture Notes in Control and Information Sciences

Edited by M. Thoma and A. Wyner

Vol. 117: K.J. Hunt
Stochastic Optimal Control Theory
with Application in Self-Tuning Control
X, 308 pages, 1989.

Vol. 118: L. Dai
Singular Control Systems
IX, 332 pages, 1989

Vol. 119: T. Başar, P. Bernhard
Differential Games and Applications
VII, 201 pages, 1989

Vol. 120: L. Trave, A. Titli, A. M. Tarras
Large Scale Systems:
Decentralization, Structure Constraints
and Fixed Modes
XIV, 384 pages, 1989

Vol. 121: A. Blaquière (Editor)
Modeling and Control of Systems
in Engineering, Quantum Mechanics,
Economics and Biosciences
Proceedings of the Bellman Continuum
Workshop 1988, June 13–14, Sophia Antipolis, France
XXVI, 519 pages, 1989

Vol. 122: J. Descusse, M. Fliess, A. Isidori,
D. Leborgne (Eds.)
New Trends in Nonlinear Control Theory
Proceedings of an International
Conference on Nonlinear Systems,
Nantes, France, June 13–17, 1988
VIII, 528 pages, 1989

Vol. 123: C. W. de Silva, A. G. J. MacFarlane
Knowledge-Based Control with
Application to Robots
X, 196 pages, 1989

Vol. 124: A. A. Bahnasawi, M. S. Mahmoud
Control of Partially-Known
Dynamical Systems
XI, 228 pages, 1989

Vol. 125: J. Simon (Ed.)
Control of Boundaries and Stabilization
Proceedings of the IFIP WG 7.2 Conference
Clermont Ferrand, France, June 20–23, 1988
IX, 266 pages, 1989

Vol. 126: N. Christopeit, K. Helmes
M. Kohlmann (Eds.)
Stochastic Differential Systems
Proceedings of the 4th Bad Honnef Conference
June 20–24, 1988
IX, 342 pages, 1989

Vol.127: C. Heij
Deterministic Identification
of Dynamical Systems
VI, 292 pages, 1989

Vol. 128: G. Einarsson, T. Ericson,
I. Ingemarsson, R. Johannesson,
K. Zigangirov, C.-E. Sundberg
Topics in Coding Theory
VII, 176 pages, 1989

Vol. 129: W. A. Porter, S. C. Kak (Eds.)
Advances in Communications and
Signal Processing
VI, 376 pages, 1989.

Vol. 130: W. A. Porter, S. C. Kak,
J. L. Aravena (Eds.)
Advances in Computing and Control
VI, 367 pages, 1989

Vol. 131: S. M. Joshi
Control of Large Flexible Space Structures
IX, 196 pages, 1989.

Vol. 132: W.-Y. Ng
Interactive Multi-Objective Programming
as a Framework for Computer-Aided Control
System Design
XV, 182 pages, 1989.

Vol. 133: R. P. Leland
Stochastic Models for Laser Propagation
in Atmospheric Turbulence
VII, 145 pages, 1989.

Vol. 134: X. J. Zhang
Auxiliary Signal Design in Fault
Detection and Diagnosis
XII, 213 pages, 1989

Vol. 135: H. Nijmeijer, J. M. Schumacher (Eds.)
Three Decades of Mathematical System Theory
A Collection of Surveys at the Occasion of the
50th Birthday of Jan C. Willems
VI, 562 pages, 1989

Vol. 136: J. Zabczyk (Ed.)
Stochastic Systems and Optimization
Proceedings of the 6th IFIP WG 7.1
Working Conference,
Warsaw, Poland, September 12–16, 1988
VI, 374 pages. 1989

Lecture Notes in Control and Information Sciences

Edited by M. Thoma and A. Wyner

Vol. 137: S. L. Shah, G. Dumont (Eds.)
Adaptive Control Strategies for Industrial Use
Proceedings of a Workshop
Kananaskis, Canada, 1988
VI, 360 pages. 1989

Vol. 138: D. C. McFarlane, K. Glover
Robust Controller Design Using Normalized Coprime Factor Plant Descriptions
X, 206 pages. 1990

Vol. 139: V. Hayward, O. Khatib (Eds.)
Experimental Robotics I
The First International Symposium
Montreal, June 19-21, 1989
XVII, 613 pages. 1990

Vol. 140: Z. Gajic, D. Petkovski, X. Shen
Singularly Perturbed and Weakly Coupled Linear Control Systems
A Recursive Approach
VII, 202 pages. 1990

Vol. 141: S. Gutman
Root Clustering in Parameter Space
VIII, 153 pages. 1990

Vol. 142: A. N. Gündeş, C. A. Desoer
Algebraic Theory of Linear Feedback Systems with Full and Decentralized Compensators
V, 176 pages. 1990

Vol. 143: H.-J. Sebastian, K. Tammer (Eds.)
System Modelling and Optimization
Proceedings of the 14th IFIP Conference
Leipzig, GDR, July 3-7, 1989
X, 960 pages. 1990

Vol. 144: A. Bensoussan, J. L. Lions (Eds.)
Analysis and Optimization of Systems
Proceedings of the 9th International Conference, Antibes, June 12-15, 1990
XII, 992 pages. 1990

Vol. 145: M. B. Subrahmanyam
Optimal Control with a Worst-Case Performance Criterion and Applications
X, 133 pages. 1990

Vol. 146: D. Mustafa, K. Glover
Minimum Entropy H_∞ Control
IX, 144 pages. 1990

Vol. 147: J. P. Zolésio (Ed.)
Stabilization of Flexible Structures
Third Working Conference,
Montpellier, France, January 1989
V, 327 pages, 1990

Vol. 148: In preparation

Vol. 149: K.H. Hoffmann, W. Krabs (Eds.)
Optimal Control of Partial Differential Equations
Proceedings of the IFIP WG 7.2 International Conference, Irsee, April 9-12, 1990
VI, 245 pages. 1991

Vol. 150: L. C. G. J. M. Habets
Robust Stabilization in the Gap-topology
IX, 126 pages. 1991

Vol. 151: J. M. Skowronski, H. Flashner, R. S. Guttalu (Eds.)
Mechanics and Control
Proceedings of the 3rd Workshop on Control Mechanics
in Honor of the 65th Birthday of George Leitmann
January 22-24, 1990, University of Southern California
IV, 497 pages. 1991

Vol. 152: J. D. Aplevich
Implicit Linear Systems
XI, 176 pages. 1991

Vol. 153: O. Hájek
Control Theory in the Plane
X, 272 pages. 1991

Vol. 154: A. Kurzhanski, I. Lasiecka (Eds.)
Modelling and Inverse Problems of Control for Distributed Parameter Systems
Proceedings of IFIP (W.G. 7.2)-IIASA Conference
Laxenburg, Austria, July 24-28, 1989
VII, 179 pages. 1991

Vol. 155: M. Bouvet, G. Bienvenu (Eds.)
High-Resolution Methods in Underwater Acoustics
V, 249 pages. 1991

Lecture Notes in Control and Information Sciences

Edited by M. Thoma and A. Wyner

Vol. 156: R. P. Hämäläinen, H. K. Ehtamo (Eds.)
Differential Games –
Developments in Modelling and Computation
Proceedings of the Fourth International Symposium
on Differential Games and Applications
August 9-10, 1990, Helsinki University of Technology,
Finland
XIII, 292 pages. 1991

Vol. 157: R. P. Hämäläinen, H. K. Ehtamo (Eds.)
Dynamic Games in Economic Analysis
Proceedings of the Fourth International Symposium
on Differential Games and Applications
August 9-10, 1990, Helsinki University of Technology,
Finland
XIII, 311 pages. 1991

Vol. 158: K. Warwick, M. Kárný,
A. Halousková (Eds.)
Advanced Methods in Adaptive Control
for Industrial Applications
X, 331 pages. 1991

Vol. 159: X. Li, J. Yong (Eds.)
Control Theory of Distributed Parameter Systems
and Applications
Proceedings of the IFIP WG 7.2 Working Conference,
Shanghai, China, May 6–9, 1990
VIII, 219 pages. 1991